TRANSLATIONAL ADMET FOR DRUG THERAPY

TRANSLATIONAL ADMET FOR DRUG THERAPY

Principles, Methods, and Pharmaceutical Applications

SOUZAN B. YANNI
DMPK Consultants, Inc.
North Carolina, USA

WILEY

Published by John Wiley & Sons, Inc., Hoboken, New Jersey
Published simultaneously in Canada

Library of Congress Cataloging-in-Publication Data:

Yanni, Souzan, author.
Translational ADMET for drug therapy : principles, methods, and pharmaceutical applications / Souzan Yanni.
p. ; cm.
Includes bibliographical references and index.
ISBN 978-1-118-83827-3 (cloth)
I. Title.
[DNLM: 1. Drug Evaluation, Preclinical. 2. Pharmacokinetics. 3. Drug-Related Side Effects and Adverse Reactions. 4. Pharmaceutical Preparations–metabolism. QV 771]
RM301.5
615.7–dc23

2015010195

Typeset in 10/12pt TimesLTStd by SPi Global, Chennai, India

Printed in the United States of America

10 9 8 7 6 5 4 3 2 1

1 2015

CONTENTS

CONTRIBUTORS

Ingrid L. Druwe PhD, Oak Ridge Institute for Science and Education, National Center for Environmental Assessment U.S. Environmental Protection Agency, Research Triangle Park, North Carolina

Gabriel A. Knudsen PhD, Laboratory of Toxicology and Toxicokinetics, Center for Cancer Research, National Cancer Institute, Research Triangle Park, North Carolina

Samuel C. Suarez PhD, Department of Entomology, North Carolina State University, Raleigh, North Carolina

PREFACE

Drug disposition implicated by absorption, distribution, metabolism, and excretion (ADME) and by toxicity (T) are always two of the most critical issues that pharmaceutical scientists and regulatory authorities focus on during the discovery and development of a new medicine for any target disease and the human population. Pharmaceutical companies spend 20–30% of their R&D budget to assess compound behavior with respect to ADMET, the factor that subsequently affects the pharmacological response and safety of any given drug in target patient populations. Preclinical and clinical tools and technologies that support the prediction of pharmacokinetics (PK) and pharmacodynamics (PD) in relation to toxicokinetics (TK) and toxicodynamics (TD) are continuously under development and improvement to reduce the cost and to increase the precision in developing a new medicine that ensures efficacy and safety of target human populations.

Because human population can vary depending on age, race, gender, disease, environment, and so forth, drug dose has to be adjusted. Updated in vitro and in vivo as well as in silico tools are developed and validated to be used in improving drug design toward the selection of the most effective and safe drug candidate. Regulatory organizations like the Food and Drug Administration and European Medicines Agency through updated guidance are reaching out to pharmaceutical companies, biotech, and contract research organizations (CROs) in order to support the research and development efforts by listing the most acceptable tools and approaches that scientists can employ to strategically design preclinical and clinical investigations. These efforts can aid in distinguishing the drug candidates that might be developed to become new therapeutic agents from those that are not and that will fail fast and fail cheap. Predictive ADMET tools such as in vitro, in vivo, and in silico models, such as physiologically based pharmacokinetic models, are tools that have the potential to enhance the selection of lead compounds, to facilitate the understanding of mechanisms underlying the disposition of drugs, to determine pharmacokinetic parameters, and to select the drug dose for first in human (FIH) investigations.

In vitro, in vivo, and in silico correlation that could translate data generated from bench to bedside or from bedside to bench is now a frequently used approach to reveal hidden adverse events, adverse drug reaction, and drug–drug interactions (DDIs) and to elucidate the mechanism of drug disposition, thus optimizing drug dose in human. Furthermore, the current in vitro and in silico technologies can now be used to predict plasma and tissue concentrations of drugs, select animal model by extrapolating across species, provide preclinical dosing regimen, assess the variability among human populations, and predict the potential DDI, thus ensuring the selection of preclinical toxicity species and drug therapy in all human populations regardless of age, gender, race, or disease.

These predictive tools have been used to calculate drug clearance. The accurate determination of drug clearance has warranted extensive translational research efforts, thus improving our ability to estimate safe and efficacious doses in different human populations, sometimes from retrospective clinical studies (from one population to other) toward the design of prospective clinical investigations in a target population.

The broad and current coverage of translational ADMET from drug discovery to drug development will serve as a handbook for scientists and managers from multidisciplinary functions within biotechnology and pharmaceutical companies and CROs to assist in designing and executing drug discovery and development programs that are conducted in compliance with regulatory guidelines. This book focuses on the most critical and emerging points that emphasize the translational ADME, also mapping the most effective approaches and technologies that are currently used to investigate ADME, PK, and toxicology prior to setup of clinical studies. The book can also be a textbook for senior graduate and medical students to be utilized as a hands-on manual in conducting in vitro, in silico, and preclinical in vivo ADMET studies.

Several case studies from drug discovery and drug development of drug candidates from varies therapeutic areas including, study design, possible data interpretations and decision-making tactics will be illustrated. These studies will demonstrate the integration of in vitro, in vivo, and in silico data to address human PK/PD/TK/TD and hence to select the safe and therapeutic dose in human and to support the design of clinical studies, investigational new drug, and new drug application submissions. Furthermore, the book will also demonstrate the strategy in translating ADME properties retrospectively from bedside to bench and from bench to bedside toward the design of a safe and effective medicine in special human populations, for example, pediatrics.

The unique features of this book will support the pharmaceutical professionals who are interested in becoming pharmaceutical managers and leaders with ADMET.

SOUZAN B. YANNI

ACKNOWLEDGEMENT

The author likes to thank Wiley Editor team especially, Jonathan Rose who presented to her the opportunity of writing this book and who provided her with access to Wiley publications and resources. Also like to thank the rest of the editorial team, specifically Ms. Sarah Brown and Ms. Kiruthika Balasubramanian for their excellent effort in revising the contents of the book and completing the process in timely fashion.

Furthermore, the author likes to thank the scientists and post doc who contributed in the writing of chapter 6 and chapter 7 of this book. Finally, the author needs to express her gratitude to all her pharmaceutical industry colleagues, mentors, and leaders who indirectly supported this book through their excellent publications, training, and guide. Lastly but not least, the author likes to thank her family that provided continuous encouragement and support towards the completion of the book, especially her husband, Professor Adel Hanna, and sons Attorney Mr. Peter Hanna and Hani Hanna.

1

TRANSLATIONAL CONCEPT AND DETERMINATION OF DRUG ABSORPTION

1.1 DRUG ABSORPTION, MECHANISM, AND ITS IMPACT ON DRUG BIOAVAILABILITY, DRUG DISPOSITION, AND DRUG SAFETY

Discovery, development, and approval of a new drug is a long process that takes on average 12–14 years and costs an average of about $1.8 billion [1]. The financial burden and time for bringing to the market a new medicine are considered as major challenges in the pharmaceutical industry. In addition, the decrease in the number of truly innovative therapeutic areas that have been approved by the regulatory authorities around the globe was a reflection of higher attrition in late-stage drug development (Phase 2 and 3), despite the advancement of new technologies. However, the high-throughput screening; structure activity relationship (SAR) using absorption, distribution, metabolism, excretion, and toxicity (ADMET) properties; and target efficacy-based molecular and cell biology in collaboration with advanced medicinal and combinatory chemistry have increased the number of drug candidates successfully reaching Phase 1 due to better preclinical characterization and improved ADMET properties. For example, the Phase 2 success rates for drug candidates have fallen from 28% in 2006 to 18% in 2009, with ~50% of success to progressing through Phase 2. The decrease in Phase 2 is mainly due to insufficient efficacy, undesired side effects, and/or poor pharmacokinetics (PK) of the newly developed drug, which account for 51% of the drug failures [2–4]. Therefore, correct prediction of the efficacy of novel drug candidates especially in the early stage preclinical phases is crucial.

Translational ADMET for Drug Therapy: Principles, Methods, and Pharmaceutical Applications, First Edition. Souzan B. Yanni.

The accurate assessment of absorbed drug dose, exposure, and disposition in in vitro and in preclinical animal models that translate to human clinical data may improve the success rate of bringing a needed medicine to the stage of reaching human patients.

For the drug to be absorbed in the intestine, several processes are involved. First, the physicochemical properties of drugs such as solubility, dissolution rate, lipophilicity, and molecular weight (MW) are major driving parameters for the drug absorption in the gastrointestinal (GI) tract, as a molecule should be in a solution to permeate the intestinal membranes, and the rate at which the molecule gets into the solution impacts its ability to get absorbed. Second, the drug has to cross several physiological parameters before it reaches the bloodstream, such as effect of pH, stomach emptying, intestinal transient time, disease state, age, diet, various GI fluids, and so forth. The sum of physiochemical and physiological parameters can either hinder or facilitate the permeability of a drug in the intestinal sections.

It is important to emphasize here, as the drug absorption will be discussed in detail, that the drug's permeability is indeed the major determinant of its ability to be absorbed in the intestine. The permeability of drugs can be either by a passive diffusion mechanism, following the "rule of 5," or by an active process driven by the intestinal transporters. Drug transporters can either promote the absorption (by uptake transporters such as OCT, OAT, PepT1, OATP) or hinder the absorption (efflux transporters such as P-gp, MRP2, BCRP).

Last, the drug absorption can be influenced by other significant factors, such as the metabolism by drug-metabolizing enzymes that are expressed mostly in the duodenal section of the small intestine. Thus, the intestinal metabolism, which may also cause drug–drug interactions (DDIs), changes the extent of oral drug absorption. As will be discussed later, effective orally absorbed drug will ensure systemic exposure. Absorption through membranes of the GI tract and metabolism by gut and hepatic metabolism are key players for drug exposure in systemic circulation—that is, oral bioavailability—before it reaches the other body organs.

1.1.1 Drug Absorption and Oral Bioavailability

In ADME processes that exert the pharmacokinetic properties of a new drug, absorption is a process by which a given extravascular dose (EV), that is, an oral dose (PO), reaches the systemic circulation. The absorption of a drug can be described by any drug dose that is administered orally, subcutaneously, intramuscularly, or any other way different from a direct injection into the vascular system. The term oral "bioavailability" (F) is a parameter that is used in pharmacokinetics to quantify the ability of a compound dosed orally to reach the systemic circulation, after surviving any first-pass extraction in the gut and liver. The systemic F can be determined from Equation (1.1):

$$F = \frac{AUC_{po} \times Dose_{iv}}{AUC_{iv} \times Dose_{po}}, \tag{1.1}$$

where AUC_{po} refers to the area under the curve (AUC) from an oral administration and AUC_{iv} refers to the AUC from intravenous (iv) administration. Accordingly, the drug becomes bioavailable when it overcomes the potential barriers to reach the systemic circulation. A compound with $F = 1$ (or 100%) indicates that a given oral dose produces an identical systemic exposure to that observed in the corresponding iv dose, indicating that it is fully absorbed and fully escaped any potential of metabolism in both the gut and liver. $F = 0.5$ (50%) indicates that in transit from the oral administration site to the systemic circulation, half of the compound is lost; in this case, the oral dose to systemic concentration relationship indicates that the oral dose must be twice that of an equivalent iv dose to achieve a similar systemic exposure.

Although there are several approaches for a drug to become bioavailable, the oral dosing route is the most convenient, well-tolerated, patient-compliant, and cost-effective route of drug administration; however, it is still a complex route of administration, as the absorption from the gut into the systemic circulation may requires consideration to avoid inter- and intrapatient variability in a compound's pharmacokinetic profile [5].

Oral administration, F_{oral}, can be described as shown in Equation (1.2):

$$F_{oral} = (f_a \cdot F_G) \times (F_H \cdot F_L), \tag{1.2}$$

where f_a is the fraction of the dose absorbed from the gut, and F_G, F_H, and F_L are the bioavailability of the compound in the intestine, liver, and lung (typically $F_L = 1$ and ignored), respectively. From Equation (1.2), it is clear that a lack of f_a or bioavailability in any one of the organs will yield $F_{oral} = 0$, and fully no systemic exposure.

As mentioned above, oral bioavailability is determined by the absorption through membranes of the GI tract and by the extent to which gut and liver are able to extract the orally administered drug (see Figure 1.1). Therefore, gut and hepatic metabolism are also key players for drug oral bioavailability. In a study with a set of 309 drugs where bioavailability, fraction absorbed (f_a), fraction escaping intestinal extraction (F_G), and fraction escaping hepatic extraction (F_H) were known, the analysis was conducted to determine which physicochemical property influences these parameters to enhance the bioavailability of a new drug candidate [6]. It was shown that f_a decreases with increasing MW (> 500), polarity (c log D > −2), polar surface area (> 125 Å^2), total H-bond donors and acceptors (> 9), and rotatable bonds (> 12). Indeed, such properties limit the capability of small organic molecules to traverse lipid membranes. Molecules with a log P ranging from 1 to 3 are considered to be highly permeable. Lipinski et al. (2001) [7] showed that particular physicochemical properties are associated with high or low oral bioavailability. They established the famous "rule of 5" that predicts that poor absorption or permeation is more likely when there are more than 5 H – bond donors, 10 H – bond acceptors, the MW is > 500 g/mole, and the calculated log P > 5 [7].

However, it was noted in the above study [6] that high lipophilicity does not necessarily have a detrimental effect on $f_{a,}$ and the analysis showed that the numbers of free rotatable bonds are negatively related, with all three processes leading to a dramatic effect on bioavailability. Also it has been noted that physicochemical properties that lead to high f_a tend to be also associated with high rates of metabolism. That means

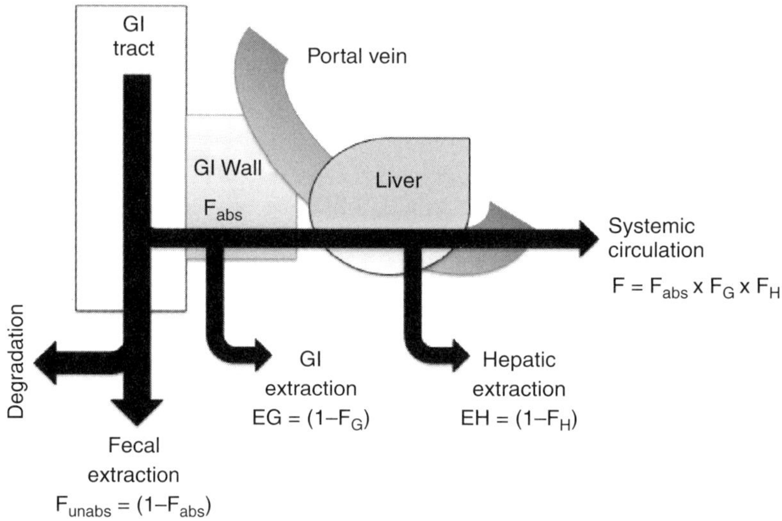

Figure 1.1 Oral drug disposition indicating the bioavailability as a function of intestinal absorption (F_{abs}), the fraction escaping intestinal metabolism (F_G), and the fraction escaping hepatic extraction (F_H), along with plasma drug bioavailability, the product of these three parameters.

enough lipophilicity is needed to ensure good membrane penetrability but too much will lead to high extraction due to metabolism in gut and liver.

In other analysis by comparing basic, acidic, and neutral drugs, the data indicated that the higher first-pass effect due to higher metabolism and relatively lower protein binding of basic drugs leads to lower bioavailability than acids or neutral drugs, although basic drugs exhibits higher f_a than the acidic and neutral drugs [6]. Indeed, the higher first-pass effect of basic molecules can be attributed to their affinity for metabolic enzymes.

1.1.2 Contribution of Intestinal Drug Transporters and Drug-Metabolizing Enzymes on Extent of Absorption and Mechanism

1.1.2.1 Intestinal Transporters As the translational investigations of drug disposition grow, along with data generated from the bench and data generated from clinical pharmacology investigations, it becomes clear that drug transporters are widely considered as a critical determinant in PK, pharmacodynamics (PD), and, most importantly, drug safety (DDI). Specifically, the intestinal transporters, as mentioned earlier, are viewed as an important determinant of oral drug absorption, bioavailability, and DDI.

Indeed, the *efflux pump ABCB1* (P-glycoprotein, P-gp, or multidrug resistance 1, MDR1) is now one of the most evaluated transporters due to the many roles it plays, for example, differential bioavailability and DDI among human populations. Because

P-gp can play a role in limiting oral absorption of particular drugs [8–11], it has emerged as a potential determinant of oral bioavailability of those drugs.

As will be discussed in the following chapters, efflux transporters are expressed in many biological membranes of body organs, including the villus tip of the apical brush border membrane of gut enterocytes. They actively cause efflux of drugs from gut epithelial cells back into the intestinal lumen (see Figure 1.2). When a drug is orally administered, intestinal efflux transports limit the amount of drug absorbed into the intestine epithelia by pump drug to gut lumina, and this process presents a significant barrier toward drug absorption. Efflux transporter is one of the adenosine triphosphate (ATP)-binding cassette (ABC) family, as well as breast cancer resistance protein (BCRP; ABCG2) and multidrug transporter proteins (MRP; ABCC), but MDR1 is the most studied transporter [11]. Although P-gp activity limits oral drug absorption for specific drugs, these efflux transporters have a detoxification protecting function against the entry of exogenous toxins to the small intestine, colon, and other nondigestive organs like CNS and testis [12–14], and its role in blocking drug absorption makes the intestinal secretion a potential mechanism for drug elimination [15]. Although it is difficult to establish SAR for MDR1 substrates (and inhibitors), some features that are shared by many MDR1 substrates include the presence of a nitrogen group, aromatic moieties, planar domains, molecular size ≥ 300 Da, presence of a positive charge at physiological pH, amphipathicity, and lipophilicity [16,17]. In the interaction between two modulators of P-gp, caution must be exercised when trying to extrapolate how the substrate/inhibitor may interact with an untested new drug, as MDR1 possesses multiple drug-binding sites and these sites are located in the middle of the lipid bilayer [18,19].

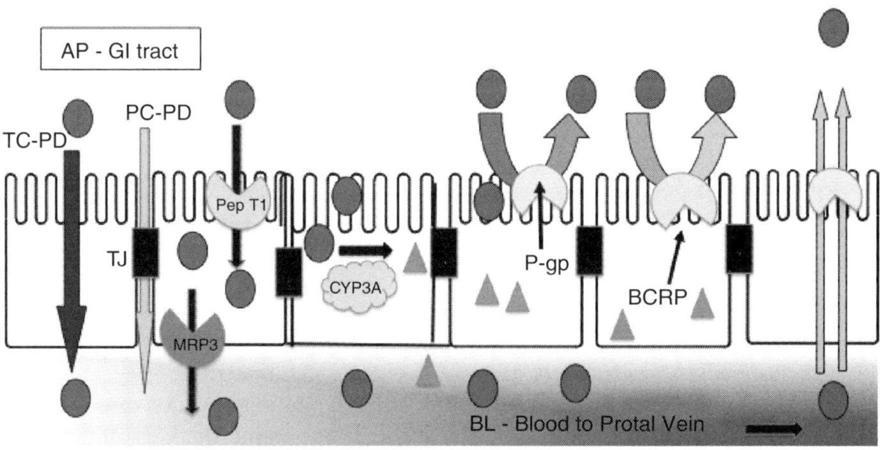

Figure 1.2 The possible mechanisms of drug absorption across the intestinal epithelia monolayers, such as transepithelial passive diffusion (TC-PD), paracellular passive diffusion (PC-PD), and active transport by uptake (PepT1) and efflux transporters (P-gp, BCRP, MRP). Furthermore, the figure indicates the interplay between drug transporters and DMEs such as CYP3A that influence the drug absorption and bioavailability.

Several studies have focused on evaluating the impact of MDR1-mediated efflux activity and its potential attenuation of the overall bioavailability of its substrates. The studies revealed that P-gp could reduce the oral bioavailability via a couple of possible mechanisms:

1. It can attenuate the rate of substrate's permeation from gut across intestinal enterocytes on apical membrane into blood, thus potentially delaying absorption time (T_{max}), reducing C_{max}, and possibly reducing total exposure (AUC).
2. It may enhance intestinal metabolic elimination (low absorbed substrate concentration below the Km of binding to P450 enzymes), thus indirectly reducing the amount of compound able to reach the bloodstream.

In clinical studies with substrates for P-gp like talinolol, the mean absorption time, AUC, and C_{max} following oral administration of MDR1 substrates are affected by efflux activity of MDR1 in the intestine [20]. Furthermore, duodenal MDR1 mRNA content was significantly correlated with the AUC and C_{max} of oral talinolol [21], and oral bioavailability of substrates such as tacrolimus and cyclosporin is known to be incomplete and variable in the clinic, as these are regulated by intestinal MDR1 and modulated by coadministered drugs, genetic polymorphisms, and disease states. Interestingly, the mRNA levels of MDR1, but not CYP3A4, correlated well with the ratio of concentration/oral dose and the oral dosage of tacrolimus [22]. In other clinical investigation with St John's wort, an inducer of intestinal MDR, and talinolol revealed that talinolol AUC decreased with a corresponding increase in intestinal MDR1 expression after long-term treatment [23]. The impact of MDR1-mediated efflux activity on drug absorption and intestinal DDI was observed in clinical studies with key prototype P-gp substrate digoxin, as the latter bioavailability is influenced by absorption mediated by intestinal P-gp only and not by first-pass metabolism. Studies of orally administered digoxin in the presence of quinidine or digoxin and rifampicin resulted in a dramatic enhancement in digoxin C_{max} and AUC [24,25]; in contrast, the treatment with the MDR1 inducer rifampicin decreased digoxin C_{max} and AUC in humans [26], as inverse correlation between intestinal MDR1 and AUC of digoxin was seen.

To conclude, the impact of efflux-mediated drug absorption of P-gp substrates can vary depending on the permeability and solubility of these drugs, for example:

1. Unlike the low-solubility low-permeability drugs, the in vivo intestinal absorption of highly soluble and highly permeable MDR1 substrates is not limited by P-gp efflux pump by the in vivo absorption dominated by their high permeability. In this case MDR1 plays a minimal role in the intestinal absorption as reported for verapamile by Cao et al. (2005) [27]. These drugs possess a relatively high-absorbed fraction and the dissolution in GI tract is not the rate-limiting step.
2. For high-solubility but low-permeability MDR1 substrates, MDR1 limits the intestinal absorption in the distal segments of the small intestine but plays a minimal role in the proximal intestinal segments because of significant lower MDR1 expression levels in this region [28].

It is important to note that MDR1 efflux activity does not always predict a compound's absorption profile. The magnitude of the effect of MDR1 efflux activity on a compound's absorption profile ultimately depends on the MDR1 activity/expression profile in combination with solubility, permeability, and metabolism.

Unlike efflux transporters, *uptake drug transporters,* known as solute carrier transporters (SLC), do not require ATP and transport the drugs according to their concentration gradient, thereby improving the intestinal absorption of a wide range of drugs. They are localized in the intestine at the apical surface of epithelial cells, and most major SLC transporters are organic anion transporter families (OATP subfamilies; gene SLCO), SLC peptide transporter family (PepT1; gene SLC15A1), and organic zwitterion/cation transporters (OCTNs; gene SLC22) [29]. The clinical significance of intestinal SLCOs and OCTNs in drug absorption is still under investigation. In contrast, the impact of PepT1 transporter on drug absorption is well defined and investigated. PepT1 is expressed primarily in the small intestine, particularly in the duodenum [30], and the substrates for proton-coupled peptide transporters are mainly cationic, anionic, or zwitterionic di- and tripeptides; the free amino acids and larger peptides are excluded and peptide bond is not a required structure for a substrate [31]. However, the transport function of PepT1 requires a proton gradient at the apical surface brush border membrane by the Na^+/H^+ exchanger of epithelial cells; the system is known as a proton-dependent cotransport system, and then the influx of protons back into the epithelial cells is coupled by PepT1 to transport its substrates [32,33]. Drugs transported by PepT1 are prodrugs (e.g., acyclovir and l-dopa [34,35]), β-lactam antibiotics (e.g., penicillins and cephalosporins [36,37]), angiotensin-converting enzyme (ACE) inhibitors (e.g., captopril) [38], and anticancer agents (e.g., bestatin [39]). In general, PepT1 has generally been characterized as a low affinity/high capacity transporter with a wide variety of compounds as substrates. The impact of PepT1 on oral drug absorption has been well established in recent years, especially with the intestinal absorption of β-lactam antibiotics [37]. The affinity of PepT1 to β-lactam antibiotics as substrates is good due to resemblance to the backbone of its physiologically occurring tripeptides.

One of major area that the human intestinal peptide transporter appears to target for increasing intestinal absorption of some small molecular weight drugs is the prodrug delivery. Because of its high capacity, broad substrate specificity, high expression in the intestinal epithelium, and low occurrence of functional polymorphisms [40,41], the intestinal peptide transporters have a significant impact on delivery of prodrugs. By using bonds that hydrolyze enzymatically in the preparation of PepT1-targeted prodrugs, it is possible to dramatically improve the systemic availability of poorly absorbed drugs, with limited systemic exposure to the intact prodrug. This general strategy of peptide transport associated with prodrug therapy [42] with valacyclovir is the most widely studied [43]. It is also used to deliver the prodrug LY544344, demonstrating the utility of PepT1-targeted non-ester prodrugs to overcome poor permeability and low bioavailability [44]. This compound exhibits near-ideal prodrug properties, with good solubility and chemical stability, extensive and reproducible absorption across species, low concentrations of circulating nontoxic prodrug, and pharmacokinetic linearity across a wide dose range [44].

1.1.2.2 The Impact of Intestinal Metabolism on Drug Absorption When a drug has been ingested, the first site capable of metabolism is the small intestine. Because both phase I metabolic enzymes (e.g., oxidative metabolic pathways) and phase II metabolic enzymes (conjugating metabolism pathways) are expressed in the intestine, metabolism in the small intestine can play an important role in the first-pass metabolism of drugs [45]. The intestinal metabolism, in animals or humans, has been extensively studied. Many difficulties have been encountered, leading sometimes to discordant results. These include (1) the low expression levels of intestinal metabolic enzymes relative to the liver; (2) intra- or interspecies variability of expression of biotransformation enzymes; (3) ethical and technical limitations for obtaining biological samples for translational studies in humans; (4) variability of sample preparation techniques; and (5) structural and functional heterogeneity of the intestine.

Several drugs such as cyclosporine [46], verapamil [47], and midazolam [48] undergo extensive intestinal first-pass metabolism, and in turn affect the intestinal bioavailability. The human intestine is divided into two parts: the small intestine, subdivided into duodenum, jejunum, and ileum; and the colon. These two parts differ in their histological structure and by their metabolism. The highest metabolic activity of the intestine is in the upper part of the small intestine, with a maximum observed at the proximal jejunum [49]. The total P450 content increases slightly between the duodenum and the jejunum, then decreases markedly at the ileum. If this heterogeneous distribution concerns phase I enzymes (i.e., CYP3A4, 2C9, or 2C19), phase II enzyme (i.e., GST, UDPGT) distribution is relatively homogeneous in small intestine but with a lower level of expression in the colon. P450 enzymes such as CYP3A and CYP2C9 were found to be the most intestinal P450 enzymes, accounting for 80% and 15% of the total immunoquantified P450s, respectively [50], but the expression of CYP3A4 in human donors varied along the length of the small intestine, decreasing from the duodenum to the distal ileum, and content is estimated to be $<1\%$ of that in the liver [50]. The variability in expression of phase I and II metabolic enzymes in small intestine among human donors has been observed more frequently compared to the liver [51], possibly due differences in diet and environment [52].

Many reports have been published to demonstrate the large degree of overlap between the broad substrate specificities of MDR1 and CYP3A4 [53], and their combined actions could contribute to the low oral bioavailability determined for many of their dual substrates. Transport studies of cyclosporin A across in vitro intestinal absorption model Caco-2 cell monolayers have shown that the actions of MDR1 and CYP3A4 may act coordinately to enhance the attenuation of apical to basolateral transport of this drug. It was observed that cyclosporin A metabolism was much greater when the compound was transported in the apical to basolateral (absorptive) direction than in the basolateral to apical (secretory) direction [54]. The rationale was that the reduction in the apical to basolateral flux of cyclosporin A, due to the P-gp efflux pump, increased the metabolism by CYP3A4 [54]. These findings that CYP3A metabolism increases with P-gp efflux have failed to be produced using PK models [55,56]. A physiologically based model showed that intestinal drug metabolism is influenced directly by drug concentrations within the enterocyte, which is governed by the degree of MDR1-mediated drug efflux and drug permeability through the enterocytes [57]. For drugs possessing high apparent permeability, increasing the efflux ratio (indicative of MDR1 substrate) predicts an increase in CYP3A drug clearance by up to 12-fold in the distal intestine.

For DDIs in intestine, as in liver, selective inhibition or induction of intestinal metabolizing enzymes either by dietary or environmental xenobiotics or by co-administered drugs has been identified as an effective method of drug interaction and a contributor to variability in oral drug bioavailability [58].

1.2 EFFECT OF PHYSIOCHEMICAL PROPERTY–RELATED FACTORS ON DRUG ABSORPTION

1.2.1 Lipophilicity, Solubility and Dissolution, and Permeability

While the physiochemical properties of drugs that are favorable and provide optimal biological activity with minimal safety liabilities have been evaluated and summarized by many medicinal chemists, the most popular is Lipinski's "rule of five." This rule is considered as the guide for drug discovery by almost all of the pharmaceutical industry. It supports the synthesis of new chemical entities [7]. The compounds with MW < 500, lipophilicity < 5 (calculated log P), total hydrogen bond acceptors < 10 (acceptors being O and N), and total hydrogen bond donors < 5 (donors being O-H and N-H groups) should have good oral absorption. It was proposed that when a compound violates two or more of these five rules, it is likely have poor intestinal absorption. Molecules within the "rule of 5" chemical space are usually absorbed through a passive process. When a drug is delivered orally, it transports from the GI fluid, primarily across the jejunum and the ileum segments of the small intestine into the portal blood system (Figures 1.1 and 1.2). This process involves transporting the drug from the GI fluid across the segments of the small intestine into the portal blood system. Before it reaches systemic circulation, it must pass through to one or more layers of membranes, either by passive/facilitated diffusion, transcellular or paracellular mechanism, active uptake or efflux, or endocytosis. Drugs cross the epithelial or mucosal cell walls, endothelial and capillary cell walls (bloodstream), and cellular plasma membranes to exert effect. These mechanisms will be discussed in more detail later in this chapter.

Ritchie et al. (2011) [59] have shown that increasing aromatic ring count influences several developability parameters such as lipophilicity and molecular weight and contributes to poor oral bioavailability. They also suggested that fused aromatic systems might have a beneficial effect relative to their nonfused counterparts. Their analysis revealed that newer drugs have increased heteroaromatic rings more than nonfused aromatic rings.

1.2.1.1 Lipophilicity *Lipophilicity* is the tendency of a compound to partition into a nonpolar lipid matrix versus an aqueous matrix. It is commonly determined by using log P from octanol/water partitioning. As has been mentioned earlier in this chapter,

TABLE 1.1 Log D Values and Impact in Drug Permeability and Oral Absorption.

Log D at pH 7.4	Impact on Drug Disposition
< 0	Poor absorption across GI tract and BBB. Potential rapid renal clearance
0–1	Balance between solubility and permeability, good oral absorption, but poor CNS permeability
1–3	Optimum oral absorption and CNS permeability. Low metabolism
3–5	Lower solubility, higher metabolic rate
> 5	Poor solubility and absorption across the GI tract, high metabolism

in order for compounds to permeate across biological membranes to be absorbed and become bioavailable, a certain balance of lipophilicity to hydrophilicity is required. The drug lipophilicity is inversely related to solubility; higher lipophilicity leads to lower aqueous solubility. However, the balance between these two properties is very critical for compounds to be absorbed from the GI tract, and the relationship between log D/log P and permeability across biological membranes is nonlinear. The permeation decreases in both the low and high end of log D values, and often a log D value of 5 is considered as an upper limit of desired lipophilicity, as indicated in Table 1.1. It has been recognized that the higher the log P, the lower the solubility [60], while neutral molecules have been shown to be more poorly soluble compared to ionizable molecules. When clog $P < 3$, the average solubility of neutral molecules approaches the average solubility of ionizable molecules. The same trend has been shown with molecular weight: as it increases, solubility decreases [60]. To generalize the relationship between log P value (lipophilicity) and solubility/permeability:

1. A drug with log P value of 0–3 is considered optimal for passive diffusion [61].
2. A log P value of < 1 suggests that a compound will have good solubility by being hydrophilic but will have poor permeability.
3. A drug with log P value > 3 indicates that a compound is highly lipophilic, may possess low solubility, and is subject to metabolism and/or biliary excretion.

Determination of the extent of ionization in discovery of new chemical entities (NCEs) can be valuable in the selection process for potential drug candidates. The ionization is determined by measuring the dissociation constant pK_a, which is indicative of compound capability to be ionized at various pH ranges, hence influences its solubility and absorption across the GI tract [62,63]. In humans, the pH of stomach is close to 2 and in the small intestine around 6, which can be affected by food.

Classically, ionization constant (pK_a) is expressed using Henderson–Hasselbalch equation, which can provide the extent of ionized versus unionized compound at a particular pH:

For acidic compounds: $pH = pK_a + \log$ ([ionized compound]/[unionized compound])

For basic compounds: $pH = pK_a + \log$ ([unionized compound]/[ionized compound])

To measure the pK_a, two methods have been used that employ cosolvents or surfactants. The cosolvent approach has been used successfully to solubilize unionized compounds. By mixing organic polar solvents like methanol, dioxin, or acetonitrile with water the solubility can be enhanced (MDM), though it was found that not all compounds would dissolve in cosolvent–water mixture. It is effective to dissolve the lipophilic compounds [62], and it can be used for compounds that are not soluble in methanol–water or other single organic cosolvent mixtures (e.g., 2-propanol, DMF, DMSO, and acetone). However, MDM also dissolves polar compounds, so it can be considered as an efficient cosolvent for pK_a determination in drug research. In an investigation by Ravichandran et al. (2011) [63], they argued that the cosolvent approach might result in erroneous pK_a determinations due to existence of two liquid boundaries, which may lead to variation in the ionization behavior of the NCE. These investigators used nonionic surfactants such as Tween 80, CremophoreEL, or Labrasol to determine pK_a values of poorly soluble compounds where the solubilization occurs by changing the ionic, hydrophobic, and amphiphilic molecule to micellar structure formed by the surfactant.

1.2.1.2 Solubility Solubility is a critical parameter for absorption because drugs have to be in solution to permeate the GI membrane, and solubility has long been recognized as a limiting factor in the absorption process. By definition, solubility is the extent to which molecules from a solid are removed from its surface by a solvent. Solubility of solid drugs in a solvent matrix reaches maximum concentration at equilibrium, and that can be optimized by modifying the structures, hence the physicochemical properties, dissolution rate, and the solvent matrix used. Aqueous solubility can be estimated by determining the ability of a drug to partition from lipid to aqueous environments, which is dependent on the ionization of the drug tested. Most drugs are weakly acidic or weakly basic compounds that cannot ionize completely in aqueous media and therfore only partly ionize. Since drug ionization is greatly dependent on the solvent pH, the above partition behavior is often considered as a function of solvent pH, and pK_a is often used as a parameter describing a compound's dissolution characteristic. In general, ionized drugs tend to exhibit far greater aqueous solubility than the unionized counterpart. As a result, the rate of solute dissolution in aqueous media can be markedly affected by the pH of the solvent. Introducing ionizable groups, reducing lipophilicity, introducing hydrogen bonding and polar groups, reducing MW, and introducing out-of-plane substitutions can improve the solubility of NCE, though the introduction of ionizable groups may impair the permeability. Interestingly, while neutral molecules and lager molecular weight drugs have been shown to be more poorly soluble compared to ionizable and small molecules, at clog $P < 3$, the average solubility of neutral and ionizable molecules, smaller or larger MW is similar [60].

As pH decreases, there is a higher solubility and in turn greater concentration of neutral molecules and lower concentration of anionic acid molecules. At basic pH it is the opposite. When drug is delivered in the GI tract and as the luminal pH changes along different sections from acidic to basic, and in the presence of food, solubility of acids and bases will vary in an opposite way, if ionization drives the solubility.

The contributions of medicinal chemistry to improve solubility, via introduction of ionizable, N-containing basic groups [64,65], or disruption of planar crystal structure, have been highlighted in various examples within drug discovery programs [66,67].

It has been estimated that up to 90% of current NCEs suffer from low solubility according to the Biopharmaceutics Classification System (BCS) [68]. Because limited solubility may compromise absorption and thus drug likeliness, it is important to assess the solubility and any potential issues as early as possible to avoid the risk of advancing of drug candidates in the development stage. Methods for measurement of solubility in the early discovery phase include kinetic and thermodynamic solubility. Kinetic solubility determination is carried out by spectrophotometry, turbidimetry, or nephelometry. Solubilities in simulated gastric fluid, simulated intestinal fluid (SIF), and fasting state SIF can be determined for selected compounds from the early discovery phase to assess its solubility in biological fluids ex vivo.

When assessing the solubility by determining the dissolution rate, which is defined as the rate at which the molecule dissolves into a solvent from a solid form, a molecule with a high dissolution rate will dissolve into solution quickly, leading to a quick absorption phase and increasing its chance to be absorbed within the GI transit time while its solubility remains constant. The dissolution rate depends on the particle size and compound physical and salt form. Reducing the particle size increases the surface area of the solid in contact with the solvent, which increases the dissolution rate.

The most frequent physical form in drug discovery is amorphous, the solid with no specific organization of molecules, unlike a crystal, which is a highly organized set of molecules. The amorphous form is often more soluble and less stable than the crystalline form. When an oral dose of poorly soluble amorphous compounds is delivered, there is potential that these compounds may precipitate in the GI tract to more stable and less soluble crystalline solid form, thus leading to lesser absorption. To increase the dissolution rate, a salt form can be developed. Salts can stay in solution in a supersaturated state and delay the compound's precipitation. However, salts of weak acid or base can precipitate because they will convert to the free acid or base, leading to reduced intestinal absorption. By introducing the use of formulations or excipients, improvement in the molecule's dissolution rate might be achieved.

1.2.1.3 Permeability As mentioned earlier, when a drug is delivered orally, it transports from the GI fluid across primarily the jejunum and the ileum segments of the small intestine into the portal blood system (Figure 1.1). This process involves transporting the drug across layers of lipid biolayer membranes, either by passive/facilitated diffusion, transcellular or paracellular mechanism, or active uptake or efflux, as shown in Figure 1.2, or by endocytosis. A drug has to cross the epithelial or mucosal cell walls, endothelial or capillary cell walls (bloodstream), and cellular plasma membranes to be absorbed, become bioavailable, and to be efficacious. Membrane permeability is not just critical for absorption of drugs across the biomembrane of the GI tract but also plays a very significant role in their distribution to all other tissues of the body (discussed in a later chapter). Most of the neutral lipophilic drugs enter a cell's lipid membrane by the transcellular passive diffusion route, but hydrophilic or charged drugs cross the intestinal epithelial cell membrane by paracellular passive diffusion through the tight junctions (TJs). Unlike the passive transcellular route, which takes place across most of the surface area of apical membrane of microvilli in the enterocytes, the paracellular route of absorption is limited, since the surface area of the TJ is about 0.01% of the small intestine surface area. However, the absorption route associated with the uptake and efflux functions of intestinal transporters is now considered as a very important mechanism of drug delivery. Those mentioned mechanisms of drug absorption are depicted in Figure 1.2 and will be discussed in more detail below.

The predominant mechanism of absorption for marketable drugs is passive diffusion [7], though other mechanisms can be involved when the physicochemical properties of drugs are beyond the "rule of 5." It has been established that transport of drugs by passive diffusion requires no energy but is driven by concentration gradient and follows Fick's law of diffusion:

$$\frac{\mathrm{dQ}}{\mathrm{dt}} = \frac{\mathrm{DAK}}{\mathrm{h}}(\mathrm{C_{GI}} - \mathrm{C_p}), \tag{1.3}$$

where $\mathrm{dQ/dt}$ = rate of diffusion, D = diffusion coefficient, K = lipid – water partition coefficient of drug in the biologic membrane, A = surface area of the membrane, h = membrane thickness, $\mathrm{C_{GI}}$ = drug concentration in the GI tract, and $\mathrm{C_p}$ = drug concentration in the plasma.

In explaining the law and its use in assessing drug permeability, these parameters are explained in more detail:

1. The term $(\mathrm{C_{GI}} - \mathrm{C_p})$ represents the difference between free drug concentrations in the lumen and free drug concentrations in the plasma. There is a high degree of drug dilution due to high GI blood flow rate after permeation of drug into the GI membranes and the relatively high drug dose given orally (in the milligram range), which thus creates a large concentration gradient between the intestinal lumen and the bloodstream.
2. K which represents the lipid–water partitioning coefficient of a drug across the theoretical GI membrane; lipophilicity (log P) drug property can significantly impact this parameter.
3. A represents the surface area of the GI membrane accessible to the drug; the larger the surface area of the GI tract, the faster the drug can permeate. The duodenum is involved most in drug permeation because it has the largest surface area in the GI tract.
4. h represents the thickness of the theoretical GI membrane and the assumption it is constant across the GI tract.
5. D represents the amount of a drug that diffuses across a membrane for a given unit area when the concentration gradient is unity.

It is worth mentioning that K, A, h, and D are constant for a given molecule with a given oral formulation and define the permeability *p* of a drug in Equation (1.4):

$$\mathrm{p} = \frac{\mathrm{DAK}}{\mathrm{h}}. \tag{1.4}$$

In addition, since the free plasma concentration is extremely low, C_p is considered negligible. Therefore, Equation (1.5) can be a simple calculation of Fick's law and used to determine the passive drug permeation through the GI membrane, which is a first-order process:

$$\frac{dQ}{dt} = P(C_{GI}). \tag{1.5}$$

Since a majority of molecules permeating through the membrane are in a neutral form, accordingly, pH of the lumen and pK_a of the molecules impact the degree of gradient concentration. As an example, without taking blood flow into account, the amount of neutral form of an acidic compound in the duodenum lumen is much higher than that in the plasma, and that further drives the gradient concentration in the direction toward greater intestinal drug permeation. However, a formal electrical charge can be highly delocalized and therefore be less of a barrier than believed, especially when lipophilicity is sufficiently high.

For the other passive diffusion mechanism through the TJs, paracellular permeation is a less frequent mechanism of intestinal absorption. TJs or zonula occludens constitute the major rate-limiting barrier toward hydrophilic drugs that are transported by paracellular mechanism. The dimensions of the paracellular space are between 10 and 50 Å, indicating the exclusion of any particles with a molecular radius exceeding 15 Å (~3.5 kDa).

For the paracellular passive diffusion mechanism, transepithelial electrical resistance (TEER) tightens the intercellular junctional complex. There are gradients in TEER values across the GI regions, less tight in duodenal than colon. When TEER data are corrected for differences in mucosal surface area (see Figure 1.3), the permeability of small intestinal and colonic epithelium is determined to be virtually identical [69,70]. However, paracellular absorption is more likely to occur in the small intestine, not due to the more leaky TJs but because of a larger mucosal surface area. Endocytosis is a constitutive process observed in most mammalian cells for the uptake of macromolecules. It requires metabolic energy and it is a slow uptake mechanism resulting in a fusion of endocytic vesicles with lysosomes containing high levels of enzymatic activity. Endocytosis may involve specific receptors, for example, vitamin B12 receptor [71]. Endocytosis of compounds, like leptin, is believed to be limited in the small intestine, and in general is not a significant mechanism for drug absorption in the intestine.

1.3 EFFECT OF GI-PHYSIOLOGICAL FACTORS AND PATIENT CONDITION ON DRUG ABSORPTION

1.3.1 Effect of pH, Intestinal Surface Area, Gastric Emptying, Transient Time, and Bile Acid

1.3.1.1 Effect of pH and Surface Area After oral dosing, the compound first encounters the buccal mucosa, where it can be absorbed, though the absorption at the buccal mucosa is negligible. The most important nonintestinal absorption site is

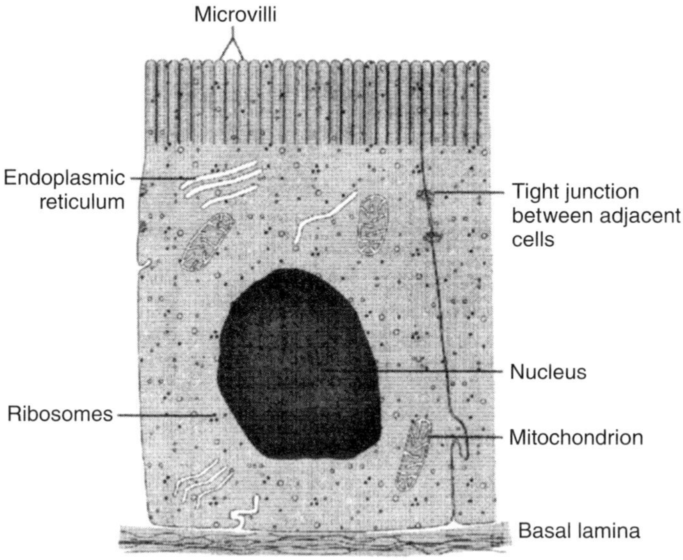

Figure 1.3 Intestinal epithelial cell (enterocyte) indicating the tight junctions and the apical surface with microvilli at the lumen givinb the appearance of a brush border. As seen, the presence of microvilli greatly increases the available absorptive surface area for the enterocyte. Adapted from Ref [72] with permission.

the stomach, which can take up nonionized, lipophilic molecules of moderate size. Compared to that of the intestines, gastric absorption is limited by the comparatively small epithelial surface area, relatively large volume, and brief amount of time that substances are in contact with the stomach epithelium. After ingestion via the esophagus, the drug arrives at the first region of the GI tract, the stomach. In the stomach, the drug is mixed with gastric acids, pepsinogen, and mucus secretions. The absorption in stomach is limited for most drugs due to the relatively small surface area ($< 0.1\ m^2$), the low blood flow perfusion rate (150 mL/min), and the rapid gastric emptying time (0.5–1 h). Although the acidic pH in the stomach can facilitate the absorption of acidic compounds in the stomach, the absorption of acidic compounds is faster in the small intestine.

The small intestine consists of three consecutive sections: duodenum, jejunum, and ileum. The pH of each section as shown in Table 1.2 increases gradually, creating a gradient from the stomach to the ileum. As has been mentioned before, the absorption in the small intestine is greater because of the larger surface area ($\sim 200\ m^2$), high blood flow perfusion rate (1 L/min), and lengthy transit time (2–4 h). To manage the intestinal drug absorption, the gastric emptying time can be the step to control the speed of drug absorption.

In the second part of the duodenum, the bile that is secreted from the gallbladder into the GI tract gets mixed with the digested drug. Bile, with its detergent-like properties, facilitates the solubilization and chemical breakdown of lipids, which explains why bile is secreted in the presence of lipids in the duodenum. The presence

TABLE 1.2 Physiological Parameters of GI Tract Regions.

Region of GI Tract	Length (cm)	Range of pH Values
Stomach	25	1.5–5.0
Small intestine:		
Duodenum	25	5.0–7.0
Jejunum	260	6.0–7.0
Ileum	395	7.0–7.4
Large intestine:		
Cecum	7	5.7–5.9
Colon	93	5.5–7.5
Rectum	55	6.5–7.0

of peptides and amino acids in the duodenum activates the secretion of pancreatic enzymes: amylases, lipases, and proteases. The pancreatic enzymes can hydrolyze some molecules that contain hydrolyzable functional groups, resulting in the deactivation or activation of the drug in the GI tract. In addition, the pancreas secretes bicarbonates ion, which neutralizes acid stomach secretions. At the ileocecal junction, the small intestine connects with the large intestine. Because of the modest surface area of the human large intestine ($0.35 m^2$, which is less than 0.15% that of the small intestine), the large intestine tends to play a minor role in absorption compared with the small intestine. Although the large intestine transit time is long (7–20 h), drug absorption is still limited mainly because of its small surface area, as mentioned, and its lack of villi.

The pH of the luminal contents (as seen in Table 1.2) is the lowest in the stomach and increases as the chyme progresses distally through the GI tract, approaching neutrality. This modification in pH as the chyme travels through the GI tract is accomplished through the secretion of various acidic and alkaline fluids. In most regions of the digestive tract, the secretions are slightly alkaline, and the luminal contents exhibit a pH of 7–8 as chyme approaches the distal small intestine. The stomach is the lone exception to this general statement. The secretion of acid by the gastric mucosa results in acidification of chyme. The pH of chyme affects the ionization state of certain molecules and, therefore, can affect absorption. The fasting versus fed, as will be explained later, can alter the intestinal pH and thus the rate of drug absorption. In addition to the pH gradient that exists along the linear axis of the GI tract, a pH gradient also exists from the center of the lumen moving radially toward the epithelial surface. The pH of the lumen is more acidic than the pH of contents at the epithelial surface, a function of the unstirred layer of water with the brush border [72] and the alkaline secretions of the intestinal epithelia [73]. This gradient may also affect the rate of uptake of endogenous and exogenous chemicals.

Many factors are involved in oral drug delivery, and the oral bioavailability of a particular drug can be a reflection of several components related to its delivery to the intestine (e.g, relative surface area, gastric emptying, pH, food). Because the surface area is a dominant factor in the passive diffusion permeability mechanism, the rate and extent of absorption is increased by expansion of the absorptive surface area.

The luminal content is also changed through the progressive absorption of fluids, electrolytes, nutrients, and xenobiotics as chyme moves through the GI tract. It should be noted that of the 8–9 L of fluid that enters the human upper digestive tract each day (~1.5 L of ingested fluid plus ~7 L of secreted digestive juice), only ~1 L enters the large intestine and only about 100 mL of water is found in the daily output of feces [72]. Surface area can be altered by physiological changes that occur in response to nutritional challenges, for example, during times of starvation; in disease states, as in diabetes (discussed later); during normal physiologic stress in pregnancy; and during human growth, though the villi of the small intestine decrease in height during old age, and that will reduce the extent of absorption.

1.3.1.2 Effect of Gastric Emptying and Intestinal Transit Time In addition to the influence of pH and surface area, there are other important parameters in initiating and enhancing the extent of drug absorption. Gastric emptying rate is an important factor because it affects the plasma concentration profile (AUC, T_{max} and C_{max}) of orally administered drugs. The significant effect of gastric emptying has been reported, using celiprolol as an example [74]. The rate and the extent of celiprolol absorption and the influence of gastric emptying on the occurrence of double peaks were demonstrated. It has been revealed that variability in gastric emptying rates due to the motility cycle can account for plasma level double peaks [75]. Furthermore, variable gastric emptying rates combined with the short plasma elimination half-life and poor gastric absorption of cimetidine cause the observed plasma level double peaks. The effect of altered gastric emptying and GI motility on the absorption of metformin in healthy subjects has been reported [76]. The results indicated that $AUC_{(0,\ infinity)}$ and percent dose excreted unchanged in urine increase with the increase in gastric emptying time and small intestinal transit times; that is, the extent of metformin absorption is improved when the GI motility is slowed. By using previously developed GI-Transit-Absorption Model, the prediction method for the plasma concentration-time profile of N-methyltyramine (NMT) was achieved [77]. By estimating the permeability of NMT at each GI segment, it is revealed that NMT is absorbed mainly from the small intestine and that permeability is at the highest level in the duodenum and jejunum. However, the contribution of these regions to the total absorption in vivo is found to be small. The substantial absorption sites in vivo were suggested to be the regions from lower jejunum to lower ileum, which have a longer residence time than duodenum and upper jejunum, thus the substantial absorption is a function of longer residence time.

Furthermore, the intestinal transit rate significantly influences the drug absorption and can make a difference in absorbability for some drugs because it determines the residence time of the drug in the absorption site.

1.3.1.3 Effect of Bile and Bile Salts Humans secrete from the gallbladder 2–22 mL of bile per kilogram body weight each day [78], at a rate of 3–45 mmol/L. Bile acids are an extensive group of molecules that share a structural similarity to cholesterol but exhibit differences with regard to substituent side groups. Bile acid in humans is moderately lipophilic and is a complex mixture of organic and inorganic materials, of which the various bile acids are the major component, composed of cholic acid, deoxycholic acid, and chenodeoxycholic acid.

Bile secreted into the GI tract may improve the bioavailability of poorly water-soluble drugs by enhancing their rate of dissolution and/or solubility. Bile salts can increase drug solubility via micellar solubilization. The increase in the rate of dissolution also may occur via a decrease in the interfacial energy barrier between solid drug and the dissolution media (via enhanced wetting), leading to an effective increase in surface area [79]. A study to examine the dissolution of low solubility–high permeability compared to high solubility–low permeability class of drugs provided results that in general low solubility–high permeability drugs depend much more on the medium, including the presence of bile salt, than high solubility–low permeability drugs [80]. The ability of sodium taurocholate (one of the bile salts secreted in the GI tract) to increase the initial dissolution rate of five steroids (hydrocortisone, triamcinolone, betamethasone, dexamethasone, and danazol) was tested [81]. The result showed that at bile salt concentrations representative of the fasted state, the wetting effects predominated over solubilization effects for all compounds [81]. At the higher bile salt concentrations, typical of the fed state, the increase in solubility was the predominant factor for the more lipophilic danazol. Furthermore, the extent to which bile salts can enhance the solubility of a drug can be predicted based on the physicochemical properties of the compound. The increase in solubility as a function of bile salt concentration can be estimated based on the partition coefficient and aqueous solubility of the compound [82].

1.3.2 Impact of Age and Disease State on Drug Absorption

1.3.2.1 Drug Absorption in Pediatric Populations Immaturity of the sphincter may lead to increased reflux of the stomach contents, which may contain drugs; drug absorption may thus be reduced [83]. Maturation of an effective antireflux barrier is not achieved until ~3 months postnatal [84]. It has been mentioned that the limited absorption from the stomach can be affected by gastric pH, which is neutral at birth but falls to $\sim 1-3$ within the 24 h after birth, followed by a gradual return to neutrality by day 8 [85] and slow decline again thereafter (e.g., $\text{pH} = 2-3$ by the age of 2–3 years) to reach adult values. At age 3 years, the amount of gastric acid excreted per kilogram body weight is similar to that excreted in adults [86]. Consequently, younger children, because of the reduced gastric acid production and secretion, have low absorption of drugs that need gastric acid for dissolution/absorption. Acid-labile drugs, such as penicillin G, ampicillin, amoxicillin, flucloxacillin, and erythromycin, are more efficiently absorbed when orally administered in the neonate and infant than in the adult [87–89]. However, such changes are unlikely to affect the absorption of non-acid-labile drugs, for which absorption will continue efficiently in the small intestine.

Gastric emptying (rate of removal of a drug from the stomach) is delayed in the neonate and infant [85,90], and emptying times of 6–8 h have been reported in neonates. This may result in delayed absorption of orally administered drugs. The T_{max} is delayed and the C_{max} is also likely to be lower, although in most cases the AUC will not be affected. The exact age at which gastric emptying time approaches adult levels is unclear.

Intestinal transit time appears to be shorter in young children, thus suggesting that sustained-release products may demonstrate incomplete absorption [91,92]. The unreliability of the sustained-release mechanisms seems to be more marked in children than in adults [93]. Therefore, extrapolation of adult data to children is particularly delicate for sustained-release formulations, as has been extensively demonstrated for theophylline [94–97].

Immaturity of secretion and activity of bile and pancreatic fluid leads to impaired fat digestion in neonates and infants in the first few months. The absorption of fat-soluble vitamins (vitamins D and E) is reduced in neonates probably because of the inadequate bile salt pool in the ileum [98]. Impaired fat digestion could be of toxicological importance when lipophilic compounds such as dichlorodiphenyltrichloroethane and structurally similar compounds are ingested, leading to a reduced uptake of these compounds in neonates and young infants. After a few months, the infant is capable of efficiently absorbing fat-soluble compounds due to a postnatal maturation of bile salt metabolism [99,100]. Ethnic/racial dietary differences can be a source of confounding for adult–pediatric comparisons pertaining to absorption. Moreover, many pediatric formulations are liquid or suspensions, as opposed to their adult-equivalent solid dosage forms, which also constitutes a source of confounding for adult–pediatric comparisons pertaining to absorption and bioavailability.

Intestinal expression of drug transporter-mediated absorption of substrates that are administrated in children was recently reported [101]. MDR1, MRP2, and OATP2B1 was determined in surgical removed small bowel samples (neonates n = 15, infants n = 3, adults n = 14). In adults, drug transporters are recognized as key determinants of variation in the pharmacokinetics of many drugs, as shown in studies using primary cell and ex vivo organ cultures, as well as clinical studies. In contrast, such data in children are scarce, and clinical studies are absent (Yanni et al. (2011) [102]). Neonatal intestinal expressions of MDR1 and MRP2 were comparable to those in adults, while intestinal OATP2B1 expression in neonates was significantly higher than in adults. This suggests that drug absorption mediated by these transporters may be subject to age-related variation in a transporter-dependent pattern.

1.3.2.2 Drug Absorption in Disease State Diabetes and its effect on altering the absorption of drugs in patients is discussed in this section. One factor that has an impact on drug absorption is drug transporters, specifically, as mentioned in this chapter, intestinal P-gp, which acts as the first barrier for drugs administered via the oral route [103]. Indeed, intestinal P-gp is likely to play a critical role in the extent of absorption of its orally delivered substrate drugs.

As has been reported, the expression and/or drug efflux activity of P-gp can be modulated not only by inherited factors, such as genetic polymorphisms [104], but also by environmental factors, such as diet [105], drugs [106], or diseases [107] like epilepsy, seizure, and diabetes.

Among these diseases, the prevalence of diabetes—classified as either type 1, characterized by dysfunction of pancreatic β cells, leading to insulin depletion, or type 2, characterized by dysfunction of the insulin receptor (insulin resistance) or by impairment in insulin secretion—is estimated as 9.2–9.8% of the adult population worldwide [108]. In addition to the modulation of P-gp in intestine, the expression and functional activity of P-gp was found to be altered in the brain, liver, and kidney under diabetic conditions [109].

Several antidiabetic drugs, such as glibenclamide [110], rosiglitazone, metformin [111], repaglinide [112], and linagliptin [113], have been found to be substrates for P-gp. Changes in the expression of intestinal P-gp due to the disease may have a critical role in the absorption process of these drugs, though few studies have been conducted to address the changes in intestinal P-gp on the PK/PD of antidiabetic drugs under diabetic conditions.

In the intestine, there is insufficient information about the features of PK/PD after oral administration of substrate drugs for P-gp. However, to design appropriate or individually targeted pharmacotherapy, it has been recommended [114] to consider the influence of changes in the expression and function of P-gp in the intestine under diabetic conditions.

Other conditions that will have an impact on drug absorption are starvation, physiologic stress in pregnancy, and during human growth decrease in absorptive surface area and, consequently, reduced efficiency in the absorption of nutrients and other materials from chyme.

1.4 EFFECT OF FOOD AND FORMULATION ON DRUG ABSORPTION

1.4.1 Effect of Food

The impact of food on intestinal absorption through bioavailability is extremely complex. Food delays gastric emptying, stimulates bile flow, changes pH of GI segments, alters the luminal metabolism, or interacts with prescribed drugs [115]. Clearly, the extent of the food's effect is variable and can be dependent on the meal content of protein, carbohydrate, or fat; volume; and fluid ingestion. Also, the effect of food can be based on the physicochemical properties of the compounds.

Food also increases blood flow to the liver, thus may change the first-pass effect and bioavailability between the fed and fasted state. The pH differences in the contents of the upper GI tract between fed and fasted states, as seen in Table 1.3, can influence the dissolution and absorption of weakly acidic and basic drugs. Elevation of gastric pH following a meal may enhance the dissolution of a weak acid in the stomach but inhibit that of a weak base. Furthermore, food inhibits the rate of gastric emptying, and prolonged retention in the stomach may increase the proportion of drug that dissolves prior to passage into the small intestine, which is the primary site of drug absorption [116]. Elevated gastric pH may afford enhanced bioavailability of acid-labile drugs such as penicillin, erythromycin, and digoxin. For ionic drugs, changing pH values can alter the fraction absorbed and thus the

TABLE 1.3 pH of the Gastrointestinal tract under Fed vs. Fasted State.

GI Tract Section	pH, Fasted	pH, Fed
Stomach	1.4–2.1	3–7
Duodenum	4.4–6.6	5.2–6.2
Jejunum	4.4–6.6	5.2–6.2
Ileum	6.8–8	6.8–8
Colon	5.5–7	5.5–7

permeability. Changing the pH can affect the dissolution of some formulations and can vary excipients–drug release.

Food has a significant effect, as it can cause drug–food interactions. As an example, if a drug chelates with ions present in the ingested meal, drug dissolution and/or absorption may be reduced. Furthermore, the meal itself may pose a physical barrier that prevents drug diffusion to the site of absorption, resulting in decreased bioavailability. Also, drug instability, as a result of acid degradation, may be exacerbated by prolonged gastric residence after food ingestion. For highly lipophylic drugs or large MW macromolecules, lymphatic uptake can be increased by the presence of a high-fat meal, thereby lowering plasma drug levels.

In general, food effect is not always predictable; however, its effect on bioavailability is the greatest when the drug is administered shortly after a meal is ingested and when meals are high in total calorie and fat content. Accordingly, the Food and Drug Administration (FDA) recommends the use of high-calorie and high-fat meals to study the effect of food on the bioavailability and bioequivalence of drugs (hamburger meal). Furthermore, qualitative prediction of food effect is often possible based solely on the BCS class of the drug, but food effect whether negative, positive, or no effect was classified for 80% of a set of 92 drugs, based simply on their dose, solubility, and permeability [117].

1.4.2 Formulation Effect

Formulation is an additive component of the dosed drug that facilitates or enhances the absorption and hence the bioavailability of the drug by ensuring drug dissolution/solubility. There are several variables that affect the dissolution rate as described by the Noyes–Whitney equation [118], as it shows the dissolution rate as a function of diffusion coefficient, surface area of particle, the thickness of the dissolution film adjacent to the dissolving surface, the saturation solubility of the drug molecule, the concentration of the dissolved solute, and the volume of the dissolution medium. Drug formulation can control the surface area and solubility but not the other variables. Clearly the most important factor achieved by formulation is the change in surface area. Increasing the surface area of a drug particle can enhance the dissolution rate of the drug. Using wetting agents that lower the surface tension of the dissolution medium and that reduce drug particle size thus increase the effective surface area for dissolution. However, due to the amount of the above surface-active agents needed to enhance in vivo drug dissolution rate, these agents are not generally used in product formulations [118]. The solubility of weak acid and weak base can be modified by using a buffer agent that slightly changes the surrounding pH. Besides buffering agents, some excipients are known to have effects on physiological conditions, such as decreasing GI transit time [119], affecting membrane permeability [120], and inhibiting efflux pumps [121].

1.4.3 The BCS in Relation to Intestinal Absorption

The BCS has been one of the most significant prognostic tools created to promote product development in recent years [122]. It is a scientific framework for classifying drug substances based on their aqueous solubility and intestinal permeability characteristics, which will substantially facilitate the drug product selection and approval process for a large group of drug candidates. The goal of the BCS is to function as a tool for developing in vitro dissolution specifications for drug products that are predictive of their in vivo performance [123]. According to the BCS, drug substances are classified as follows:

Class 1: high solubility–high permeability: generally very well-absorbed compounds. Effect of drug transporters is minimum

Class 2: low solubility–high permeability: exhibit dissolution rate-limited absorption. Effect of efflux transporters is predominante

Class 3: high solubility–low permeability: exhibit permeability rate-limited absorption. Effect of absorptive transporters is predominante

Class 4: low solubility–low permeability: very poor oral bioavailability. Effect of absorptive and efflux transporters could be both important

There some limitations in these classified system, such as:

1. A drug substance is considered *highly soluble* when the highest dose strength is soluble in ≤ 250 ml water over a pH range of 1–7.5.
2. A drug substance is considered *highly permeable* when the extent of absorption in humans is determined to be $\geq 90\%$ of an administered dose, based on mass balance or in comparison to an intravenous reference dose.
3. A drug product is considered to be *rapidly dissolving* when $\geq 85\%$ of the labeled amount of drug substance dissolves within 30 minutes using USP Apparatus 1 or 2 in a volume of ≤ 900 ml buffer solution.

The pH-solubility profile of the test drug substance should be determined in aqueous media with a pH in the range of 1–7.5 using the traditional shake-flask method as well as acid or base titration methods. A sufficient number of pH conditions should be evaluated to accurately define the pH-solubility profile. Concentration of the drug substance in selected buffers (or pH conditions) should be determined using a validated stability-indicating assay that can distinguish the drug substance from its degradation products.

The permeability class of a drug substance can be determined in human subjects using mass balance, absolute bioavailability, or intestinal perfusion approaches:

1. Pharmacokinetic studies in humans: mass balance studies or absolute bioavailability studies.
2. Intestinal permeability methods: the following are the methods that can be used to determine the permeability of a drug substance from the GI tract: (a) in vivo intestinal perfusion studies in humans; (b) in vivo or in situ intestinal perfusion studies using suitable animal models; (c) in vitro permeation studies using excised human or animal intestinal tissues; or (d) in vitro permeation studies across a monolayer of cultured epithelial cells.
3. Instability in the GI tract: determining the extent of absorption in humans based on mass balance studies using total radioactivity in urine does not take into consideration the extent of degradation of a drug in the GI fluid prior to intestinal membrane permeation.

For immediate-release formulations of Class I drugs, during the dissolution tests, verify that the drug is rapidly released from the dosage form under mild aqueous conditions.

For Class II drugs, to establish a strong correlation between the results of dissolution tests and the in vivo absorption rate, it is essential to reproduce the conditions existing in the GI tract following administration of the dosage form. Adequate comparison of formulations for Class II drugs requires dissolution tests with multiple sampling times in order to characterize the release profile. Use of more than one dissolution medium might be applied.

Class III drugs are defined as being rapidly dissolved as Class I drugs, then the formulation can be used to release the drug under mild aqueous conditions within a predetermined time. The duration of the dissolution test should be at least as stringent for Class III drugs. For Class I drugs, the contact time between the dissolved drug and intestinal epithelia needs to be maximized, thus increasing the bioavailability.

As for Class IV drugs, which are generally considered poorly absorbed, special attention should be given to the formulation to avoid further deterioration in the rate and extent of drug absorption caused by poor formulation.

The BCS is a scientific framework for classifying drug substances based on their aqueous solubility and intestinal permeability and it takes into account three major factors that govern the rate and extent of drug absorption from immediate-release solid oral dosage forms: dissolution, solubility, and intestinal permeability. The BCS classification can be used to request a waiver from the FDA for in vivo bioavailability and/or bioequivalence studies for immediate-release solid oral dosage forms.

1.5 TRANSLATIONAL APPROACHES TO DETERMINE DRUG ABSORPTION IN CLINICAL STUDIES

1.5.1 Cellular Intestinal Model

The use of epithelial cell cultures for studies that predict the extent of oral drug absorption and mechanism of drug transport has been widely introduced in drug discovery and development programs within the pharmaceutical industry. Cell lines originated from human tumor, which possesses properties and most of the morphological and functional characteristics of enterocytes, were used since the use of primary intestinal epithelial cells is limited due to the rapid loss of their differentiated characteristics during culture. Most commonly used cell lines are Caco-2, HT-29, and T84 cells. Other cell lines, such as HCT8, HRT18, or SW1116, express only a partially intestinal differentiated phenotype [124]. Although these cell lines have some recognized limitations, the simplicity of their use as in vitro models provides a clear advantage over their disadvantages. Caco-2 cells, which derived from a human colorectal adenocarcinoma, express the majority of the morphological and functional characteristics of small intestinal absorptive cells, including the spontaneous formation of polarized membranes, formation of TJs, expression of drug transporters, and phase I and phase II drug-metabolizing enzymes. Upon cultivation as monolayers on semipermeable membrane of Transwell format for 21 days, the differentiated Caco-2 cells form the apical compartment and basolateral compartment, which correspond to the intestinal lumen side and the serosal side, respectively. However, because Caco-2 cells are originated from colon and tumors that do not resemble the in vivo physiological environment of the small intestine, extrapolation of the data to the in vivo situation may be difficult.

The typical experimental setup for a Caco-2 permeability experiment is to measure the apparent rate of permeability (P_{app}) of the compound in both an apical to basolateral (A to B) and a basolateral to apical (B to A) direction after incubation at 37°C for a given period of time. The relative rates of permeability from A to B (influx) and B to A (efflux) directions and ratio (B to A/A to B) can be indicative of the mechanism of permeability. This is a well-established technique in academia and industry for high-quality permeability measurements, but the culture of the cell line is expensive and manually intensive when used as a screen for permeability [125,126].

On the other hand, MDCK (Madin–Darby canine kidney) is a dog kidney cell line that has also been applied to permeability measurements [127–129]. Although this cell line requires 3 days of culture compared to 15–21 days for most Caco-2 models, it is still relatively expensive and requires cell culture expertise. It is also less physiologically relevant, as this is a canine kidney cell line and generally the permeability measured is used to estimate intestinal absorption in humans.

1.5.2 In Vitro Artificial Membrane

In order to meet the need for higher throughput and less cost to measure the permeability, the parallel artificial membrane permeability assay (PAMPA) is utilized.

The artificial membranes are designed to mimic the GI membranes [130–133]. The amount of compound that moves from the donor chamber to the acceptor chamber after a period of incubation is measured, along with the amount left in the donor chamber. These measurements and other parameters characteristic of the experimental setup are used to determine the passive diffusion permeability, P_e, and the retention factor, R_f [132]. PAMPA is performed in a 96-well plate and suitable for automation setting; accordingly PAMPA has been used as a primary screen for measuring the intrinsic permeability for thousands of NCEs per week prior to selecting leads to be tested in the cell model assay [130,133]. Because PAMPA does not provide information about whether a compound is a substrate for active influx or efflux transporters, further mechanistic information could be determined using the Caco-2 cell line. Also, PAMPA does aid the chemist, as the output can be a simple measure of permeability that allows for clearer SARs and easier interpretation when attempting to design a more permeable series.

As the FDA guideline recommended, the process to assess drug absorption using in vitro tools is as follows:

1. Examine bidirectional transport of the test compounds in Caco-2 or MDR1-MDCK cell monolayers.
2. Select likely substrates based on an efflux ratio > 2.
3. Confirm P-gp substrate activity using specific inhibitors, determining K_i or $[I]/IC_{50}$ of the test compounds.
4. Select compounds for in vivo interaction studies with a P-gp substrate such as digoxin if efflux ratio > 2 and in step 3, compounds possesses K_i or $[I]/IC_{50} > 0.1$.

1.5.3 Non–In Vitro Models: In Situ and In Vivo

To predict the human absorption in more accurate fashion before first injection in humans (FIH), studies using in situ models and in vivo models were applied. Although in situ models are seldom used in a drug discovery setting, in situ single-pass perfusion of the rat intestine (sometimes mice model) can provide useful mechanistic information. The drug concentration, usually radiolabeled, in the intestine is known and controlled and the subsequent barriers that a compound has to cross to reach the portal blood circulation are identical in the in situ and in vivo situation. The model dimension certainly offers reliable data and justifies the use of an in situ model despite the technical challenge in its setting, labor intensity, and low throughput.

The in vivo pharmacokinetic studies to give a sense of the absorption profile of their molecules are always the most applicable approach. These studies depend on using radiolabeled drugs and including as one element the preclinical mass balance study. Rat is normally used to determine the fraction absorbed and the bioavailability, and hence to understand any obstacles that may hinder these parameters in humans. By comparing rat absorption with human absorption for approximately a hundred drugs, a similarity of percentage of dose absorbed was presented [134]. Consequently, the authors recommended that the in vivo absorption information in rats could be used as a translational method to predict the extent of intestinal absorption in humans. In addition, by finding a moderate correlation in the expression of all known transporters in both human and rat duodenum, the mechanism of human absorption can be determined from rat data. However, there was no correlation in the expressions of drug-metabolizing enzymes, indicating a disconnection in the bioavailability between rat and human [135]. It is worth mentioning that although dog was commonly used as a nonrodent preclinical species for drug discovery and development in humans, there was poor correlation between fraction absorbed based on a retrospective study in human and dog for approximately 40 drugs [136].

Drug development studies that determine drug absorption by using radiolabeled drug are generally conducted by introducing radiolabel of ^{3}H or ^{14}C. The drug is administered orally (PO) and intravenously (IV) to intact or bile duct–cannulated animals or intact humans. The urine, bile (in animals only), plasma, and feces are collected until a full recovery of radioactivity has occurred. The fraction absorbed (f_a) can then be assessed using several approaches.

1. Calculating the percentage of cumulative excretion of radioactive drug-related material in urine and bile from bile-duct animals following oral administration:

$$f_a = \frac{\%\ \text{total dose excreted in urine and bile}}{100}$$

2. Measuring the amount of unchanged drug in feces after oral administration from intact animals:

$$f_a = \frac{\%\ \text{dose of parent drug excreted in feces}}{100}$$

3. Comparing the amounts of total radioactivity excreted in the urine after PO and IV dose:

$$f_a = \frac{\%\ \text{dose excreted in urine after PO dose}}{\%\ \text{dose excreted in urine after IV dose}}$$

4. Comparing the exposure (AUC) of total radioactivity after PO dose and IV dose:

$$f_a = \frac{\text{AUC total radioactivity after PO dose}}{\text{AUC total radioactivity after IV dose}} \times \frac{\text{Dose IV}}{\text{Dose PO}}$$

By employing these approaches, the fraction absorbed and mechanisms of excretion for several drugs were assessed [137–141].

In a more recent investigation, Augustijns et al. (2014) [142] reported the use of simulated media as an attractive approach for the prediction of oral absorption, especially for drugs that have poor aqueous solubility. Solubility assessment in human intestinal fluid (HIF) can be considered as the most optimum approach, but unfortunately, HIF is not commercially available and its collection is impractical as well as requires institutional review board approval because the use of biological samples in an industry setting is regulated by strict legality rules. The use of simulated media was found to be a practical approach and found to mimic the HIF. Augustijns [142] and his colleagues demonstrated a correlation between simulating intestinal fluid for fasting and fed state (FASSIF and FESSIF, respectively) with HIF for fasting and fed state (FaHIF and FeHIF, respectively) for hundreds of drugs as a simple but effective way of getting adequately accurate estimates of intestinal solubility.

REFERENCES

1. Collier R. Drug development cost estimates hard to swallow. Can Med Assoc J 2009; 180:279–280.
2. Schuster D, Laggner C, Langer T. Why drugs fail—a study on side effects in new chemical entities. Curr Pharm Des 2005;11:3545–3559.
3. Arrowsmith J. Trial watch: phase III and submission failures: 2007–2010. Nat Rev Drug Discov 2011;10:87.
4. Arrowsmith J. Trial watch: phase II failures: 2008–2010. Nat Rev Drug Discov 2011;10:328–329.
5. Rowland M, Tozer TN. Clinical pharmacokinetics: concepts and applications, 3rd rev. ed. Media, PA: Lippincott Williams and Wilkins, 1995.
6. Varma MV, Obach RS, Rotter C, et al. Physicochemical space for optimum oral bioavailability: contribution of human intestinal absorption and first-pass elimination. J Med Chem 2010;53(3):1098–1108.
7. Lipinski CA, Lombardo F, Dominy BW, et al. Experimental and computational approaches to estimate solubility and permeability in drug discovery and development settings. Adv Drug Delivery Rev 2001;46(1–3):3–26.
8. Chan LM, Lowes S, Hirst BH. The ABCs of drug transport in intestine and liver: efflux proteins limiting drug absorption and bioavailability. Eur J Pharm Sci 2004;21(1):25–51.
9. Cvetkovic M, Leake B, Fromm MF, et al. OATP and P-glycoprotein transporters mediate the cellular uptake and excretion of fexofenadine. Drug Metab Dispos 1999;27(8):866–871.
10. Drescher S, Glaeser H, Murdter T, et al. P-glycoprotein-mediated intestinal and biliary digoxin transport in humans. Clin Pharmacol Ther 2003;73(3):223–231.
11. Murakami T, Takano M. Intestinal efflux transporters and drug absorption. Expert Opin Drug Metab Toxicol 2008;4(7):923–939.
12. Borst P, Schinkel AH. What have we learnt thus far from mice with disrupted P-glycoprotein genes. Eur J Cancer 1996;32A(6):985–990.
13. Borst P, Schinkel AH, Smit JJ, et al. Classical and novel forms of multidrug resistance and the physiological functions of P-glycoproteins in mammals. Pharmacol Ther 1993; 60(2):289–299.
14. Cordon-Cardo C, O'Brien JP, Casals D, et al. Multidrug-resistance gene (P-glycoprotein) is expressed by endothelial cells at blood-brain barrier sites. Proc Natl Acad Sci USA 1989;86(2):695–698.

15. Gramatte T, Oertel R, Terhaag B, et al. Direct demonstration of small intestinal secretion and site-dependent absorption of the beta-blocker talinolol in humans. Clin Pharmacol Ther 1996;59(5):541–549.
16. Etievant C, Schambel P, Guminski Y, et al. Requirements for P-glycoprotein recognition based on structure-activity relationships in the podophyllotoxin series. Anticancer Drug Des 1998;13(4):317–336.
17. Seelig A. A general pattern for substrate recognition by P-glycoprotein. Eur J Biochem 1998;251(1,2):252–261.
18. Martin C, Berridge G, Higgins CF, et al. Communication between multiple drug binding sites on P-glycoprotein. Mol Pharmacol 2000;58(3):624–632.
19. Loo TW, Clarke DM. Location of the rhodamine-binding site in the human multidrug resistance P-glycoprotein. J Biol Chem 2002;277(46):44332–44338.
20. Spahn-Langguth H, Baktir G, Radschuweit A, et al. P-glycoprotein transporters and the gastrointestinal tract: evaluation of the potential in vivo relevance of in vitro data employing talinolol as model compound. Int J Clin Pharmacol Ther 1998;36(1):16–24.
21. Bernsdorf A, Giessmann T, Modess C, et al. Simvastatin does not influence the intestinal P-glycoprotein and MPR2, and the disposition of talinolol after chronic medication in healthy subjects genotyped for the ABCB1, ABCC2 and SLCO1B1 polymorphisms. Br J Clin Pharmacol 2006;61(4):440–450
22. Masuda S, Uemoto S, Goto M, et al. Tacrolimus therapy according to mucosal MDR1 levels in small-bowel transplant recipients. Clin Pharmacol Ther 2004;75(4):352–361.
23. Schwarz UI, Hanso H, Oertel R, et al. Induction of intestinal P-glycoprotein by St John's wort reduces the oral bioavailability of talinolol. Clin Pharmacol Ther 2007;81(5):669–678.
24. Angelin B, Arvidsson A, Dahlqvist R, et al. Quinidine reduces biliary clearance of digoxin in man. Eur J Clin Invest 1987;17(3):262–265.
25. de Lannoy IA, Silverman M. The MDR1 gene product, P-glycoprotein, mediates the transport of the cardiac glycoside, digoxin. Biochem Biophys Res Commun 1992;189(1):551–557.
26. Greiner B, Eichelbaum M, Fritz P, et al. The role of intestinal P-glycoprotein in the interaction of digoxin and rifampin. J Clin Invest 1999;104(2):147–153.
27. Cao X, Yu LX, Barbaciru C, et al. Permeability dominates in vivo intestinal absorption of P-gp substrate with high solubility and high permeability. Mol Pharm 2005;2(4):329–340.
28. Dahan A, Amidon GL. Segmental dependent transport of low permeability compounds along the small intestine due to P-glycoprotein: the role of efflux transport in the oral absorption of BCS class III drugs. Mol Pharm 2009;6(1):19–28.
29. Shitara Y, Horie T, Sugiyama Y. Transporters as a determinant of drug clearance and tissue distribution. Eur J Pharm Sci 2006;27(5):425–446.
30. Liang R, Fei YJ, Prasad PD, et al. Human intestinal H+/peptide cotransporter. Cloning, functional expression, and chromosomal localization. J Biol Chem 1995;270(12):6456–6463.
31. Brandsch M. Transport of drugs by proton-coupled peptide transporters: pearls and pitfalls. Expert Opin Drug Metab Toxicol 2009;5(8):887–905.
32. Ogihara H, Saito H, Shin BC, et al. Immuno-localization of H+/peptide cotransporter in rat digestive tract. Biochem Biophys Res Commun 1996;220(3):848–852

33. Adibi SA. The oligopeptide transporter (Pept-1) in human intestine: biology and function. Gastroenterology 1997;113(1):332–340.
34. Balimane PV, Tamai I, Guo A, et al. Direct evidence for peptide transporter (PepT1)-mediated uptake of a nonpeptide prodrug, valacyclovir. Biochem Biophys Res Commun 1998;250(2):246–251.
35. Tamai I, Nakanishi T, Nakahara H, et al. Improvement of L-dopa absorption by dipeptidyl derivation, utilizing peptide transporter PepT1. J Pharm Sci 1998;87(12):1542–1546.
36. Ganapathy ME, Prasad PD, Mackenzie B, et al. Interaction of anionic cephalosporins with the intestinal and renal peptide transporters PEPT 1 and PEPT 2. Biochim Biophys Acta 1997;1324(2):296–308.
37. Terada T, Saito H, Inui K. Interaction of beta-lactam antibiotics with histidine residue of rat H+/peptide cotransporters, PEPT1 and PEPT2. J Biol Chem 1998;273(10): 5582–5585.
38. Zhu T, Chen XZ, Steel A, et al. Differential recognition of ACE inhibitors in Xenopus laevis oocytes expressing rat PEPT1 and PEPT2. Pharm Res 2000;17(5):526–532.
39. Gonzalez DE, Covitz KM, Sadee W, et al. An oligopeptide transporter is expressed at high levels in the pancreatic carcinoma cell lines AsPc-1 and Capan-2. Cancer Res 1998;58(3):519–525.
40. Brandsch M, Knutter I, Leibach FH. The intestinal H+/peptide symporter PEPT1: structure-affinity relationships. Eur J Pharm Sci 2004;21(1):53–60.
41. Zhang EY, Emerick RM, Pak YA, et al. Comparison of human and monkey peptide transporters: PEPT1 and PEPT2. Mol Pharm 2004;1(3):201–210.
42. Han HK, Amidon GL. Targeted prodrug design to optimize drug delivery. AAPS PharmSci 2000;2(1):E6.
43. Balimane PV, Tamai I, Guo A, et al. Direct evidence for peptide transporter (PepT1)-mediated uptake of a nonpeptide prodrug, valacyclovir. Biochem Biophys Res Commun 1998;250(2):246–251.
44. Perkins EJ, Abraham T. Pharmacokinetics, metabolism, and excretion of the intestinal peptide transporter 1 (SLC15A1)-targeted prodrug (1 S,2 S,5 R,6 S)-2-[(2′S)-(2-amino) propionyl]aminobicyclo[3.1.0.]hexen-2,6-di carboxylic acid (LY544344) in rats and dogs: assessment of first-pass bioactivation and dose linearity. Drug Metab Dispos 2007;35(10):1903–1909.
45. Lin JH, Chiba M, Baillie TA. Is the role of the small intestine in first-pass metabolism overemphasized? Pharmacol Rev 1999;51(2):135–158.
46. Kolars JC, Awni WM, Merion RM, et al. First-pass metabolism of cyclosporin by the gut. Lancet 1991;338(8781):1488–1490.
47. Fromm MF, Busse D, Kroemer HK, et al. Differential induction of prehepatic and hepatic metabolism of verapamil by rifampin. Hepatology 1996;24(4):796–801.
48. Paine MF, Shen DD, Kunze KL, et al. First-pass metabolism of midazolam by the human intestine. Clin Pharmacol Ther 1996;60(1):14–24.
49. Schwenk M. Mucosal biotransformation. Toxicol Pathol 1988;16:138–146.
50. Paine MF, Hart HL, Ludington SS, et al. The human intestinal cytochrome P450 "pie." Drug Metab Dispos 2006;34(5):880–886.
51. Paine MF, Khalighi M, Fisher JM, et al. Characterization of interintestinal and intraintestinal variations in human CYP3A-dependent metabolism. J Pharmacol Exp Ther 1997;283(3):1552–1562.

52. Kaminsky LS, Zhang QY. The small intestine as a xenobiotic-metabolizing organ. Drug Metab Dispos 2003;31(12):1520–1525
53. Benet LZ, Izumi T, Zhang Y, et al. Intestinal MDR transport proteins and P-450 enzymes as barriers to oral drug delivery. J Control Release 1999;62(1,2):25–31.
54. Gan LS, Moseley MA, Khosla B, et al. CYP3A-like cytochrome P450-mediated metabolism and polarized efflux of cyclosporin A in Caco-2 cells. Drug Metab Dispos 1996;24(3):344–349.
55. Johnson BM, Chen W, Borchardt RT, et al. A kinetic evaluation of the absorption, efflux, and metabolism of verapamil in the auto perfused rat jejunum. J Pharmacol Exp Ther 2003;305(1):151–158.
56. Tam D, Sun H, Pang KS. Influence of P-glycoprotein, transfer clearances, and drug binding on intestinal metabolism in Caco-2 cell monolayers or membrane preparations: a theoretical analysis. Drug Metab Dispos 2003;31(10):1214–1226.
57. Badhan R, Penny J, Galetin A, et al. Methodology for development of a physiological model incorporating CYP3A and P-glycoprotein for the prediction of intestinal drug absorption. J Pharm Sci 2009;98(6):2180–2197.
58. Lown KS, Bailey DG, Fontana RJ, et al. Grapefruit juice increases felodipine oral availability in humans by decreasing intestinal CYP3A protein expression. J Clin Invest 1997;99(10):2545–2553.
59. Ritchie TJ, Macdonald SJ, Young RJ, et al. The impact of aromatic ring count on compound developability: further insights by examining carbo- and hetero-aromatic and aliphatic ring types. Drug Discov Today 2011;16:164–171.
60. Gleeson MP. Generation of a set of simple, interpretable ADMET rules of thumb. J Med Chem 2008;51(4):817–834.
61. Kwon Y. Absorption. In: Handbook of essential pharmacokinetics, pharmacodynamics and drug metabolism for industrial scientists. New York: Springer, 2001, 35–72.
62. Völgyi G. Development of methods for determination of physico-chemical parameters (pK_a, logP) of water-insoluble compounds in early phase of drug discovery [PhD thesis]. Budapest: Semmelweis University, Department of Pharmaceutical Chemistry, 2007.
63. Ravichandran V, Devarajan V, Masilamani K. Determination of ionization constant (pK_a) for poorly soluble drugs by using surfactants, a novel approach. Scholars Research Library 2011;3:183–192.
64. Grey R, Pierce AC, Bemis GW, et al. Structure-based design of 3-aryl-6-amino-triazolo[4,3-b]pyridazine inhibitors of Pim-1 kinase. Bioorg Med Chem Lett 2009; 19(11):3019–3022.
65. Hall A, Billinton A, Bristow AK, et al. Discovery of brain penetrant, soluble, pyrazole amide EP1 receptor antagonists. Bioorg Med Chem Lett 2008;18(14):4027–4032.
66. Palmer BD, Trumpp-Kallmeyer S, Fry DW, et al. Tyrosine kinase inhibitors. 11. Soluble analogues of pyrrolo- and pyrazoloquinazolines as epidermal growth factor receptor inhibitors: synthesis, biological evaluation, and modeling of the mode of binding. J Med Chem 1997;40(10):1519–1529.
67. Oslob JD, Heumann SA, Yu CH, et al. Water-soluble prodrugs of an Aurora kinase inhibitor. Bioorg Med Chem Lett 2009;19(5):1409–1412.
68. Benet LZ, Broccatelli F, Oprea TI. BDDCS applied to over 900 drugs. AAPS J 2011; 13(4);519–547.

69. Madara JL, Trier JS. The functional morphology of the mucosa of the small intestine. In: Physiology of the gastrointestinal tract, ed. Johnson LR. New York: Raven, 1994, 1577–1622.

70. Blikslager AT, Moeser AJ, Gookin JL, et al. Restoration of barrier function in injured intestinal mucosa. Physiol Rev 2007;87(2):545–564.

71. Russell-Jones GJ. The potential use of receptor-mediated endocytosis for oral drug delivery. Adv Drug Delivery Rev 2001;46(1–3):59–73.

72. DeSesso JM, Jacobson CF. Anatomical and physiological parameters affecting gastrointestinal absorption in humans and rats. Food Chem Toxicol 2001;39:209–228.

73. Flemstrom G, Kivilaasko E. Demonstration of a pH gradient at the luminal surface of rat duodenum in vivo and its dependence on mucosal alkaline secretion. Gastroentrology 1983;84:787–794.

74. Lipka E, Lee ID, Langguth P, Spahn-Langguth H, Mutschler E, Amidon GL. Celiprolol double-peak occurrence and gastric motility: nonlinear mixed effects modeling of bioavailability data obtained in dogs. J Pharmacokinet Biopharm 1995;23(3):267–286.

75. Oberle RL, Amidon GL. The influence of variable gastric emptying and intestinal transit rates on the plasma level curve of cimetidine; an explanation for the double peak phenomenon. J Pharmacokinet Biopharm 1987;15(5):529–544.

76. Marathe PH, Wen Y, Norton J, Greene DS, Barbhaiya RH, Wilding IR. Effect of altered gastric emptying and gastrointestinal motility on metformin absorption. Br J Clin Pharmacol 2000;50(4):325–332.

77. Kimura T, Iwasaki N, Yokoe JI, Haruta S, Yokoo Y, Ogawara KI, Higaki K. Analysis and prediction of absorption profile including hepatic first-pass metabolism of N-methyltyramine, a potent stimulant of gastrin release present in beer, after oral ingestion in rats by gastrointestinal-transit-absorption model. Drug Metab Dispos 2000; 28(5):577–581.

78. Dressman JB, Yamada K. Animal models for oral drug absorption. In: Pharmaceutical bioequivalence, ed. Welling PG, Tse FLS, Dighe SV. New York: Marcel Dekker, 1991, 235–266.

79. Charman WN, Porter CJH, Mithani S, Dressman JB. Physicochemical and physiological mechanisms for the effects of food on drug absorption: the role of lipids and pH. J Pharm Sci 1997;86(3):269–282.

80. Galia E, Nicolaides E, Horter D, Lobenberg R, Reppas C, Dressman JB. Evaluation of various dissolution media for predicting in vivo performance of class I and II drugs. Pharm Res 1998;15(5):698–705.

81. Bakatselou V, Oppenheim RC, Dressman JB. Solubilization and wetting effects of bile salts on the dissolution of steroids. Pharm Res 1991;8(12):1461–1469.

82. Mithani SD, Bakatselou V, TenHoor CN, Dressman JB. Estimation of the increase in solubility of drugs as a function of bile salt concentration. Pharm Res 1996; 13(1):163–167.

83. Radde IC. Mechanisms of drug absorption and their development. In: Textbook of pediatic clinical pharmacology, ed. McLeod SM, Radde IC. Littleton, MA: PSG Publishing, 1995,17–31.

84. Boix-ochoa J, Canals J. Maturation of the lower esophagus. J Pediatr Surg 1976; 11(5):749–756.

85. Morselli PL. Clinical pharmacokinetics in neonates. Clin Pharmacokin 1976;1:81–98.

86. Stewart CF, Hampton EM. Effect of maturation on drug disposition in pediatric patients. Clin Pharm 1987;6:548–564.
87. Huang NN, High RH. Comparison of serum levels following the administration of oral and parenteral preparations of penicillin to infants and children of various age groups. J Pediatr 1953;6:57–68.
88. Olive G. Pharmacocinétique et biotransformation des médicaments chez l'enfant. Louvain Med 1991;110:565–569.
89. Brown RD, Campoli- Richards M. Antimicrobial therapy in neonates, infants and children. Clin Pharmacokin 1989;17:105–115.
90. Butler DR, Kuhn RJ, Chandler MHH. Pharmacokinetics of anti-infective agents in paediatric patients. Clin Pharmacokinet 1994;26(5):374–395.
91. Gilman JT. Therapeutic drug monitoring in the neonate and paediatric age group—problems and clinical pharmacokinetic implications. Clin Pharmacokin 1990; 19:1–10.
92. Grand RJ, Watkins JB, Torti FM. Development of the human gastrointestinal tract: a review. Gastroenterology 1976;70:790–810.
93. Pedersen S. Effects of food on the absorption of theophylline in children. J Allergy Clin Immunol 1986;78:704–709.
94. Pedersen S, Steffensen G. Absorption characteristics of once-a-day slow-release theophylline preparation in children with asthma. J Pediatr 1987;110:953–959.
95. Pedersen S, Moeller-Petersen J. Influence of food on the absorption rate and bioavailability of a sustained release theophylline preparation. Allergy 1982;37:531–534.
96. Rogers RJ, Kalisher A, Wiener MB, Szefler SJ. Inconsistent absorption from a sustained-release theophylline preparation during continuous therapy in asthmatic children. J Pediatr 1985;106:496–501.
97. Isles AF, Newth CJL. Pharmacokinetics of a sustained-release theophylline preparation in infants and preschool children with asthma. J Allergy Clin Immunol. 1985;75:377–381.
98. Radde IC. Mechanisms of drug absorption and their development. In: Textbook of pediatric clinical pharmacology, ed. McLeod SM, Radde IC. Littleton, MA: PSG Publishing, 1985, 17–43.
99. Boehm G, Braun W, Moro G, Minoli I. Bile acid concentrations in serum and suosenal aspirates of healthy preterm infants: effects of gestational and postnatal age. Biol Neonate 1997;71:207–214.
100. Heubi JE, Balistreri WJ, Suchy FJ. Bile salt metabolism in the first year of life. J Lab Clin Med 1992;100:127–136.
101. Mooij MG, Schwarz UI, de Koning deB, et al. Ontogeny of human hepatic and intestinal transporter gene expression during childhood: age matters. Drug Metab Dispos 2014;42:1268–1274.
102. Yanni SB, Smith PB, Benjamin DK Jr, Augustijns PF, Thakker DR, Annaert PP. Higher clearance of micafungin in neonates compared with adults: role of age-dependent micafungin serum binding. Biopharm Drug Dispos 2011;32(4):222–232.
103. Ikemura K, Urano K, Matsuda H, Mizutani H, Iwamoto T, Okuda M. Decreased oral absorption of cyclosporine A after liver ischemia-reperfusion injury in rats: the contribution of CYP3A and P-glycoprotein to the first-pass metabolism in intestinal epithelial cells. J. Pharmacol Exp Ther 2009;328:249–255.

104. Zhang Y, Jiang XH, Hu YQ, Li ZR, Su L, Wang ZG, Ma G. MDR1 genotypes do not influence the absorption of a single oral dose of 600mg valacyclovir in healthy Chinese Han ethnic males. Br J Clin Pharmacol 2008;66:247–254.

105. Okura T, Ozawa T, Ito Y, Kimura M, Kagawa Y, Yamada S. Enhancement by grapefruit juice of morphine antinociception. Biol Pharm Bull 2008;31:2338–2341.

106. Kobori T, Kobayashi M, Harada S, Nakamoto K, Fujita-Hamabe W, Tokuyama S. RhoA affects oral morphine analgesia depending on functional variation in intestinal P-glycoprotein induced by repeated etoposide treatment. Eur J Pharm Sci 2012; 47:934–940.

107. Garcia MG, Alaniz LD, Cordo Russo RI, Alvarez E, Hajos SE. PI3K/Akt inhibition modulates multidrug resistance and activates NF-kappaB in murine lymphoma cell lines. Leuk Res 2009;33:288–296.

108. Whiting DR, Guariguata L, Weil C, Shaw J. IDF diabetes atlas: global estimates of the prevalence of diabetes for 2011 and 2030. Diabetes Res Clin Pract 2011;94:311–321.

109. Kameyama N, Arisawa S, Ueyama J, Kagota S, Shinozuka K, Hattori A, Tatsumi Y, Hayashi H, Takagi K, Wakusawa S. Increase in P-glycoprotein accompanied by activation of protein kinase Calpha and NF-kappaB p65 in the livers of rats with streptozotocin-induced diabetes. Biochim Biophys Acta 2008;1782:355–360.

110. Cygalova LH, Hofman J, Ceckova M, Staud F. Transplacental pharmacokinetics of glyburide, rhodamine 123, and BODIPY FL prazosin: effect of drug efflux transporters and lipid solubility. J Pharmacol Exp Ther 2009;331:1118–1125.

111. Hemauer SJ, Patrikeeva SL, Nanovskaya TN, Hankins GD, Ahmed MS. Role of human placental apical membrane transporters in the efflux of glyburide, rosiglitazone, and metformin. Am J Obstet Gynecol 2010;202:383.e1–387.e7.

112. Li C, Choi DH, Choi JS. Effects of efonidipine on the pharmacokinetics and pharmacodynamics of repaglinide: possible role of CYP3A4 and P-glycoprotein inhibition by efonidipine. J Pharmacokinet Pharmacodyn 2012;39:99–108.

113. Fuchs H, Runge F, Held HD. Excretion of the dipeptidyl peptidase-4 inhibitor linagliptin in rats is primarily by biliary excretion and P-gp-mediated efflux. Eur J Pharm Sci 2012; 45:533–538.

114. Kobori T, Harada S, Nakamoto K, Tokuyama S. Functional alterations of intestinal P-glycoprotein under diabetic conditions. Biol Pharm Bull 2013;36(9):1381–1390.

115. Lentz KA. Current methods for predicting human food effect. AAPS J 2008;10(2): 282–288.

116. Charman WN, Porter CJH, Mithani S, Dressman JB. Physicochemical and physiological mechanisms for the effects of food on drug absorption: the role of lipids and pH. J Pharm Sci 1997;86(3):269–282.

117. Gu CH, Li H, Levons J, et al. Predicting effect of food on extent of drug absorption based on physicochemical properties. Pharm Res 2007;24(6):1118–1130.

118. Abdou HM. Dissolution, bioavailability and bioequivalence. Easton, PA: Mack Publishing, 1989.

119. Adkin DA, Davis SS, Sparrow RA, Huckle PD, Wilding JR. The effect of mannitol on the oral bioavailability of cimetidine. J Pharm Sci 1995;84:1405–1409.

120. Del Estal JL, Alvarez AI, Villaverde C, Coronel P, Fabra S, Prieto JG. Effect of surfactants on albendazole absorption. J Pharm Biomed Analysis 1991; 9:1161–1164.

121. Woodcock LK, Linsenmeyer E, Chojnowski G, Kriegler AB, Nink V, Webster LK, et al. Reversal of multidrug resistance by surfactants. Br J Cancer 1992;66:62–68.

122. Amidon GL, Lennernas H, Shah VP, Crison JR. A theoretical basis for a biopharmaceutics drug classification: the correlation of in vitro drug product dissolution and in vivo bioavailability. Pharm Res 1995;12(3):413–420.

123. Waiver of In Vivo Bioavailability and Bioequivalence Studies for Immediate-Release Solid Oral Dosage Forms Based on a Biopharmaceutics Classification System, 2000. [Online]. Available: http://www.fda.gov/cder/ guidance/index.htm.

124. Sambruy Y, Ferruzza S, Ranaldi G, et al. Intestinal cell culture models: applications in toxicology and pharmacology. Cell Biol Toxicol 2001;17:301–317.

125. Artursson P, Borchardt RT. Intestinal drug absorption and metabolism in cell cultures: Caco-2 and beyond. Pharm Res 1997;14(12):1655–1658.

126. Bu HZ, Poglod M, Micetich RG, et al. High-throughput caco-2 cell permeability screening by cassette dosing and sample pooling approaches using direct injection/on-line guard cartridge extraction/tandem mass spectrometry. Rapid Commun Mass Spectrom 2000;14(6):523–528.

127. Cho MJ, Adson A, Kezdy FJ. Transepithelial transport of aliphatic carboxylic acids studied in Madin Darby canine kidney (MDCK) cell monolayers. Pharm Res 1990;7(4): 325–331.

128. Cho MJ, Thompson DP, Cramer CT, et al. The Madin Darby canine kidney (MDCK) epithelial cell monolayer as a model cellular transport barrier. Pharm Res 1989;6(1): 71–77.

129. Tran TT, Mittal A, Gales T, et al. Exact kinetic analysis of passive transport across a polarized confluent MDCK cell monolayer modeled as a single barrier. J Pharm Sci 2004;93(8):2108–2123.

130. Di L, Kerns EH, Fan K, et al. High throughput artificial membrane permeability assay for blood-brain barrier. Eur J Med Chem 2003;38(3):223–232.

131. Kansy M, Senner F, Gubernator K. Physicochemical high throughput screening: parallel artificial membrane permeation assay in the description of passive absorption processes. J Med Chem 1998;41(7):1007–1010.

132. Wohnsland F, Faller B. High-throughput permeability pH profile and high-throughput alkane/water log P with artificial membranes. J Med Chem 2001;44(6):923–930.

133. Zhu C, Jiang L, Chen TM, et al. A comparative study of artificial membrane permeability assay for high throughput profiling of drug absorption potential. Eur J Med Chem 2002;37(5):399–407.

134. Zhao YH, Abraham MH, Le J, et al. Evaluation of rat intestinal absorption data and correlation with human intestinal absorption. Eur J Med Chem 2003;38(3):233–243.

135. Cao X, Gibbs ST, Fang L, et al. Why is it challenging to predict intestinal drug absorption and oral bioavailability in human using rat model. Pharm Res 2006;23(8):1675–1686.

136. Chiou WL, Jeong HY, Chung SM, et al. Evaluation of using dog as an animal model to study the fraction of oral dose absorbed of 43 drugs in humans. Pharm Res 2000;17(2):135–140.

137. Prakash C, Kamel A, Miao Z. Radiochemical tracers: essential tools for the drug metabolism studies. In: Synthesis and application of isotopically labelled compounds, ed. Dean DC, Filer CN, McCarthy KE. New York: Wiley, 2004, 115–120.

138. Penner N, Klunk LJ, Prakash C. Human radiolabeled mass balance studies: objectives, utilities and limitations. Biopharm Drug Dispos 2009;30(4):185–203.

139. Prakash C, Johnson KA, Schroeder CM, et al. Metabolism, distribution, and excretion of a next generation selective estrogen receptor modulator, lasofoxifene, in rats and monkeys. Drug Metab Dispos 2008;36(9):1753–1769.

140. Barbhaiya RH, Dandekar KA, Greene DS. Pharmacokinetics, absolute bioavailability, and disposition of [14C]nefazodone in humans. Drug Metab Dispos 1996;24(1):91–95.

141. Shukla UA, Marathe PH, Pittman KA, et al. Pharmacokinetics, absolute bioavailability, and disposition of [14C]nefazodone in the dog. Drug Metab Dispos 1993;21(3):502–507.

142. Augustijns P, Wuyts B, Hens B, Annaert P, Butler J, Brouwers. A review of drug solubility in human intestinal fluids: implications for the prediction of oral absorption. J Eur J Pharm Sci 2014;57:322–332.

2

DISTRIBUTION: PRINCIPLE, METHODS, AND APPLICATIONS

2.1 INTRODUCTION: DRUG DISTRIBUTION IN RELATION TO DRUG DISPOSITION IN HUMANS

When oral drug is delivered, it is absorbed through the intestinal epithelia and translocates across the basolateral (BL) membrane to the portal-vein blood, as discussed in the previous chapter. A drug transports to hepatocytes, where it is either metabolized, excreted to bile, or transported to systemic circulation. There, drug distributes to various tissues (e.g., muscle and fat) or organs (e.g., brain). It has been described that drug distribution is an integral part of absorption (A), distribution (D), metabolism (M), and excretion (E), which encompasses the complete fate of a drug when it enters the body and exerts its own therapeutic effect [1].

A complete circulation of the blood typically takes about 1 min, and after this time, a drug will begin to distribute into tissues. Factors that can affect the ability of the drug to distribute into tissues include water and fat solubility. A drug that is more water soluble will tend to stay in the blood or interstitial spaces, whereas a fat-soluble drug may concentrate in fatty tissues. Drugs can also be limited in their ability to penetrate into tissues depending on their ability to cross barriers, such as the blood-brain barrier (BBB) or placental barrier. Drug binding to plasma proteins can also affect its ability to distribute. Bound drug is not able to distribute into tissues; only the fraction unbound (free) moves into tissues and determines the pharmacodynamics (PD) or efficacy of the drug [2]. The extent of protein binding can vary among human

Translational ADMET for Drug Therapy: Principles, Methods, and Pharmaceutical Applications, First Edition. Souzan B. Yanni.

populations based on age, gender, and disease state, and the alteration in protein binding could potentially alter the drug exposure, thus clearance and its safety.

Like drug absorption, many factors can influence the ability of a drug to distribute in the body, such as physiochemical and physiological factors [3], including the role of drug transporters, as discussed in the previous chapter, that have an important role in drug tissue distribution. There are several methods employed to investigate drug distribution; the most common is in vivo, in animal and human pharmacokinetics (PK) studies. However, with the development of new in vitro and in silico predictive tools, it is now common to predict drug distribution in the drug discovery stage.

For drug development, radiolabeled studies in preclinical animal models are conducted to determine absorption, distribution, metabolism, and excretion (ADME) and potential toxicity prior to Phase 1 clinical investigation in man. In this preclinical investigation, quantitative measurement of drug distribution is recommended and generally conducted for each new drug candidate using a technique referred as "quantitative whole-body autoradiography" (QWBA) [4]. In addition to QWBA, a traditional mass balance study (MBS) is required to investigate the quantitative distribution of drugs in ADME or toxicity animal models before clinical studies in humans are initiated [5]. With the new advancement of the inclusion of imaging as one of the tools to ensure full understanding of drug efficacy and safety, the use of positron emission tomography (PET) becomes common in human investigations after treating subjects with radiolabeled drugs to monitor the distribution of drugs in various body organs such as the brain [6].

As described above, drug distribution is one important parameter that plays a role in the overall PK of drugs. Its accurate assessment is critical in understanding and explaining drug effects and safety; this will be discussed in this chapter. The factors that influence the ability of a drug to distribute in the body, including physiochemical and physiological factors and the contribution of transporters, as well as the extent of protein binding among human populations, will also discussed in the coming sections.

Drug distribution refers to the reversible transfer of a drug from one location to another within the body [7]. The compound is distributed to different tissues of the body after entering into the systemic circulation (i.e., blood). In most cases, the volume of distribution at steady state (V_{dss}) is used to describe the distribution of a compound in the body. Distribution is generally uneven because of differences in blood perfusion, tissue binding, regional pH, and permeability of cell membranes. The distribution of a drug to specific tissue and rate of distribution are important parameters in determining drug pharmacological action. The acidic drugs such as warfarin and aspirin are highly protein bound and thus have a small apparent volume of distribution (V_d) [8]. The basic drugs such as amphetamine [9] and meperidine are extensively taken up by tissues and thus have an apparent V_d larger than the V_d of acidic drugs. All these topics will also discussed in more detail in the next section of this chapter.

2.2 INFLUENCE OF DRUG-RELATED PHYSIOCHEMICAL FACTORS ON DRUG DISTRIBUTION

For a drug to diffuse and distribute into systemic circulation and body organs to exert the PK/PD effect, it has to cross several barriers, such as the cell membrane of all body organs and tissues. As discussed in the previous chapter, there are well-identified mechanisms for a drug molecule to cross cell membranes: (1) it diffuses directly through the membrane lipid bilayer (transcellular passive diffusion mechanism); (2) it may diffuse through aqueous pores formed by special proteins in the membrane called tight junctions (TJs) (paracellular passive diffusion mechanism); (3) it may bind to transcellular carrier proteins (drug transporters) in the membrane (active transport mechanism); or (4) it can translocate across the cell membrane by pinocytosis [10].

Diffusion through the lipid bilayer (trancellularly or paracellularly) and transport via carrier-mediated mechanisms (active transport) represent the most common mechanisms of drug distribution, and thus its disposition [10]. The fourth mechanism of crossing drugs through the membrane, via pinocytosis, is commonly the mechanism in which large molecular weight (MW) proteins such as insulin can translocate the cell membrane, as pores in membranes are too small (0.4 nm). With recent advances in technologies supported by pharmaceutical interest during the last two decades, considerable knowledge on the transport of drugs and mechanisms has become available; most important are identifying the drug transporters for xenobiotics and understanding their role in PK/PD and safety of drugs [11].

As also discussed in the previous chapter, there are drug-related factors that influence the passive diffusion mechanism of drug molecules through the lipid membrane, such as *lipophilicity*, *pH*, pK_a, and *ionization*. The solubility of drug in the membrane lipid bilayer is a major determinant of the permeability of a drug across the cell membrane, hence its disposition and PK. The absorption from the intestine, translocation to the BBB, and rate of renal elimination are to a large extent dependent on drug solubility in the lipid component of cell membrane, which is determined by the partition coefficient of the drug between the membrane lipid phase and the aqueous phase and its diffusion coefficient. Nearly all drug-like molecules are weak acids or bases, meaning that they contain at least one site that can reversibly disassociate or associate a proton (a hydrogen ion) to form a negatively charged anion or a positively charged ion. Molecules that disassociate protons are acids, and those that associate protons are bases. Drug in solution is always in equilibrium with some fraction protonated and the rest deprotonated according to Equation (2.1):

$$HA \leftrightarrow H^+ + A^- \text{ or } HB+ \leftrightarrow H^+ + B. \tag{2.1}$$

By varying the availability of protons, that is, the acidity of the media (pH), the balance of the equilibrium can be shifted. The pK_a value is defined and determined by the Henderson–Hasselbalch equation [Equation (2.2)], and this will assess the extent of ionization of weak acids and bases at different pH, as shown below:

$$\text{Weak base:}\quad pK_a = pH + \log \frac{[\text{ionized form}]}{[\text{nonionized form}]}$$

$$\text{Weak acid:}\quad pK_a = pH + \log \frac{[\text{nonionized form}]}{[\text{ionized form}]}. \tag{2.2}$$

Alternatively, the pK_a of a site can be thought of as the pH at which the protonated and deprotonated fractions are equal. If the pH is higher than the pK_a, the site is mostly deprotonated, and if the pH is lower than the pK_a, the site is mostly protonated. The pH of physiological body content can control its ionization forms, ionized or nonionized. The pH determines the extent of drug ionization and in turn that can determine the lipid solubility of drug and its distribution in the body by crossing the cell membranes.

Based on the structure, the ionized form, whether the drug is a weak base or acid, is not lipid soluble and permeability will be rate limited (unless it is facilitated by a specific transporter mechanism). On the other hand, the nonionized form will be lipid soluble and drugs permeate rapidly. In Table 2.1, known acidic and basic drugs with a wide range of pK_a values are shown. When a pH difference exists between body compartments, then the extent of ionization of a drug would influence its ability to distribute within the body, with the ratio of ionized to nonionized drug being determined by the pK_a of the compound and the pH of the compartment fluid. The impact

TABLE 2.1 pK_a of Some Acids and Base.

Drug	pK_a
A-Bases	
Chloroquine	10.8 (strong base)
Amphetamine	9.9
Ephedrine	9.6
Propranolol	9.5
Chlorpromazine	9.3
Erythromycin	8.8
Dopamine	8.4
Morphine	7.9
Codeine	5.8
Diazepam	3.3
B-Acids	
Levodopa	2.3 (strong acids)
Penicillin	2.7
Probenacid	3.4
Aspirin	3.5
Warfarin	5.0
Phenytoin	8.3

of pH and pK_a of a drug on relative distribution of drug at different body compartments is explained below for a weak base such as erythromycin (pK_a 8.8) and a weak acid such as aspirin (pK_a 3.5).

Erythromycin drug concentration was seen at relatively highest concentration in gastric compartment pH of 2.0, followed by in urine pH 6.0, then in plasma at pH 7.4 in plasma, and a trace amount in urine at pH 8.0. In contrast, Aspirin was found to be at the highest concentration in urine pH 8.0, followed by plasma (pH 7.4), then urine pH 6.0, and a trace amount in the stomach pH 2.0. The distribution profile of those two drugs indicates that the nonionized drug will cross membranes and reach equal concentrations in each compartment, whereas the ionized drug will not cross at all. As a result, the total drug concentration in each compartment will be different. Acidic drug will be trapped as an ion in the compartment with the highest pH, and a basic one will be trapped at low pH. Large concentration gradients can be built between compartments. This is theoretically logical; however, in reality, the constant flux of fluids between compartments reduces the concentration gradients because fluids in body compartments are not static.

It is important to state that partition between different pH environments is not a major determinant of the site of absorption of drugs from the gastrointestinal (GI) tract, but the enormous absorptive area of the villi of the small intestine compared with the smaller absorptive area of the stomach is the major dominant contributing factor. For this reason, weak acids and bases are well absorbed from the intestine, while strong acids and bases are not, as discussed by studies such as that published by Schanker et al. (1957) [4], which showed that strong acids, with pK_a of <2, and strong bases, with $pK_a > 10$, are not well absorbed in human and rat. Unlike GI distribution, urinary excretion and brain distribution are influence by the pH of the environment. If the urine becomes acidic, then the excretion of weak acids is decreased while the excretion of weak bases is increased. On the contrary, if urine becomes basic, the opposite effects become true. Strategically, by altering the pH of the urine, elimination of overdose aspirin is increased (see Table 2.2). In addition to altering urine pH, the distribution of weak acids from the central nervous system (CNS) into the plasma can be increased by administration of sodium bicarbonate, the agent that increases

TABLE 2.2 Comparison between the Distribution of Strong Acid and Base in Body Compartments with Different pH. Values are listed in [12].

Compartment	Relative Concentrations	
	Weak Acid (Aspirin)	Weak Base (Erythromycine)
Gastric juice (pH 2.0)	0.4	6×10^6
Plasma (pH7.4)	100	100
Urine (pH 6.0)	4	2520
Urine (pH 8.0)	400	25

plasma pH. In overdose weak acids like phenobarbitone, administration of sodium bicarbonate that increases the pH of urine and plasma would increase the elimination and decrease its CNS toxicity. On the other hand, plasma pH decreases by administration of acetazolamide, as it concentrates the drug in the CNS, thus increasing its neurotoxicity [12].

2.3 INFLUENCE OF PHYSIOLOGICAL FACTORS ON DRUG DISTRIBUTION

Distribution of a drug occurs through the blood via the circulatory system from blood to other tissues, and that occurs at various rates. The various regions of the systemic circulation are perfused with blood through innumerable branching pathways, which are effectively arranged in parallel [13], flowing from the heart to all areas of the body through arteries, to capillaries through organs and skin, then return to the heart through the blood veins. Several factors determine the distribution pattern of drug with time, including the delivery of drug to tissue by blood, ability to cross tissue membrane, binding with blood and tissues, and partitioning into fat. Tissue uptake continues toward equilibrium of the diffusible form between tissue and blood that perfuses it [7].

The first stage of drug distribution is dependent on the cardiac output and regional blood flow to the various organs. Drug distribution is rapid to highly perfused organs, where the rate of blood flow is higher and these organs receive most of the drug. In contrast, in non-well-perfused tissues such as adipose tissues, the distribution is generally slower and drug content in these compartments is markedly lower. As discussed earlier, distribution of drug from systemic circulation to tissues is dependent on the physiochemical properties of the drug, such as Pk_a and pH, but also dependent on the lipid solubility and molecular size in relation to physiological factors such as binding to plasma proteins, rate of blood flow, cardiac output, body volume, and, more important, susceptibility to passively diffuse or to undergo active transport. Ionic and cellular biochemical conditions as well as biological barriers such as BBB and placental barrier may affect the extent of distribution. Like absorption, distribution can be rate limited by either perfusion or permeability. A perfusion-rate limitation becomes key when there is no barrier to distribution by tissue membrane, such as in the case of small lipophilic drugs diffusing across membranes of the body. Perfusion is expressed as unit blood volume per unit time per volume of tissue. Table 2.3 illustrates the perfusion rate of tissues, which varies from 0.025 to ~ 10 mL/min/mL for fat or resting muscle to lung, respectively. It is worth mentioning that there is a direct correlation between tissue perfusion rate and the time required to distribute a drug to tissue.

The drug exists in two forms in blood, the free form, which is active and available for biotransformation and excretion, or the bound form, which is usually pharmacologically inactive. The main physiological compartments, where drugs are distributed, are the plasma, interstitial fluid, intracellular fluid, fat, and specific organs like liver and kidney, as well as muscle, bone, and the CNS [12].

TABLE 2.3 Blood Flow, Perfusion Rate, and Relative Size of Different Organs and Tissues in Standard 70-kg Human.

Organ	% of Body Volume	Blood Flow, mL/min	% of Cardiac Output	Perfusion Rate mL/min/mL
Blood	7	5000	100	–
Bone	16	250	5	0.02
Brain	2	700	14	0.5
Fat	20	200	4	0.03
Heart	0.4	200	4	0.6
Kidney	0.5	1100	22	4
Liver	2.3	1350	27	0.8
Portal	1.7	1050	21	–
Arterial	–	300	6	–
Lungs	1.6	5000	100	10
Muscle	43	750	15	0.025
Skin	11	300	6	0.04
Spleen	0.3	77	1.5	0.4
Thyroid gland	0.03	50	1	2.4
Total body	100	5000	100	0.071

2.3.1 Effect of Body Water Content, Perfusion, and Diffusion on Drug Distribution

2.3.1.1 Effect of Body Water Drug distribution or V_d can be determined by assessing organ compartments and content of body water. The V_d relates the amount of drug in the body to the plasma concentration as follows:

$$C_p = C_{p^\circ}\ e^{-kt}, \tag{2.3}$$

where C_{p° equals the initial plasma concentration and is a function of the distribution of the drug in the body, while e^{-kt} is the elimination parameter. By definition, V_d is the ratio between dose in mg and plasma drug concentration in mg/mL. When drug binds to plasma proteins, which limits its distribution, or binds to tissue protein, which increases its distribution, these cases can have an impact on the V_d. The body water ranged from 60% to 75% of the total body weight with the average being around 44 L, found in the intracellular fluid (61%), followed by interstitial fluid (27%), and to a lesser extent blood plasma (7%) and blood cell (5%) [7]. As PK parameters were determined, including V, they can provide information on the extent of distribution between body water and tissues. Furthermore, assessment of drug distribution between body water compartments can be achieved based on their volume, lipid membranes, binding to plasma proteins, and pH gradients that are encountered. When a drug enters the body either by oral, intravenous (iv), or topical administration, it is distributed into intracellular fluid, interstitial fluid, blood plasma, or blood cell compartments or is sequestered within a cellular site. As drug is absorbed, it passes through cell linings of the absorbing organs into the interstitial fluid. The interstitial and intracellular fluids are in equilibrium with water and electrolytes, which move slowly into and out of cells in contrast to fast-moving blood. By entering local tissue cells, drug may have two passes, either entering blood capillaries of the systemic circulation system or entering the lymphatic system. If a drug is administered orally, it enters transcellular fluids through an epithelium of cerebrospinal fluid (CSF), aqueous humor of the eye, or renal tubules. It is also distributed into GI secretion, synovial fluid in joints, pleural fluids, peritoneal fluids, bile, saliva, glandular contents, and the urine in the bladder. If the drug is administered intravenously to the bloodstream, it directly enters the plasma compartment.

2.3.1.2 Effect of Perfusion and Diffusion on Drug Distribution In addition to all factors that determine drug distribution, like the extent of plasma protein binding-drug free fraction, drug transporter-mediated permeability, and body water, tissue perfusion rate or diffusion of the drug across barrier membranes can alter the distribution of the drug unbound fraction. Simply, diffusion is the mechanism that rules the distribution of unbound-free drug (small MW). When bound to albumin, the MW of the complex is so high (>68, 000 kD) that it limits diffusion [13]. Rate of diffusion is described by the equation for Fick's law of diffusion [12]:

$$\text{Rate of diffusion} = \frac{-\text{DKA}\,(\text{Cp} - \text{Ct})}{\text{h}}, \tag{2.4}$$

where Cp is the concentration of drug in plasma, Ct is the concentration of drug in tissue, A is the surface area of the membrane, K is the lipid–water partition coefficient, D is the diffusion constant and is inversely proportional to the weight of the drug, and h is the thickness of the membrane. The negative sign denotes the net transfer of drug from inside the capillary lumen into tissues and extracellular spaces. According to this relationship, the diffusion is spontaneous and a temperature-independent process. As has been tested by Yanni et al. (2010) [14], the passive diffusion mechanism is distinguished from transporter-mediated mechanism by measuring the rate of drug accumulation in tissues (or in the hepatocytes in vitro) during an uptake study at 4°C versus 37°C [14]. Perfusion can be rate limited for drugs that are polar and well lipid soluble; as soon as drug enters the blood circulation, it is free to distribute completely and instantly across membranes without diffusion barriers but is limited by the rate of blood flow to the organ.

It is worth mentioning that when perfusion is low, it limits the distribution. In this case, the distribution equilibrium may take longer to achieve, especially when the partition coefficient is high. High permeability coefficient results in high concentration of drug at the tissue before equilibrium takes place. For a polar drug with the distribution diffusion rate limited, the rate of distribution into tissue is a function of the permeability coefficient of the drug, the concentration gradient, the distance that the drug travels, the surface area for diffusion, the temperature, and the MW of the drug, since smaller, lighter molecules will diffuse faster than larger, heavier ones. Capillary permeability can also have a profound effect on drug distribution.

2.4 PLASMA PROTEIN BINDING

Additional factors can influence the ability of a drug to distribute, including the tendency of the drug to bind to plasma proteins. Plasma makes up ~55% of the total blood volume and is composed of water containing trace amounts of salts, minerals, and nutrients. Plasma contains a variety of proteins, such as immunoglobulin and fibrinogen (for clotting); however, one of its main constituents is albumin, which is ~60% of the total plasma protein [15].

Among serum proteins, human serum albumin (HSA) and globulins (α, β,γ) and $\alpha - 1 -$ acid glycoprotein (AGP) play important roles in protein binding of many drugs, which is important in determining drug distribution throughout the body [16]. In humans, albumin concentration tested in a pool of plasma from 100 subjects was found to be ~50 mg/mL.

Albumin is basic, and therefore preferentially binds to acidic (negatively charged drugs at physiological pH) and neutral compounds, while AGP, which is an acidic component, can bind to basic drugs. It is known that many drugs circulate through the bloodstream bound to albumin, as it is a good carrier for acidic drugs. In addition to binding to serum albumin, drugs bind to other plasma proteins nonspecifically, to a lesser extent. Binding of drugs to albumin enables their circulation throughout the body, but the binding may only allow free drug to become available for surrounding tissues, and that reduces the extent of drug distribution [16]. In the plasma, in addition to binding to albumin or AGP, drugs may bind to lipoproteins. Since protein binding can have a large impact on drug distribution properties, the lipid solubility of the drug can also affect its distribution.

It has been observed that higher protein binding to intestinal tissue, reducing the fraction unbound in the gut, will effectively increase the V_d of the enterocyte compartments and delay the absorption to portal and systemic circulation, thus changing a high solubility–high permeability drug PK profile in reference to T_{max} from <1 h to >3 h [17].

2.4.1 Effect of Biomedical Conditions: Disease State and Pregnancy

Based on what has been mentioned, protein binding can be considered as the factor that inversely determines the amount of drug in the central blood compartment (V_d), thus a higher degree of protein binding will produce a lower V_d. There are medical conditions that alter the levels of albumin and alpha acid glycoprotein (AAG) in plasma and thus can affect the degree of plasma protein binding (Table 2.4) [17–21]. From the table and as has been reported, disease, stress, and biomedical conditions such as pregnancy can cause conformational changes in plasma or proteins in tissues that may interfere with binding and reduce or increase the plasma proteins. Renal failure, starvation, and liver disease were found to reduce plasma proteins, while heart infarction or liver morbidities can increase AAG levels and have been implicated in changing the plasma binding of basic drugs, and consequently changing the distribution and PK/PD of these drugs. During pregnancy, albumin levels

TABLE 2.4 Physiological and Medical Conditions Associated with Altered Plasma Albumin and AAG.

Decreased Albumin	Increased Albumin	Decreased AAG	Increased AAG
Acute viral hepatitis	Dehydration	Oral contraceptive	Acute myocardial
Advanced age	Psychosis	Advanced age	Burn injury
Young age	Optic neuritis/retinitis	Infarction	Cancer
Cirrhosis		Cirrhosis	Chronic pain syndrome
Hyperthyroidism		Sever liver disease	Inflammatory disease
Starvation			Pneumonia
Pregnancy		Pregnancy	Renal translation
Renal failure		Nephrotic syndrome	Surgery
Surgery			Trauma injury
Stress and trauma			

decrease, resulting in significant increase in unbound fractions for drugs such as phenobarbitone, salicylic acid, dexamethasone, diazepam, valproic acid, phenytoin, and propranolol [22,23]. Safety on the systemic exposure of those drug-unbound fractions raised a concern to the fetus via the maternal placenta following maternal treatment. To minimize any drug-mediated toxicity to the fetus, a lower dose may be suggested.

2.4.2 Protein Binding as a Function of Age

In pediatrics, it has been reported by Benedetti et al. (2005) [24] that the drug distribution is altered throughout life due to the change in extent of plasma protein binding with age. Protein binding was found to be different in young pediatric patients than adults for some drugs like micafungin [25,26]. Table 2.5 shows that plasma proteins in young infants may be not present, present but at low concentration, or a different type of protein than those in adult plasma. In addition, the permeability (uptake or secretion from cells of body organs) across the cell membrane is known to have a high impact on the changes of drug distribution across a child's growth. Membrane

TABLE 2.5 Changes in Plasma Proteins Component with Child Development.

Plasma Component	Age Group in Child Development		
	Neonate(<28 Days)	Infant (1 month–1 Year)	Child (1–8 Years)
Total protein	↓	↓	< = >
Plasma albumin	↓	< = >	< = >
Fetal albumin	Present	Absent	Absent
Plasma globulin	↓	↓	< = >
α_1-acid glycoprotein	↓	Unknown	< = >

Notes: < = > Equivalent to adult level.
↓ below adult level.

permeability is particularly high in immature neonates. Myelinization, which counteracts drug passage, continues after birth [27]. The International Conference on Harmonisation of Technical Requirements for Registration of Pharmaceuticals for Human Use (ICH) guideline (2000) emphasizes that penetration of therapeutic products into the CNS should be considered as a critical matter for preterm newborn infants. Likewise, the BBB in term newborn infants is not fully mature, and drugs easily penetrate the epithelial cell membrane to the CNS, causing unwanted high exposure and thus toxicity. CNS permeability is decreased in neonates and infants as a function of age as is evident, for example, by the diminution of the brain/plasma ratio of anticonvulsants [28]. Although the distribution of a drug is the result of simple passive diffusion along concentration gradients, tissue transporters that are capable of producing a biologic barrier can also contribute.

It is important to note that changes in the unbound drug fraction do not always result in proportional changes in clearance or V_d. Important contributors to the change of drug concentration will also be the extent of binding, type of clearance, distribution volume, administration route, and concomitant changes in intrinsic (cellular) clearance due to induction or inhibition of hepatic enzymes [29,30].

In drug discovery, during the lead optimization processes and candidate selection, plasma protein binding studies play an important role in the process of candidate drug selection and development. The evaluation processes deal with not only identifying the ability of drug to bind to plasma protein but also determining the amount of drug that binds to a given protein concentration, that is, the concentration of free drug, and the drug's affinity for the binding sites of the protein molecule. Plasma protein binding studies allow the determination of the amount of unbound (free) drug concentrations in the blood. A low fraction unbound (fu) value such as < 0.1 indicates that a compound is highly bound to plasma proteins. Plasma protein binding data should be considered with other in vitro data to determine the potential disposition, PK/PD, and safety in a clinic. Low protein binding may be more favorable than high protein binding, as this will be implicated with higher free fraction, hence recommended lower drug dose than results in potential dose-dependent toxicity. However, factors such as solubility of the drug and nonspecific binding need to be considered when interpreting data.

2.5 ROLE OF DRUG TRANSPORTERS IN DRUG DISTRIBUTION

Drug transporters are expressed in all body organs, in particular in those tissues responsible for drug disposition and clearance such as liver, kidney, intestine, and brain. Giacomini et al. (2010) [31]; Ho and Kim (2005) [32]; and Chandra and Brouwer (2004) [33] have reported the expression and localization of the major transporters to carry on their function of drug distribution and excretion in liver, kidney, intestine, and brain.

Most drugs are distributed across cellular membranes with the assistance of transporter proteins that have been expressed in every living cell at various levels and are distributed differentially throughout tissues and barrier membranes where they possess different functions either to support tissue accumulation or excretion [34–37]. It is thought that transporters may play an important role not only in distributing drug molecules to their effector site but also in distribution to almost every organ and tissue in the body in a concentration-dependent fashion [34], thus contributing to drug pharmacological and toxicological effects. Because the binding to drug transporters is a concentration-dependent process (a saturable kinetic), it may lead to nonlinear PK, including distribution.

2.5.1 Drug Distribution as a Function of Efflux Drug Transporters

Efflux drug transporters, most importantly P-glycoprotein (P-gp), have been recognized as an important contributor to drug transport across membranes, drug disposition, and their underlying role in multidrug resistance (MDR) [38]. Hundreds of clinical-relevant investigations reveal the significant role of P-gp on ADME and safety [31]; thus it becomes a key element in every preclinical and clinical drug investigation. To exert its function, P-gp and other efflux transporters require the direct binding and hydrolysis of adenosine triphosphate (ATP); consequently they are referred to as superfamily of ABC (ATP-binding cassette) transporters and have at least 49 members identified in humans. Those 49 transporters are divided into seven families, the most well investigated of which are ABCB1, ABCC1, ABCC2, and ABCG2 [39]. Another important family of transporters that play a critical role in drug distribution are the solute carrier (SLC) transporters of 50–55 gene families and at least 362 members [40,41] responsible for drug uptake. These major uptake transporters include the organic anion transporting polypeptides (OATPs) and organic cation transporters (OCTs) [33]. Unlike the ABC transporters, the SLC transporters do not directly utilize an energy source [31]. Together the ABC and SLC transporters are increasingly being implicated in regulation of the passage of endogenous and exogenous chemicals across cellular and tissue membranes, thereby playing a pivotal role in drug distribution and subsequent disposition, efficacy, and toxicity as they all have been implicated in clinically relevant drug–drug interactions (DDIs) [31]. In addition to their association with DDI, the drug transporters mediate a differential distribution of drugs through body compartments within a human population as a result of the occurrence of single-nucleotide polymorphisms (SNP) within the transporter molecules [42].

The expression and localization of ABC and SLC transporters in the AP and BL sides of epithelial and endothelial cells, respectively (except in the intestine, where ABC and SLC are located in the AP side), protect and detoxify as they modulate the distribution of drugs to different body compartments [41]. ABCB1 (PGP) is highly expressed in most barrier tissues, such as intestinal epithelia, the BBB, testis, placenta, and endothelial cells of human cardiac vasculature. Additional transporters of importance include the ABCC family, which consists of the MRPs and ABCG2 (BCRP) and are also expressed in many of the epithelia and endothelia that constitute important barrier membranes [43]. Furthermore, the apparent V_d -mediated uptake of hepatocyte drug transporters was evident when cerivastatin uptake by OATP was associated with DDI with cyclosporine, which inhibited cerivastatin uptake. A recent review by Nakanishi and Tamai (2014) [44] summarized the effect of various SLC transporters in up-regulating cancer cells, and among them, OATPs have been proposed to have pathophysiological significance in cancer. The review focused on hormone-dependent breast and prostate cancers with all the compelling evidence accumulated over the last two decades on the significant role of SLC transporters, including OATPs, upon interacting with clinically important anticancer agents and their influence on agents ADME and PK. Therefore, a better understanding of not only the pathophysiological roles but also the pharmacological effects of OATPs in cancer cells should be helpful in combating human malignant tumors. This work will be discussed in detail in Section 2.7.

2.6 TRANSLATIONAL METHODS AND APPROACHES IN DETERMINING DRUG DISTRIBUTION

2.6.1 In Vitro Methods for Determination of Protein Binding

There are several in vitro methods used in the pharmaceutical industry to measure extent of drug binding to plasma or cellular compartments, including equilibrium dialysis, which is one of the most common, easy to set up, and data are easy to interpret by pharmaceutical scientists. There are other in vitro methods used to determine protein binding, such as ultrafiltration, ultracentrifugation, and chromatographic separation (see Figure 2.1). In addition to these methods, there are much less commonly used approaches, such as dynamic dialysis, exclusion chromatography, and circular dichroism. The methods that allow high throughput are 96-well ultrafiltration and 96-well equilibrium dialysis [45–47]. The method of chromatographic separation, for example, using an HSA-immobilized column, separates multiple test compounds on the same column. This allows relative ranking by calculation of percent binding for each compound. In addition to ease of setup and use, it is relatively inexpensive. With the 96-well ultrafiltration method (an automated and rapid method for analyzing multiple compounds), percent binding is calculated. Nonspecific binding of the analyte to the plastic housing or ultrafiltration membrane surface can affect the quality of the data; however, by using Teflon materials, the issue of binding to film and apparatus is less critical. Consequently, the 96-well equilibrium dialysis method is the "gold standard" means of protein binding analysis [46]. Protein binding studies are traditionally performed with ^{14}C-labeled compounds, but unlabeled drug using LC-MS/MS detection has been also used. With the equilibrium dialysis method, the procedure is performed in a 96-well system. There is a membrane in each of the 96 wells with buffer solution added to one side of the membrane and an equal volume of plasma containing the test compound added to the other side of the membrane, as shown in Figure 2.1. The plate is incubated at 37°C with rotation for 4–24 h to allow the incubations to reach equilibrium; most drugs achieve equilibrium at 4–6 h. Pre-assessment of drug stability in plasma is always recommended to better select time for equilibrium if the analyte is not stable in plasma under long incubation conditions. Samples are removed from either side of the membrane for analysis by LC-MS/MS

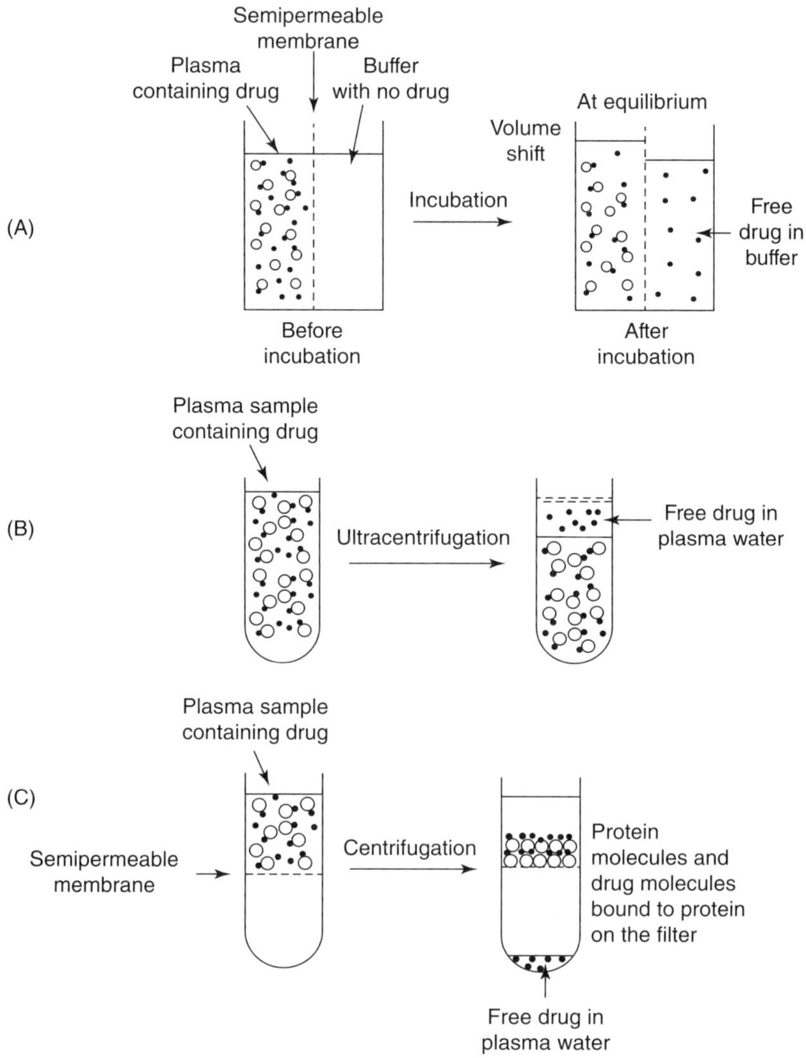

Figure 2.1 Methods to determine plasma protein binding. (A) Equilibrium dialysis, (B) ultracentrifugation, and (C) ultrafiltration.

(or counted by liquid scintillation counter if radiolabeled drug is used). The equation to calculate free fraction is as follows:

$$\text{fu} = \frac{\text{Drug concentration in buffer}}{\text{Drug concentration in plasma}}, \quad (2.5)$$

where drug concentration in buffer is the unbound compound after dialysis, and drug concentration in plasma is the bound drug remaining in the plasma compartment postdialysis [46]. This equation assumes that error added as a result of volume shift is negligible since potential error in the 96-well format is minimized.

The percentages of drug unbound (free) and bound to protein are calculated as follow:

$$\%\ \text{Free} = \text{fu} \times 100, \quad \text{and}\ \%\ \text{bound} = (1 - \text{fu}) \times 100. \tag{2.6}$$

Recovery of test compound at the end of the study should theoretically be 100%, and can be determined from the percentage of the ratio of total drug concentration in both buffer and plasma compartments after the dialysis with total drug concentration in both buffer and plasma compartments before the dialysis. Any deviation from 100% may indicate binding of the compound to the dialysis equipment or a solubility issue of the test drug. Protein binding studies are conducted as a routine part of the ADME process as mentioned, using radiolabeled drug or nonlabeled using LC-MS/MS for sensitivity and selectivity matters.

In translational setting, plasma protein binding using equilibrium binding can be investigated as follow:

1. Interindividual variability among subjects can be examined by testing six plasma samples from the same subjects that the Phase 1-PK clinical investigations used, or from similar genders, ethnicity, age, or disease state that the clinical investigation is intended for [26].
2. Minimize inter-/intratechnical errors by testing the same plasma sample 3 times within the same assay and on three different occasions [26].
3. To validate the assay, both high and low binding compounds can be tested as control in the same assay when new compound will be tested.

2.6.2 In Vivo Protein Binding Studies in Preclinical Animals and Humans

2.6.2.1 Using Radiolabeled Drugs Drug distribution studies have been performed in preclinical species by applying techniques such as QWBA and MBS where radiolabeled drugs are used. Generally, for MBS, a single oral or intravenous dose of radiolabeled drug is given to a preclinical species such as rat. Blood samples are taken at time intervals from animals for PK analyses, while urine, feces, and tissues are removed at specific time points to determine tissue distribution in whole body or in excreta by elimination [48]. In addition, QWBA is applied to determine drug distribution, also by using radiolabeled drug in a live animal such as rat [49]. It is a sensitive and quantitative method. The main advantage of QWBA is that it allows quantitative mapping of all body organs as a function of time of dosing to assess the distribution of a drug in individual tissues and throughout the body and study duration [49]. However, it is not possible to distinguish whether the radioactivity detected is from the drug or from its metabolites, as it is in the case of MBS, where radiolabeled detector in conjunction with LC-MS/MS and liquid scintillation counter are applied to distinguish the distribution of parent compared to metabolites. It is an invasive technique and analysis is postmortem, therefore a large number of animals are required per each time point.

In the mass balance and tissue distribution of the drug, either in animals or humans, the total dose used in the radiolabel studies is usually a pharmacologically relevant dose. If a specific dose has not yet been established, a total dose of radioactivity could be 100 μCi or less, providing that exposure of a specific tissue to the radiolabeled drug does not exceed radiation limits to avoid local radiation damage [50–52]. The most commonly used radioisotope is ^{14}C. It is highly stable and therefore no chemical decomposition of the compound occurs; however, the possibility of compromised safety to radioactivity exposure and its impact on overall human health make it risky [52]. In addition, the cost to synthesize and purify the radiolabeled compound under good manufacturing practice guidelines, the time required for this process, and the further expense of correct handling and disposal of radioactive material make the use of radiolabeled drugs in human ADME investigation less common. These classic approaches were used in a study with paclitaxel followed by gemcitabine given to patients with non-small-cell lung cancer where V_d and other PK parameters were generated to optimize their treatment regimens in Phase 1 trials [53]. Furthermore, plasma distribution of antiretrovirals in HIV-1-infected patients was investigated in a routine manner as indicated above, but the distribution of drug in the fat compartment was also analyzed after collection of adipocyte by-product of lipostructure plastic surgery in patients [54].

2.6.2.2 Applying Advanced Translational Tools for Determining Drug Distribution in Humans In addition to the use of traditional methods in assessing tissue distribution of a test compound in patients after oral or iv dosing by analyzing plasma, feces, and urine levels of parent drug and metabolites, the noninvasive use of PET, magnetic resonance imaging (MRI), and in silico modeling is now considered a significant tool for ADME investigation with the development of a new drug.

PET and imaging are some of the most modern technologies, injected noninvasively, in patients with a biologically active radiolabeled drug (radiotracer) at targeted sites within the body. Radiation emitted from the sites of accumulation is detected in a three-dimensional image of the radiotracer of drug distribution. It is very sensitive and allows determination of absolute radioactivity concentrations within tissues [55]. PET technology has been used effectively in cancer, especially in localization of brain tumors and to monitor the distribution of pharmacotherapy, as in applying temozolomide. By applying the PET technology, temozolomide concentration profiles of different dosing regimens in brain tumor were assessed, allowing the optimization of dosing and treatment schedules [56].

In addition to PET, in vivo noninvasively are two other imaging methods to measure drug disposition, MRI and magnetic resonance spectroscopy (MRS). MRI can create an image that depends on the rates of magnetization for each point, while MRS produces spectra that depict individual changes in the concentrations of parent drug and metabolite only qualitatively and does not to measure absolute concentrations [55,57], though it has low spatial resolution compared to PET, and it provides average drug concentrations from large regions, for example, whole brain.

For the in silico model, as discussed in chapter 8, physiologically based pharmacokinetic (PBPK) models have also been used to predict ADME parameters [58,59]. PBPK uses mathematical models to predict PK parameters and to present quantitative structure-activity relationship models of clinical PK for clearance and V_d.

2.6.3 Assess Drug Distribution from Transporter Studies

There are several established in vitro methods to evaluate the role of uptake and efflux transporters in the distribution of a drug. While the in vitro methods used to evaluate the uptake transporters were discussed in chapter 1, in this section, the methods used to assess the role of efflux transporters will be discussed.

2.6.3.1 Use of Membrane Vesicles Membrane vesicles are usually prepared from control and transfected cells, which express the transporters such as ABC efflux transporters. The most commonly used system is baculovirus-transfected Sf9 insect cells. When conducting the study, drugs are incubated with the membrane vesicles that are prepared from the cell system that expresses the transporters in the presence or absence of ATP (-ve control). Drugs that are substrate of the transporter accumulate in the vesicles in the presence of ATP; without ATP no accumulation is observed. The vesicular transport assay is a direct measure of transporter activity. Following incubation, vesicles can be separated from the incubation solution by rapid filtration through filters or nitrocellulose membranes. The compound trapped inside the vesicle is retained on the filter and quantitatively determined by bioanalytical methods (LC-MS/MS, radioactivity, fluorescence) depending on drug label.

In general, the advantages of using membranes and vesicles are the simplicity, that they can be used in a high throughput manner, and that they are a good system to investigate a single transporter. The disadvantage of these technologies is that there can be a high background of nonspecific binding of the compound to cell membranes in vesicle or cellular systems. However, these assays when used with other in vitro methods help the investigator to understand, more thoroughly, the relevant PK properties of new drugs in relation to specific transporters [60].

2.6.3.2 Use Cultured-Cell Based Assay Unlike the membrane vesicles, cultured cells are utilized to study both uptake and efflux of test compounds and determine the transport mechanism as discussed in chapter 1. Several well-developed cell cultures are routinely used depending on the parameters under investigation. To determine the distribution of drug in the intestine after oral administration, the Caco-2 cell line or MDCK-transfected cells with drug transporters (with uptake or efflux transporters) are used. Use of transfected cell lines such as MDCK-MDR1 allows focus on a single transporter and could help identify drug permeability issues early on in drug development.

On the other hand, to determine the distribution of drug in the brain, the BBB is used.

Cells isolated from intact tissue (primary cells), such as hepatocytes isolated from liver tissue [61,62], or those isolated from whole kidney such as proximal tubular cells

(PTCs) [61], can also be used to evaluate the role of transporters in the hepatobiliary disposition or kidney disposition and distribution of drugs, respectively. Preparation of primary cell cultures is more technically difficult and requires specific skills to ensure their use is valid. In the case of hepatocytes, the sandwich-cultured hepatocytes model [14] was developed to help maintain more in vivo–like properties such as cell polarity and shape and bile canaliculi reformation. Also, the PTC model that can be isolated and maintained in Transwell format was used to determine the transport [61], accumulation, and excretion of drug in kidney [62]. The different primary cell models such as suspension, monolayer, sandwich-cultured format (as in hepatocytes), or Transwell format as in PTC require more technical skills, use of specific chemical inhibitors, or use of primary cells from transporter-deficient animals (or genetically polymorphic human subjects) in order to assess mechanisms underlying the transporter-mediated drug disposition [14].

For brain, the in vitro cell model approach to determine the permeability of drug across the BBB was assessed in vitro using MDCK cell-expressed P-gp for efflux transport, through the transcellular membrane for passive diffusion, or paracellular through resisting (high transepithelial electrical resistance value) TJ between the cells, as those exist in the BBB. However, the epithelial cells such as Caco-2 express more uptake and efflux transporters, which makes them much better suited for studies investigating different transporters within the BBB [63].

Furthermore, estimation of unbound drug concentrations, drug distribution, and influence of efflux transporter P-gp was achieved based on cumulative work of Pollack and his collaborators [64–67]. For example, when the role of P-gp in determining the net brain uptake of fexofenadine was tested by Zhao et al. (2009) [64], the following models were used. The investigators conducted in vivo experiments with subcutaneous administration of fexofenadine in mdr1a (−/−) mice and in mdr1a (+/+) mice. The results indicated that brain penetration was 48-fold higher in mdr1a (−/−) mice than in mdr1a (+/+) mice. Using an in situ brain perfusion technique, the P-gp efflux ratio of fexofenadine at the BBB was only ∼ 4. PK modeling based on experimental results indicated that the apparent fexofenadine P-gp efflux ratio was time dependent due to low passive permeability at the BBB. Fexofenadine brain penetration after terfenadine administration was approximately 25- to 27-fold higher than after fexofenadine administration in both mdr1a (+/+) and mdr1a (−/−) mice, consistent with terfenadine metabolism to fexofenadine in murine brain tissue. By using tissue homogenate to determine the impact of metabolism, the results suggested that the hydroxylation of terfenadine is the rate-limiting step in fexofenadine formation and regional brain metabolism seems to be an important factor in terfenadine brain disposition and, consequently, fexofenadine brain exposure.

In addition to this work, brain slice and brain homogenate were also used to describe distribution of unbound drugs from the intracellular to extracellular in brain tissue [68].

2.7 IMPACT OF DRUG DISTRIBUTION IN DRUG DISPOSITION DDI IN CLINIC

Understanding the interplay of ADME with the disposition of a drug is an essential matter for drug development programs at the preclinical and clinical stage as well as for the clinicians responsible for prescribing an individualized drug treatment for each patient. Furthermore, it is essential that clinicians can use the drug optimally in all groups of patients but have full understanding that subgroups of patients may be at risk during therapy, due to their genetic polymorphism, age (pediatrics or geriatrics), diet, coadministered drugs, and renal- or hepatic-impaired functions. This is particularly important for drugs with a low therapeutic index, such as warfarin. While in hepatic-impaired patients drugs that are known to be cleared by liver are excluded otherwise, dose adjustment is made by monitoring changes in liver enzymes through the review of blood chemistry, for example, with drugs such as Plavix and tamoxifen.

The critical determination of PK is crucial to enable optimal and safe use of drugs in patients and for new product development. By measuring drug concentration in tissues and body fluids over time and fitting them to a mathematical model of the body, PK parameters including V_d can be determined. There are several ways to determine V_d from clinical data, including (1) noncompartmental model using WinNonlin, (2) one-compartment model, or (3) two- compartment model.

For a *one-compartment model*, it is a simple analysis assuming that all the body compartments are in rapid equilibrium with a central compartment, usually the blood, and that the concentration of a drug is constant throughout this compartment as indicated in Figure 2.2(A). In this simple model, the body is considered to be a single well-stirred compartment, and it is assumed that if a drug is injected iv into this compartment, the drug will instantaneously distribute throughout it. Thus, the concentration of drug at zero time (C_o) can be calculated; however if C_o is known, then V_d can be calculated from Equation (2.7):

$$C_o = \frac{D}{V_d}, \tag{2.7}$$

where D is the drug dose. V_d does not represent the volume of an anatomical compartment but links the total amount of drug in the body to its concentration in blood plasma. If the drug only distributes into plasma water (about 3 L in man), then its plasma concentration will be higher than if it distributes into all the extracellular water (about 14 L) or indeed total body water (about 40 L).

Although V_d is a mathematical parameter, without absolute physiological meaning, it gives important information about a compound:

1. A very high V_d value indicates that the drug is localized or sequestered in a storage site such as fat or bone.
2. If the V_d is low, it indicates that the drug is retained in the plasma

In Table 2.6 various drugs with their corresponding V_d are listed.

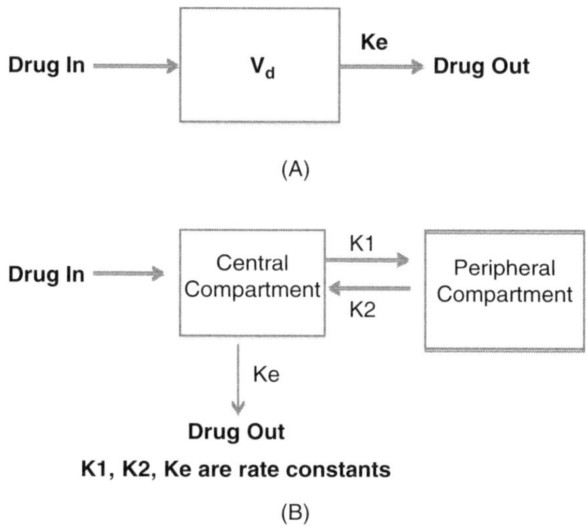

Figure 2.2 Determination of V_d from in vivo data. (A) One-compartment model with well-perfused central compartment, and (B) two-compartment model with central and peripheral compartments. K1, K2, and Ke are rate constants.

TABLE 2.6 Values of Volume of Distribution for Selected Drugs.

Drug	Vd L/Kg
Heparine	0.05–0.1
Insulin	0.05–0.1
Warfarin	0.1–0.2
Atenolol	0.1–0.2
Theophylline	0.4–0.7
Paracetamol	1–2
Lidocaine	1–2
Morphine	2–5
Propranolol	2–5
Digoxin	2–5
Chloroquine	200

A *two-compartment model*, which introduces a separate peripheral compartment to represent the tissues, in contact with the central blood compartment, resembles the real body situation more closely, as indicated in Figure 2.2(B).

As expected, the analysis of a two-compartment model is more complex than that for a one-compartment model and where the drug is not usually given intravenously. There, the plasma concentration profile will be a composite of absorption in addition to distribution and elimination. For a one-compartment model, the absorption rate constant (K_a) and the elimination rate constant K_{el} can be calculated by extrapolation of the appropriate part of the log dose curve. However, in the two-compartment model, absorption, distribution, and elimination are occurring simultaneously. The administered dose of the drug may not be the same as the dose that was absorbed and available systemically. This is due partly to incomplete absorption from the GI tract and to first-pass metabolism mainly by the liver. The extent of first-pass metabolism by the liver can be extensive. For example, approximately 70% of the drug propranolol is metabolized before it reaches the systemic circulation when given orally. The dose of a drug actually absorbed can be quantified by measuring the area under the curve, which is a measure of the total body load of a drug.

Indeed, the accurate measurement of drug distribution will positively impact on the PK/PD of therapy in patients. As discussed, the influence of drug uptake efflux transporters on the distribution of drugs and metabolites across body compartments and organs can also contribute to drug efficacy and safety. The compelling evidence summarized by Nananishi and Tamai [44] showing OATPs are up-regulated in cancer cells presents the case that OATPs can be considered potential targets for chemotherapy. There is a requirement of OATP-mediated active transport of steroid hormones that trigger breast and prostate cancers, through plasma membranes. Based on this recent review, it has been recommended that suppression of OATP, which is up-regulated in cancer cells, can be a therapeutic strategy to block the cell entry of sulfate conjugates of steroid hormones, thus diminishing tumor growth. Also, an efficient anticancer-specific drug delivery system would be required to deliver OATP inhibitors, which would be selected based on the expression profiles of OATPs in the targeted cancer.

This concept of interference of drug transporter in drug therapy, as in the case of cancer, was used to gain therapeutic value during World War II when penicillin was still in short supply. Doctors maximized its efficacy by coadministration of probenecid, which competitively inhibits its secretion from the renal tubules [69].

It has been commonly shown that DDI may occur when interference with plasma protein binding takes place, when one drug interferes with carrier-mediated transport that is responsible for hepatobiliary and renal disposition or metabolism of another drug, for example, statin drugs that depend on hepatic OATP family for the uptake from blood to hepatocytes. Several cases of severe side effects caused by increasing plasma concentration of statins were reported, when various drugs such as cyclosporin A were coadministered. Cyclosporin A is known to inhibit OATP1B1-mediated uptake of statin drugs and the degree of inhibition of OATP1B1 in humans may lead to severe side effects such as myopathy and rhabdomyolysis [70–72]. When in vitro inhibition parameter K_i of potential inhibitor to OATPs was compared to the unbound drug concentration of inhibitor in blood circulation, several OATP inhibitors were identified, such as rifampicin, clarithromycin, and rifamycin SV in addition to cyclosporin A. These drugs would interact with pitavastatin in vivo, causing an increase in the plasma concentration, becoming more efficacious in cholesterol-lowering activity but leading to unwanted side effects.

REFERENCES

1. Caldwell GW, Yan Z, Tang W, et al. ADME optimization and toxicity assessment in early- and late-phase drug discovery. Curr Top Med Chem 2009;9(11):965–980.
2. Wang J. Comprehensive assessment of ADMET risks in drug discovery. Curr Pharm Des 2009;15(19):2195–2219.
3. Song N-N, Zhang S-U, Liu C-X. Overview of factors affecting oral drug absorption. Asian J Drug Metabolism Pharmacokinetics 2004;4(3):167–176.
4. Schanker LS, Shore PA, Brodie BB, Hogben CAM. Absorption of drugs from the stomach I. The rat. J Pharmacol Exp Ther 1957;120(4):528–539.
5. Penner N, Klunk LJ, Prakash C. Human radiolabeled mass balance studies: objectives, utilities and limitations. Biopharm Drug Dispos 2009;30(4):185–203.
6. Kitson SL, Mansi L. Editorial review of 2014. Curr Radiopharm 2014;7(1):1–4.
7. Rodgers T, Rowland M. Mechanistic approaches to volume of distribution predictions: understanding the processes. Pharm Res 2007;24(5):918–933.
8. Rowland M, Riegelman S. Determination of acetylsalicylic acid and salicylic acid in plasma. J Pharm Sci 1967;56(6):717–720.
9. Rowland M. Amphetamine blood and urine levels in man. J Pharm Sci 1969;58(4): 508–509.
10. Hidalgo IJ. Assessing the absorption of new pharmaceuticals. Curr Top Med Chem 2001;1(5):385–401.
11. Benet LZ, Izumi T, Zhang Y, et al. Intestinal MDR transport proteins and P-450 enzymes as barriers to oral drug delivery. J Control Release 1999;62(1,2):25–31.
12. Wallace GC, Ramsden DB, Grant MH. General principles of drug distribution. In: Encyclopedia of drug metabolism and interactions, ed. Lyubimov AV. Hoboken, NJ: Wiley, 2012, 3:1–47.
13. Levitt DG. Quantitative relationship between the octanol/water partition coefficient and the diffusion limitation of the exchange between adipose and blood. Clin Pharmacol 2010;10:1–13.
14. Yanni SB, Augustijns PF, Benjamin DK Jr, Brouwer KL, Thakker DR, Annaert PP. In vitro investigation of the hepatobiliary disposition mechanisms of the antifungal agent micafungin in humans and rats. Drug Metab Dispos 2010 38:1848–1856.
15. Vogel S. Vital circuits: on pumps, pipes and the workings of circulatory systems. New York: Oxford University Press, 1992.
16. Banker MJ, Clark TH. Plasma/serum protein binding determinations. Curr Drug Metab 2008;9:854–859.
17. Heikkinen AT, Baneyx G, Caruso A, Parrott N. Application of PBPK modeling to predict human intestinal metabolism of CYP3A substrates—an evaluation and case study using GastroPlus. Eur J Pharm Sci. 2012;47(2):375–386.
18. Otagiri M. Study on binding of drug to serum protein. Yakugaku Zasshi 2009;129(4): 413–425.
19. Tozer TN. Implications of altered plasma protein binding in disease states. In: Benet Lz, editor. Pharmacokinetic basis for drug treatment. New York: Raven Press; 1984. pp. 173–193.
20. Piaky KM. Disease-induced changes in the plasma binding of basic drugs. Clin Pharmacokinet 1980;3:246–262.

21. Verbeek RK, Cardinal JA, Wallace SM. Effect of age and sex on the plasma binding of acidic and basic drugs. Eur J Clin Pharmacol 1984;27(1):91–97.

22. Perucca E, Crema A. Plasma protein binding of drugs in pregnancy. Clin Pharmacokinet 1982;7(4):336–352.

23. Perucca E, Ruprah M, Richens A. Altered drug binding to serum proteins in pregnant women: therapeutic relevance. J R Soc Med 1981;6:422–426.

24. Benedetti MS, Whomsley R, Baltes EL. Differences in absorption, distribution, metabolism and excretion of xenobiotics between the paediatric and adult populations Expert Opin Drug Metab Toxicol 2005;1(3):447–471.

25. Heresi GP, Gerstmann DR, Reed MD, van den Anker JN, Blumer JL, Kovanda L, Keirns JJ, Buell DN, Kearns GL. The pharmacokinetics and safety of micafungin, a novel echinocandin, in premature infants. Pediatr Infect Dis J 2006;25(12):1110–1115.

26. Yanni SB, Smith PB, Benjamin DK Jr, Augustijns PF, Thakker DR, Annaert PP. Higher clearance of micafungin in neonates compared with adults: role of age-dependent micafungin serum binding. Biopharm Drug Dispos 2011;32(4):222–232.

27. Olive G: Pharmacocinétique et biotransformation des médicaments chez l'enfant. Louvain Med 1991;110:565–569.

28. Daood M, Tsai C, Ahdab-Barmada M, Watchko JF. ABC transporter (P-gp/ABCB1, MRP1/ABCC1, BCRP/ABCG2) expression in the developing human CNS. Neuropediatrics 2008;39(4):211–218.

29. Tesseromatis C, Alevizou A. The role of the protein-binding on the mode of drug action as well the interactions with other drugs. Eur J Drug Metab Pharmacokinet 2008;33(4):225–230.

30. Meijer DK, Van der Sluijs P. The influence of binding to albumin and alpha 1-acid glycoprotein on the clearance of drugs by the liver. Pharm Weekbl Sci 1987;9(2):65–74.

31. Giacomini KM, Huang SM, Tweedie DJ, et al. Membrane transporters in drug development. Nat Rev Drug Discov 2010;9:215–236.

32. Ho RH, Kim RB. Transporters and drug therapy: implications for drug disposition and disease. Clin Pharmacol Ther 2005;78:260–277.

33. Chandra P, Brouwer KL. The complexities of hepatic drug transport: current knowledge and emerging concepts. Pharm Res 2004;21:719–735.

34. Deoon Lee EJ, Lean CB, Limenta LM. Role of membrane transporters in the safety profile of drugs. Expert Opin Drug Metab Toxicol 2009;5(11):1369–1383.

35. Szakacs G, Varadi A, Ozvegy-Laczka C, Sarkadi B. The role of ABC transporters in drug absorption, distribution, metabolism, excretion and toxicicty (ADME-Tox). Drug Discov Today 2008;13(9/10):379–393.

36. Jung N, Taubert D. Organic cation transporters and their roles in antiretroviral drug disposition. Expert Opin Drug Metab Toxicol 2009;5(7):773–787.

37. Hagenbuch B, Gao B, Meier PJ. Transport of xenobiotics across the blood-brain barrier. News Physiol Sci 2002;17:231–234.

38. Sharom FJ The P-glycoprotein multidrug transporter. Essays Biochem 2011;50(1):161–178.

39. Tsybovsky Y, Palczewski K. Expression, purification and structural properties of ABC transporter ABCA4 and its individual domains. Protein Expr Purif 2014;97:50–60.

40. Fredriksson R, Nordstrom KJ, Stephansson O, et al. The solute carrier (SLC) complement of the human genome: phylogenetic classification reveals four major families. FEBS Lett 2008;582(27):3811–3816.

41. He L, Vasiliou K, Nebert DW. Analysis and update of the human solute carrier (SLC) gene superfamily. Hum Genomics 2009;3(2):195–206.

42. Shitara Y, Horie T, Sugiyama Y. Transporters as a determinant of drug clearance and tissue distribution. Eur J Pharm Sci 2006;27:425–446.

43. Grover A, Benet LZ. Effects of drug transporters on volume of distribution. AAPS J 2009;11(2):250–261.

44. Nakanishi T, Tamai I. Putative roles of organic anion transporting polypeptides (OATPs) in cell survival and progression of human cancers. Biopharm Drug Dispos 2014. doi:10.1002/bdd.1915.

45. van Liempd S, Morrison D, Sysmans L, et al. Development and validation of a higher-throughput equilibrium dialysis assay for plasma protein binding. J Lab Autom 2011;16(1):56–67.

46. Banker MJ, Clark TH, Williams JA. Development and validation of a 96-well equilibrium dialysis apparatus for measuring plasma protein binding. J Pharm Sci 2003;92(5):967–974.

47. Kariv I, Cao H, Oldenburg KR. Development of a high throughput equilibrium dialysis method. J Pharm Sci 2001;90(5):580–587.

48. Gad SC. Preclinical development handbook: ADME and biopharmaceutical properties. Hoboken, NJ: Wiley, 2009, 1:1103–1132.

49. Richter WF, Starke V, Whitby B. The distribution pattern of radioactivity across different tissues in quantitative whole-body autoradiography (QWBA) studies. Eur J Pharm Sci 2006;28(1–2):155–165.

50. Roffey SJ, Obach RS, Gedge JI, et al. What is the objective of the mass balance study? A retrospective analysis of data in animal and human excretion studies employing radiolabeled drugs. Drug Metab Rev 2007;39(1):17–43.

51. EMA. Toxicokinetics: a guidance for assessing systemic exposure in toxicology studies. ICH topic S3A, CPMP/ICH/384/95. EMA, 1995.

52. U.S. Department of Health and Human Services. Food and Drug Administration Guidance for Industry M3. Nonclinical safety studies for the conduct of human clinical trials for pharmaceuticals. FDA, 2010.

53. Fogli S, Danesi R, De Braud F, De Pas T, Curigliano G, Giovannetti E, Del Tacca M. Drug distribution and pharmacokinetic/pharmacodynamic relationship of paclitaxel and gemcitabine in patients with non-small-cell lung cancer. Ann Oncol 2001;12:1553–1559.

54. Dupin N, Buffet M, Marcelin AG, Lamotte C, Gorin I, Ait-Arkoub Z, Tréluyer JM, Bui P, Calvez V, Peytavin G. HIV and antiretroviral drug distribution in plasma and fat tissue of HIV-infected patients with lipodystrophy. AIDS 2002;16:2419–2424.

55. Port RE, Wolf W. Noninvasive methods to study drug distribution. Invest New Drugs 2003;21:157–168.

56. Rosso L, Brock CS, Gallo JM, Saleem A, Price PM, Turkheimer FE, Aboagye EO. A new model for prediction of drug distribution in tumor and normal tissues: pharmacokinetics of temozolomide in glioma patients. Cancer Res 2009;69(1):120–127.

57. Singh M, Waluch V. Physics and instrumentation for imaging in-vivo drug distribution. Adv Drug Deliv Rev 2000;41:7–20.

58. Gombar VK, Hall SD. Quantitative structure-activity relationship models of clinical pharmacokinetics: clearance and volume of distribution. J Chem Inf Model 2013 53(4):948–957.

59. Madan AK, Dureja H. Prediction of pharmacokinetic parameters. Methods Mol Biol 2012;929:337–357.

60. Marchetti S1, Pluim D, van Eijndhoven M, van Tellingen O, Mazzanti R, Beijnen JH, Schellens JH. Effect of the drug transporters ABCG2, Abcg2, ABCB1 and ABCC2 on the disposition, brain accumulation and myelotoxicity of the aurora kinase B inhibitor barasertib and its more active form barasertib-hydroxy-QPA. Invest New Drugs 2013;31(5):1125–1135.

61. Verhulst A, Sayer R, De Broe ME, D'Haese PC, and Brown CDA. Human proximal tubular epithelium actively secretes but does not retain rosuvastatin. Mol Pharmacol 2008;74:1084–1091.

62. Kusuhara H and Sugiyama Y. In vitro–in vivo extrapolation of transporter-mediated clearance in the liver and kidney. Drug Metab Pharmacokinet 2009;24:37–52.

63. Alavijeh MS, Chishty M, Qaiser MZ, Palmer AM. Drug metabolism and pharmacokinetics, the blood-brain barrier, and central nervous system drug discovery. J Am Soc Exp Neurother 2005;2(4):554–571.

64. Zhao R, Kalvass JC, Yanni SB, Bridges AS, Pollack GM. Fexofenadine brain exposure and the influence of blood-brain barrier P-glycoprotein following fexofenadine and terfenadine administration. Drug Metab Dispos 2009;37(3):529–535.

65. Padowski JM, Pollack GM. Influence of time to achieve substrate distribution equilibrium between brain tissue and blood on quantitation of the blood-brain barrier P-glycoprotein effect. Brain Res 2011;1426:1–17.

66. Zhao R, Pollack GM. Regional differences in capillary density, perfusion rate, and P-glycoprotein activity: a quantitative analysis of regional drug exposure in the brain. Biochem Pharmacol 2009;78(8):1052–1059.

67. Kalvass JC, Olson ER, Cassidy MP, Selley DE, Pollack GM. Pharmacokinetics and pharmacodynamics of seven opioids in P-glycoprotein-competent mice: assessment of unbound brain EC50,u and correlation of in vitro, preclinical, and clinical data. J Pharmacol Exp Ther 2007;323(1):346–355.

68. Friden M, Ducrozet F, Middleton B, Antonsson M, Bredberg U, Hammarlund-Udenaes M. Development of a high-throughput brain slice method for studying drug distribution in the central nervous system. Drug Metab Dispos 2009;37(6):1226–1233.

69. Walker WF, Hunter RB. Oral administration of Procaine penicillin with and without benemid p(DI-n-propylsulphamyl) benzoic acid. J Lancet 1951;258(6673):104–106.

70. Evans M, Rees A. Effects of HMG-CoA reductase inhibitors on skeletal muscle: are all statins the same? Drug Saf 2002;25(9):649–663.

71. Shitara Y, Itoh T, Sato H, Li AP, Sugiyama Y. Inhibition of transporter-mediated hepatic uptake as a mechanism for drug-drug interaction between cerivastatin and cyclosporin A. J Pharmacol Exp Ther 2003;304(2):610–616.

72. Hirano M, Maeda K, Shitara Y, Sugiyama Y. Drug–drug interaction between pitavastatin and various drugs via OATP1B1. Drug Metab Dispos 2006;34(7):1229–1236.

3

METABOLISM: PRINCIPLE, METHODS, AND APPLICATIONS

3.1 INTRODUCTION: AN OVERVIEW ON DRUG METABOLISM IN RELATION TO CLEARANCE—MEDIATED BY PHASE I, PHASE II, AND PHASE III DRUG-METABOLIZING ENZYMES

Drug metabolism occurs in almost all body organs and comprises a diverse set of chemical reactions within four general categories: oxidation, reduction, conjugation, and hydrolysis. Liver is the body organ that has the greatest metabolic capacity, and consequently drug metabolism has been extensively studied in liver. Hepatic drug metabolism, in particular, is important in the clinical action of drugs because it is often the main means by which drugs are cleared from the body. Consequently, hepatic enzymes play an important role in drug disposition, clearance, and drug–drug interaction (DDI), hence the efficacy and safety of drugs that are cleared by the liver. Hundreds of studies have been conducted during the last 30 years to identify the factors implicated in drug metabolism and metabolic clearance; these include:

- Expression and activity of drug-metabolizing enzymes (DMEs) and drug transporters
- Mg microsomal protein per gram liver and liver size relative to body weight
- Liver blood flow
- Extent of plasma protein binding

Translational ADMET for Drug Therapy: Principles, Methods, and Pharmaceutical Applications, First Edition. Souzan B. Yanni.

Drug metabolism is divided into three phases: phase I (mainly oxidation), phase II (conjugation), and phase III (transport/elimination). The role of the three phases in the disposition of drugs is depicted in Figure 3.1. The role of DMEs in metabolic clearance, renal clearance, or hepatic clearance will be discussed in more detail in this chapter and following chapters. Figure 3.2(A) presents the percentage of metabolic clearance catalyzed by phase I/II in comparison to renally or biliary clearance catalyzed by phase III mediated by transporters. Figure 3.2 clearly indicates that clearance by phase I/II is the predominant elimination pathway of drugs analyzed (~ 70%), followed by renal clearance (~ 25%), while drugs that are predominantly eliminated by biliary clearance are < 5% of drugs analyzed.

Phase I drug metabolism is metabolic reactions with enzymes that catalyze: (a) hydroxylation (aliphatic, aromatic, or olephinic); (b) epoxidation (aliphatic or aromatic); (c) dealkylation of function groups containing O−, N−, or S−; (d) deamination, oxidation (N− or S−); (e) reduction (nitro, disulfide, ketoaldehyde, or olefin); (f) and hydrolysis (amide, ester, carbamate, or epoxide) as shown in Figures 3.3 and 3.4. The phase I reactions are known to introduce or unmask functional groups within a molecule to increase its solubility [1–5].

Phase II is metabolic reactions, mainly conjugation, including glucuronidation, sulfation, methylation, acetylation, amino acid conjugations (glycine, glutamic acid, and taurine), and glutathione (GSH) conjugation [6–9].

In phase III, which is the transporter-mediated process, including P-glycoprotein (P-gp) [10], multidrug resistance-associated protein (MRP) [11], and organic anion transporting polypeptide 2 (OATP2) [12] transporters are expressed in many tissues

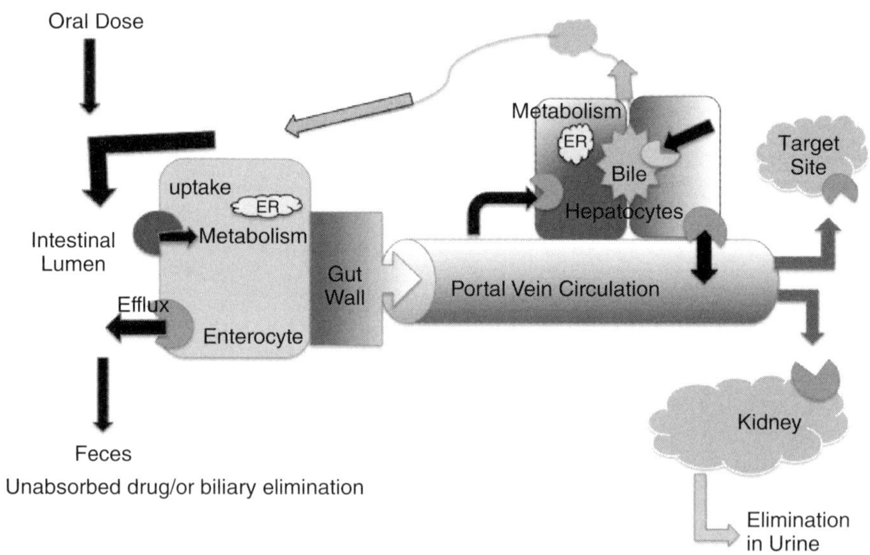

Figure 3.1 Role of drug metabolism by phase I/II DME at endoplasmic reticulum (ER) and by phase III mediated by membrane-bound drug transporters (efflux or uptake functions) in the disposition and elimination of drugs in bile and urine.

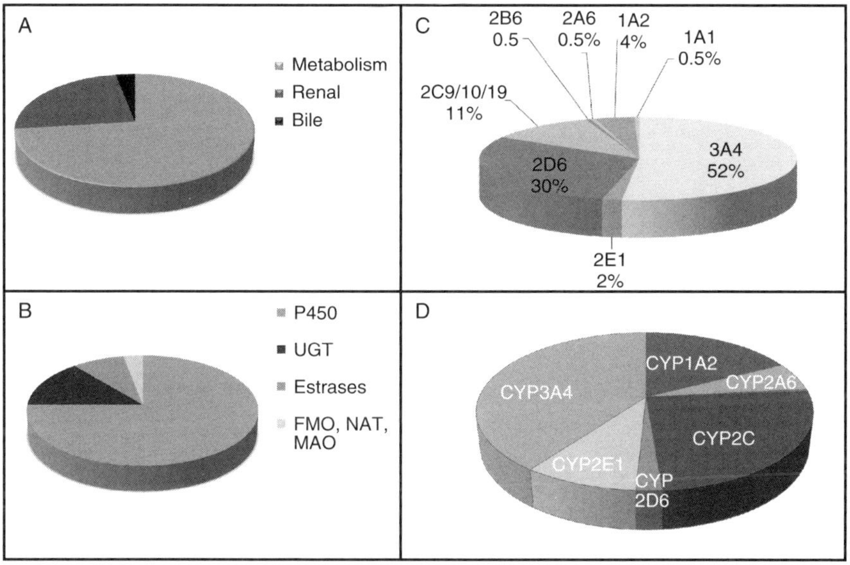

Figure 3.2 Contribution of metabolism, P450, CYP3A to metabolism relative to other elimination pathways, other enzymes, and pther P450s of hundreds of drugs on the market (data listed in various publications). (D) represents the % of each P450 enzyme to total P450 protein content in hepatocytes.

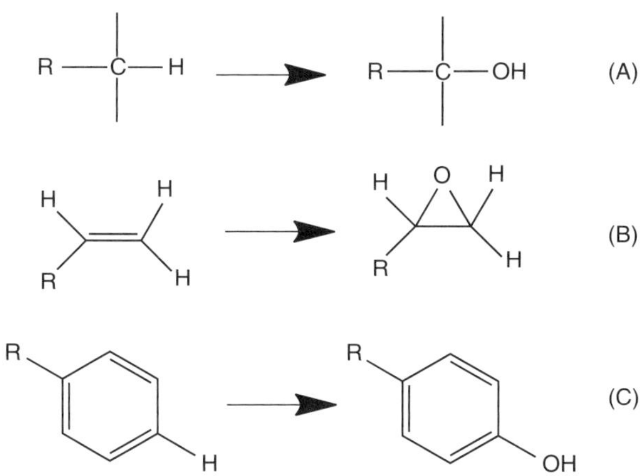

Figure 3.3 Examples of reactions catalyzed by cytochrome P450-mediated phase I. The metabolism is via NADPH-mediated oxidative and reductive reactions covering various classes of drugs and structures such as (A) aliphatic, (B) olephinic, and (C) aromatic.

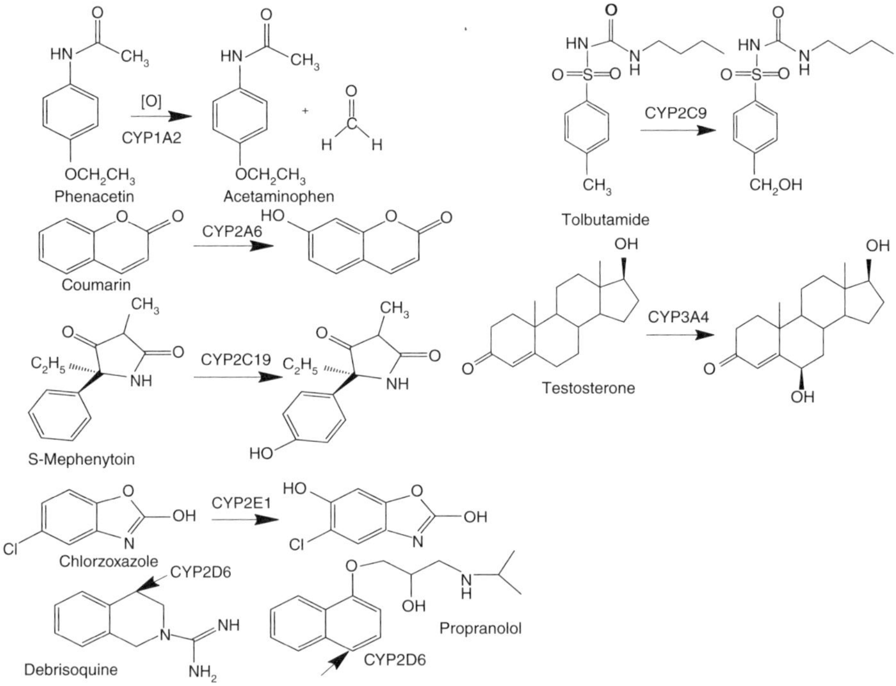

Figure 3.4 Examples of P450-mediated reactions for known drugs and probe substrates catalyzed by CYP1A2, CYP3A4, CYP2A6, CYP2C9, CYP2C19, CYP2E1, and CYP2D6.

such as the liver, intestine, kidney, and brain, where they provide a formidable barrier against drug penetration and play crucial roles in drug absorption, distribution, and excretion [10, 13–15].

The major function of phase II and phase III drug metabolism is the detoxification and subsequent transport/elimination of drugs and other xenobiotics [16]. Consistent with this function, conjugation generally produces a metabolite that is more polar, larger in molecular weight (MW), and charged at physiological pH [17], characteristics that make the metabolite more amenable as a substrate for transport proteins for ultimate excretion into the urine or bile [18]. Most phase II conjugates are inactive and nontoxic [19]; however, there are exceptions where bioactivation or reactive metabolites formed as a result of conjugation [20]. In general, most phase II reactions generate metabolites with higher hydrophilic property and obviously with lower volume of distribution (V_{dss}) and higher excretion from the body. The rate and extent of metabolism of a drug determines the dose of drug and the duration of drug effect. Overall, the role of drug metabolism is to clear the xenobiotics from the body, so that the metabolites tend to be more polar and soluble than the parent drug, making them easier to be excreted. With the aid of drug transporters, parent drug penetrates into the metabolizing organs and passages of ionic metabolites across cell membranes (biliary canaliculi of hepatocytes or apical and baselateral membranes of proximal tubular cells in kidney) into the excreta, bile, or urine, respectively [21]. As lipophilic drug is oxidized, hydrolyzed, or reduced by phase I reaction, to form epoxide, hydroxyl, sulfhydryl, or primary amine, it forms electrophiles or nucleophiles that undergo conjugations by phase II enzymes (glucuronidation, acetylation, sulfation, or GSH conjugation) that form the hydrophilic form of the drug (metabolite). Finally, in phase III, the conjugated xenobiotics may be further processed, before being recognized by efflux transporters and pumped out of cells. Figure 3.2(B) shows the comparison between metabolism reactions catalyzed by oxidative P450 pathways, other oxidative pathways (e.g., flavin-containing monooxygenase [FMO]), and conjugation pathways. Clearly, oxidative metabolism by P450 represents the most common pathway for metabolic clearance of most drugs on the market.

When we discuss translational absorption, distribution, metabolism, and excretion (ADME) and, in particular, drug metabolism, it is necessary to identify the enzymes mediating the metabolism of drug candidates, as it may reveal several aspects of drug disposition during actual drug therapy in various patient populations. The extent of drug metabolism and potential metabolite profile can be influenced by genetic variation, DDIs, time-dependent inhibition, autoinduction, and the effect of disease, age, diet, and ethnicity. Because of the influence of these factors on drug metabolism in humans, the clinical drug development programs of new drug candidates have covered the characterization of the metabolic pathways, identification of major circulating metabolites, and determination of the enzymes that produce these metabolites. Accordingly, several types of clinical studies of metabolism are mandated for new drug registration, including determination of the major pathways of clearance (CL), characterization of DDIs based on metabolic phenomena, and assessment of the extent of excretion of drug-derived materials from the body [18, 22].

As mentioned earlier, most of the metabolic reactions of drugs occur by enzymes through oxidation, reduction, conjugation, or hydrolysis. These reactions are summarized in Table 3.1 by indicating significant characteristics that distinguish these reactions from each other, cofactor, location of enzyme at the subcellular level, and its role in the overall metabolic pathways [21].

Oxidation (reduction) involves insertion of oxygen, removal of hydrogen, or removal of electrons. Reduction reactions involve exactly the opposite changes. Oxidative (reductive) reactions are catalyzed by cytochrome P450 (P450), FMO, peroxidases (cyclooxygenases [prostaglandin H synthase], lactoperoxidase, myeloperoxidase), catalase, amine oxidases (monamine oxidase [MAO], diamine oxidase [DAO], polyamine oxidase [PAO]), dehydrogenases (alcohol dehydrogenase, aldehyde dehydrogenase, carbonyl reductase), and xanthine oxidase. For other phase I reactions, reductive reactions include the reduction of carbonyl groups (like aldehydes and ketones), reduction of quinones, reduction of nitro groups, azo groups, and disulfides, and reduction of halogen-substituted carbons. Last, there are phase I reactions that involve hydrolysis of carboxy esters, thioesters, amide, carbonates, carbamates, sulfates, phosphates, oxides/hydration. These reactions are catalyzed by enzymes like esterases, peptidases, proteases, phosphatases, nucleases, phosphodiesterases, and epoxide hydrolase [23].

TABLE 3.1 Phase I oxidative Reactions and Reductions.

Phase I Reaction	Specification
Cytochrome P450 (CYP) typical reaction: oxygenases, oxidases, and dehydrogenases	$R - H + O_2 + NADPH + H^+ \rightarrow R - OH + H_2O + NADP^+$ Required cofactors: O_2, NADPH, CPR, and cytochrome b_5. Tissue expression: liver and most other tissues. Localization: endoplasmic reticulum. Substrate binding to enzyme is rate-limiting step. Enzyme is subject to inhibition and induction
Flavin-containing monooxygenase (FMO)	$R_3N + O_2 + NADPH + H^+ \rightarrow R_3N^+ - O^- + H_2O + NADP^+$ FAD. Required cofactors: O_2 and NADPH. Tissue expression: liver and other tissues. Endoplasmic reticulum. Substrate binding to enzyme is not rate limiting. No inhibition or induction
Aldehyde oxidase (AOX)	$RCH_2NH_2 + H_2O + O_2 \rightarrow RCH = O + NH_3 + H_2O_2$. Molybdenum, [2Fe–2S] centers and flavin. Required cofactors: O_2. Tissue expression, liver and other tissues. Localization: cytosol
Monoamine oxidase (MAO)	$RCH_2NH_2 + H_2O + O_2 \rightarrow RCH = O + NH_3 + H_2O_2$. Prosthetic group: flavin. Required cofactors: O_2. Tissue expression: many tissues. Localization: mitochondria
Xanthine oxidase (XOR)	Xanthine $+ H_2O + O_2 \rightarrow$ uric acid $+ H_2O_2$. Prosthetic group: molybdenum, [2Fe–2S] centers and flavin. Required cofactors: O_2. Tissue expression: liver. Localization: cytosol
Alcohol dehydrogenase (ADH)	$R - CH_2 - OH + NAD^+ \rightarrow R - CH = O + NADH + H^+$. Prosthetic group: none. Required cofactors: NAD^+. Tissue expression: liver. Localization: cytosol
Aldehyde dehydrogenase (ALDH)	$R - CH = O + NAD^+ + H_2O \rightarrow R - COOH + NADH + H^+$. Prosthetic group: none. Required cofactors: NAD^+. Tissue expression: liver. Subcellular localization: cytosol
Aldo-keto reductase (AKR). Quinone reductase (NQOI). Quinone + NADPH $\rightarrow$ hydroquinone + $NADP^+$. Prosthetic group: flavin	$RCH = O$ or $R_2C == O + NADPH + H^+ \rightarrow RCH_2OH$ or $R_2CHOH + NADP^+$. Prosthetic group: none. Cofactors: NADPH. Tissue expression: liver and others, cytosol

3.2 COMMON PHASE I, II, AND III DRUG METABOLISM REACTIONS

3.2.1 Phase I Drug Metabolism

The major and most significant phase I reactions in drug dispositions are those that are catalyzed by P450 enzymes [Figure 3.2(B)]. Human P450 is a superfamily of membrane-associated proteins found either in the inner membrane of mitochondria or in the endoplasmic reticulum (ER) cellular compartments that can be isolated from the microsomal subcellular fractions of all living organs. They are responsible for the metabolism of various endogenous body circulating compounds such as steroids, prostaglandins, and bile acids and bilirubin, as well as exogenous compounds and xenobiotics, including drugs, carcinogens, pesticides, pollutants, and food toxicants [24, 25]. These enzymes are cellular colored proteins, which contain heme pigments that absorb light at a wavelength of 450 nm when exposed to carbon monoxide; for these characteristics, these enzymes are named as cytochrome P450 [26, 27]. Liver is the major body organ where the expression and function of P450 enzymes are detected. In addition to liver, P450 enzymes are in extrahepatic tissues such as intestine, kidney, lung, heart, brain, and skin. Human P450s can either metabolize only one or sometimes multiple substrates. These characteristics account for their central importance in medicine [28].

The Human Genome Project has identified 57 human genes coding for the various P450 enzymes. Currently, more than 10,000 P450 gene members have been identified in humans and other living species, including bacteria and yeast [28–30]. The content of major P450 isozymes in liver is determined by several investigators who have indicated that CYP3A, found to be 40%, followed by CYP2C enzymes are the major P450 enzymes expressed in hepatocytes and measured in liver microsomal subcellular fraction as % of P450 protein. The CYP3A subfamily is the most abundant (~ 28% of total P450 content) and important DME. Human CYP3A has been shown to catalyze the metabolism for the majority of marketed drugs, as indicated in Figure 3.3(C). In human, CYP3A has four known family members, CYP3A4, CYP3A5, CYP3A7, and CYP3A43. CYP3A4 is expressed in liver, stomach, lung, intestine, brain, skin, and renal tissues. CYP3A4 expression levels are higher in both liver and small intestine, where it metabolizes a large number of therapeutic popular drugs. CYP3A4 has been well studied for its induction or inhibition because of drug–drug, drug–herbal, and drug–food interactions. Drugs such as terfenadine, cisapride, and astemizole cause ventricular arrhythmias when CYP3A4 inhibitors ketoconazole or erythromycin are taken along with these drugs. In the small intestine, CYP3A4 plays a major role in the first-pass metabolism (presystemic clearance) of xenobiotics. Catalytic activity of CYP3A4 decreases longitudinally along the small intestine. Generally, the CYP3A4 concentrations in intestine are 10–50% lower than in liver [31]. CYP3A4 has large active sites that can interact with two drugs simultaneously [32]. Popular drugs such as testosterone and midazolam both can interact in two distinct sites of CYP3A4 called the steroid and benzodiazepine sites, respectively. Although both CYP3A4 and CYP3A5 are expressed in liver and intestine, CYP3A5 is the predominant form expressed in extrahepatic tissues. CYP3A5 expression is polymorphic; five allelic variants of CYP3A5 have been reported [33]. There is some ambiguity in reports of the relative rate of drug and steroid metabolism by CYP3A4 and CYP3A5 [34], but an extensive study of recombinant enzymes with 14 compounds showed CYP3A4 was subject to greater inhibition than CYP3A5 [35]. CYP3A7 and CYP3A43 both play a minor role in drug metabolism. CYP3A7 is a fetal enzyme and is involved in the activation of drugs to teratogenic metabolites and CYP3A43 expressed in liver but has very low or restricted activity (0.2–5%) [36].

P450-mediated phase I reactions cover various classes of drugs and structures such as aliphatic, olephinic, and aromatic (Figures 3.3 and 3.4) via NADPH-mediated oxidative and reductive reactions. As mentioned earlier, these reactions introduce (or unmask) a functional group such as hydroxyl (−OH), carboxylic acid ($-CO_2H$), amine ($-NH_2$), or sulfhydryl (−SH) within a molecule that can enhance its water solubility by becoming hydrophilic drug product. This can occur through direct introduction of the functional group (e.g., aromatic and aliphatic hydroxylation) or by modifying existing functionalities (e.g., oxidative hydrolysis of the esters and amides, oxidative N−, O−, and S-dealkylation, and reduction of aldehydes and ketones as shown in Table 3.1 [30]. As a result, more hydrophilic (water soluble) and polar entities are formed, which are eliminated from the body. In general, metabolism leads to compounds that are generally pharmacologically inactive and relatively nontoxic. However, metabolic biotransformation of drugs at times can lead to the formation of metabolites with pharmacological activity or toxicity [37, 38].

P450 enzymes are responsible for the elimination and metabolism of 75% of drugs on the market, while CYP3A is responsible for more than 50% of these drugs and other developed xenobiotics [39] (Figure 3.2).

3.2.1.1 Oxidation Reaction There four types of oxidation reactions [40]: hydroxylation reactions, epoxidation reactions, heteroatom oxidation, and dehydrogenation reactions.

Hydroxylation reactions replace a hydrogen atom (C–H) with hydroxyl group o to become (C–OH). Hydroxylation of an aliphatic carbon or an aromatic ring is one of the most common drug metabolism reactions. The other common biotransformation reaction is hydroxylation at the α carbon to heteroatoms, which results in oxidative cleavage of the molecule. An example of this reaction is depicted in Figure 3.4. Although hydroxylation is the most common NADPH-dependent P450 oxidative reaction, hydrolysis can also be mediated by non-NADPH-dependent P450 oxidative reactions, such as with hydrolases.

The oxidase activity of P450s involves one electron transfer from reduced P450 to molecular oxygen with the formation of superoxide anion radical and H_2O_2 as shown in the two equations below:

$$\mathrm{NADPH} + \mathrm{O_2} \rightarrow \mathrm{O_2}^{-} + \mathrm{NAD(P)}^{+}$$

$$\mathrm{2NADPH} + \mathrm{2H^{+}} + \mathrm{O_2} \rightarrow \mathrm{H_2O_2} + \mathrm{NAD(P)}^{+}$$

The reductase activity of P450s involves direct electron transfer to reducible substrates such as quinones and proceeds readily under anaerobic conditions. The P450 oxidation catalytic cycle is a complex multistep process [23] as indicated below:

1. P450 enzyme (Fe^{3+}) first binds to a substrate RH to form $Fe^{3+} - RH$. A lowering of the redox potential is generated, which makes the transfer of an electron favorable from its redox partner, NADH or NADPH. This is accompanied by a change in the spin state of the hem iron at the active site.
2. The next step in the cycle is the first reduction of the $Fe^{3+} - RH$ to $Fe^{2+} - RH$ by an electron transferred from NAD(P)H via an electron-transfer chain.
3. An O_2 molecule binds rapidly to the $Fe^{2+} - RH$ to form $Fe^{2+} - O^{-}{}_2 - RH$, which slowly forms a more stable complex $Fe^{3+} - O^{-}{}_2 - RH$.
4. The following cycle is a second reduction of $Fe^{3+} - O^{-}{}_2 - RH$ to $Fe^{3+} - O_2{}^{2-} - RH$ via the electron donors either NADPH or cytochrome b5; this is the rate-determining step of the reaction.
5. The $Fe^{3+} - O_2{}^{2-} - RH$ reacts with two protons from the surrounding solvent, breaking the $O - O$ bond, forming water, and leaving an $(Fe - O)^{3+} - RH$ complex.
6. The Fe-ligated O atom is transferred to the substrate forming a hydroxylated form of the substrate ($Fe^{3+} - XOH$).
7. The last step involves the release of product from the active site of the enzyme, which returns to its initial state.

Epoxidation reactions introduce an oxygen atom into carbon-carbon double, triple bond. Aromatic ring can be subjected to CYP-mediated epoxidation, as indicated in Figure 3.3. Epoxidation results in the formation of unstable products, which hydrolyze by epoxidegydrolyses to form diols or react with nucleophilic groups in macromolecules to initiate toxicological effects. Epoxides can also be further biotransformed to stable metabolites [41, 42].

Heteroatom oxidation: add an oxygen atom to nitrogen or sulfur. Aromatic amines and secondary or tertiary amines are subjected to N-oxidation, which is mediated by a large spectrum of enzymes including P450s (such as CYP3A4) and FMOs as in the case of voriconazole N-oxide formation [43]. This reaction is catalyzed by CYP2C9, CYP2C19, CYP3A4, FMO3/FMO1 (Figure 3.5).

Dehydrogenation reactions: replace two hydrogen atoms with double bond [40]. Dehydrogenation reactions occur by abstraction of a hydrogen atom by the $(Fe - O)^{3+}$ species to form a carbon-centered radical. Abstraction of another hydrogen atom results in double bond formation.

3.2.2 Phase II Conjugation Biotransformation Reactions

3.2.2.1 UDP-Glucuronosyltransferase (UGT) Glucuronidation reactions are catalyzed by a family of enzymes known as UGTs. These reactions involve the conjugation of an acceptor molecule with glucuronic acid. The source of glucuronic acid is a cosubstrate known as UDP glucuronic acid (UDPGA). The acceptor molecule functional group is typically −OH, −OOH, −NH, and even −CH and −SH. In general, glucuronide-conjugation reaction increases the solubility of xenobiotic compounds so that they are more readily excreted into the bile and urine.

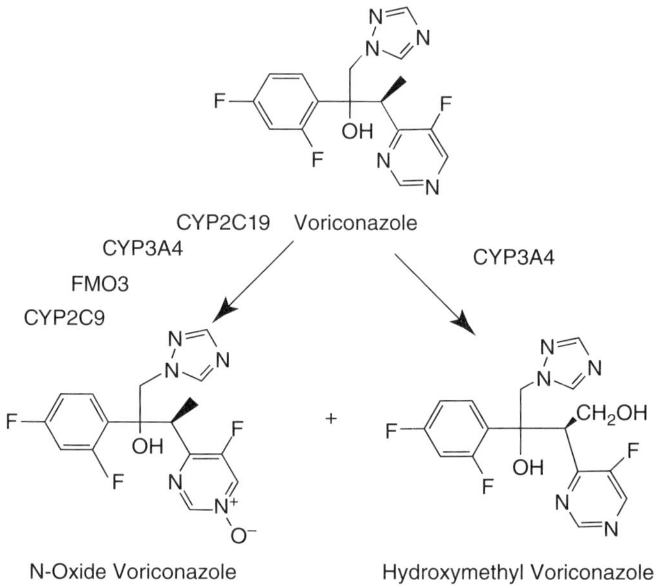

Figure 3.5 N-oxidation reaction of phase I mediated by several enzymes including P450s, such as CYP3A4 and FMOs. Antifungal drug voriconazole N-oxidation reaction is catalyzed by CYP2C9, CYP2C19, CYP3A4, and FMO3/FMO1.

Many drugs are glucuronidated by the UGT enzymes, including acetaminophen, codeine, morphine, S-naproxen, oxazepam, and zidovudine [44]. Relative to P450 metabolism, glucuronidation generally occurs more rapidly as seen in the glucuronidation of flavonoid galangin that occurs 11–31 times faster (depending on the site of glucuronidation) than CYP450-mediated metabolism [45]. When metabolism of a drug by P450s first is followed by UGT, the oxidation by P450 becomes rate limiting in the clearance of the compound. Consequently, the changes in phase I metabolism may influence the extent of glucuronidation. As with most phase II enzymes, UGTs are also capable of biotransformation of important endogenous substrates such as bilirubin, bile acids, steroids, and glycolipids.

The concentration of UDPGA in human liver ranges from 201 to 349 μmol/kg liver [46], which makes the glucuronidation reaction a predominant phase II metabolic pathway and clearance in humans. As indicated in Table 3.2, UGTs are present in liver, lung, skin, intestine, kidney, brain, and other tissues. They are membrane-bound enzymes with a MW in the range of 50–60 kDa and are located in the ER of the cell [47, 48]. Unlike the CYP450 enzymes, which have their active sites facing the cytosol, the active site of UGTs is facing the luminal side of the ER [49]. Thus, UGT conjugation occurs within the ER [50]. The UGTs consist of four families, UGT1, UGT2, UGT3, and UGT8, though UGT1 and UGT2, with about 19 human UGTs, are relatively well characterized. The UGT isoforms in the two subfamilies UGT1 and UGT2 are found to be the only two families involved in ~ 87% of hepatic UGT metabolism of drugs, including, UGT1A1,

TABLE 3.2 Phase II Conjugative Reactions and Hydrolysis.

Phase II Reactions	Specifications
UDP-glucuronosyl transferase (UGT)	R–OH or Ar–OH + UDPGA → alkyl or aryl β–d–glucuronide + UDP. Prosthetic group: none. Required cofactors: UDPGA. Tissue expression: liver and intestine. Subcellular localization: endoplasmic reticulum
Sulfotransferase (SULT)	Ar–OH + PAPS → Ar–OSO_3H + adenosine–3′, 5′-diphosphate. Prosthetic group: none. Required cofactors: PAPS. Tissue expression: liver and intestine. Subcellular localization: cytosol
Glutathione S-transferase (GST)	R–X + GSH → R–SG + HX (X is a leaving group such as halogen). Prosthetic group: none. Required cofactors: GSH. Tissue expression: liver, kidney, and other tissues. Subcellular localization: cytosol
N-acetyl transferase (NAT)	Ar–NH_2 + AcSCoA → Ar–$NHCOCH_3$ + CoASH. Prosthetic group: none. Required cofactors: AcSCoA. Tissue expression: liver and other tissues. Subcellular localization: cytosol
Catecholamine O-methyl transferase (COMT)	A catechol + SAM → a guaiacol + S–adenosyl–l–homocysteine. Prosthetic group: none. Required cofactors: SAM. Tissue expression: liver and other tissues. Subcellular localization: both cytosol and endoplasmic reticulum.
Hydrolases:	
Carboxylesterase (CES)	RCOOR′ + H_2O → RCOOH + R′OH. Prosthetic group: none. Required cofactors: none. Tissue expression: liver and many other tissues. Subcellular localization: cytosol and endoplasmic reticulum
Epoxide hydrolase (EH)	Epoxide + H_2O → trans–1,-2-diol. Prosthetic group: none. Required cofactors: none. Tissue expression: liver and other tissues. Subcellular localization: endoplasmic reticulum

UGT1A4, UGT1A9, and UGT2B7 [51]. Examples of UGT reactions are shown in Figure 3.6

Inhibition of UGT activity was reported in vitro by several drugs, such as tacrolimus, cyclosporine, and diclofenac (K_i values range from 0.033 to 7.9 μM). That DDIs involving glucuronidation seem to be less prevalent than those identified for CYP450s might be due to higher substrate K_m values (> 300 μM) compared to those of CYP450s (K_m typically around 3 μM), thus UGTs rarely saturate their own UGT-mediated metabolism, and also due to the participation of multiple UGTs

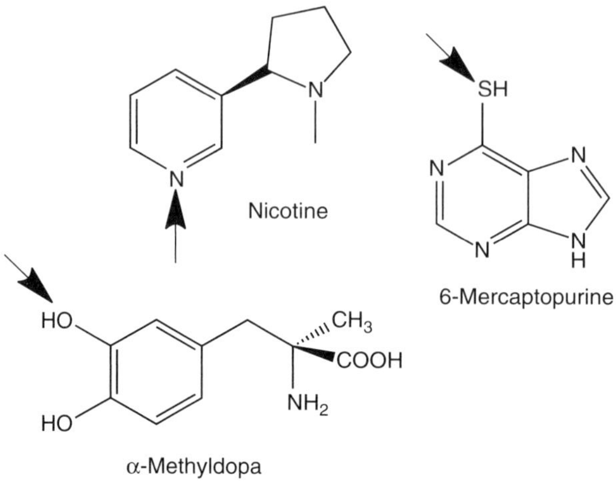

Figure 3.6 Examples of function group that can be subject to methylation-mediated phase II conjugation reactions.

in the metabolism [52]. Furthermore, K_i values for most UGT inhibitors are usually > 10 μM, making glucuronidation-mediated DDIs that result in toxicity rare, but they have been observed [53]. For example, lamotrigine coadministered with valproic acid increases the incidence of skin rash, which is a known side effect of lamotrigine [54]. For P450 reactions, there are also number of intrinsic and extrinsic factors that affect the extent of glucuronidation in humans, including age, as in the case of gray baby syndrome, cigarette smoking, diet, disease state, ethnicity, genetic polymorphism, and alteration in hormonal level [55].

UGT enzyme expression and activities are variable among human populations due to the genetic polymorphism, most commonly associated with the UGT1A1, UGT1A6, and UGT2B7 isoforms [56]. Of these three isoforms, the most relevant polymorphism-mediated drug metabolism is the one related to UGT1A1. Hereditary diseases such as Gilbert's and Crigler–Najjar syndromes occur due to UGT1A1 polymorphism, which reduces the individual's ability to metabolize bilirubin, causing hyperbilirubinemia [57]. In Gilbert's syndrome, it results in only minor jaundice in roughly 3–13% of the population and is less severe as expression of UGT1A1 is ~ 30% of normal [52]. For Crigler–Najjar, bilirubin can reach lethal levels [58] and is characterized as two types; type I is more severe as subjects have no expression of UGT1A1 (bilirubin levels > 340 μmol/L), and type II is less severe in that patients have some UGT1A1 activity at levels that are 10% of normal (bilirubin levels between 150 and 340 μmol/L), which is treatable with phenobarbital that induces UGT1A1 activity [59].

Glucuronidation is considered a reversible process; once glucuronide metabolites enter the intestine, they are exposed to enzymes of the intestinal microflora such as β-glucuronidase and become deconjugated back to the parent drug. With the process of enterohepatic recirculation, the parent drug is often reabsorbed in the lower intestine.

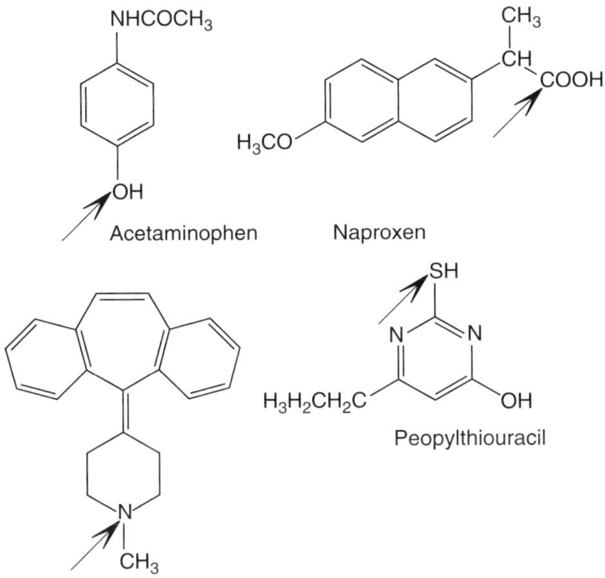

Figure 3.7 Examples of function group that can be subject to glucuronidation reactions catalyzed by phase II, UGTs.

3.2.2.2 Other Conjugation Reactions: Sulfonyltransferase, Glutathione-S-Transferases, Methyl Transferases, and N-Acetyl Transferases In addition to UGTs, phase II conjugation reactions involve the attachment of a sulfate, GSH, methyl, and acetyl moiety catalyzed by a group of enzymes known as sulfonyl transferase (SULT), glutathione S-transferase (GST), methyltransferase (MT), and N-acetyltransferase (NAT), respectively (see Table 3.2), though, as described earlier, glucuronidation is the most common conjugation, which accounts for ~ 35% of all conjugation reactions. SULTs, GSTs, and NAT are found to be responsible for conjugative metabolic reactions to a lesser extent, by 20%, 15%, and 10%, respectively [60]. As in phase I reactions, conjugating enzymes are also involved in important biosynthetic and biochemical pathways of endogenous biomarkers. Metabolic reactions catalyzed by these phase II conjugative pathways are depicted in Figures 3.6 and 3.7.

Sulfonation reaction is catalyzed by a large family of enzymes known as SULTs. The SULTs catalyze the transfer of a sulfonate (SO_3^-) group from 3′-phosphoadenosine 5′-phosphosulfate (PAPS) to a −COH, an −NOH, or an −NH group (as indicated in Table 3.2). Sulfonation can either be direct—for example, the sulfonation of troglitazone—or occur after oxidation via phase I metabolism, as, for example, in the sulfonation of hydroxyphenolbarbitol. The capacity of sulfonation is limited by the quantity of PAPS available in vivo for conjugation. With sulfonation relatively low, PAPS in hepatocytes is 10 times lower than UDPGA; it was reported to be as low as 23 μmol/kg [61]. Also, glucuronidation typically compensates the sulfonation pathway when the latter becomes saturated. Thus, sulfonation and glucuronidation are balancing each other. Similar to glucuronidation, sulfate conjugation is a reversible process and sulfate conjugates may be desulfated by sulfatases present in the intestinal microflora. Hence, sulfate conjugates can also undergo enterohepatic recirculation. The SULTs are expressed in most tissues, including liver, kidney, lung, skin, breast, gastrointestinal, brain, and platelets [62, 63], with the highest levels in the small intestine and liver. SULTs are mostly isolated from the cytosolic subcellular fraction or are membrane-bound of Golgi apparatus of cells. SULTs that are membrane-bound are not involved in drug metabolism but catalyze the conjugation of endogenous macromolecules such as proteins and peptides. The cytosolic SULTs catalyze the conjugation of drugs and other xenobiotics as well as lower MW endogenous compounds such as neurotransmitters, bile acids, and steroids. Several investigators have reported the role of sulfonation in the metabolism of several drugs; most known is its role in the metabolism of acetaminophen and toxicity [62]. It is worth noting that due to the ontogeny of SULTs, the hepatotoxicity-mediated acetaminophen treatment in pediatrics is diminished compared to these toxicities observed in adults [64].

DDI and polymorphisms have been identified in several human SULT genes. Few or no clinically relevant DDIs involving SULTs have been reported [65], while common genetic polymorphisms of SULT1A1 and SULT1A2 isoforms have been investigated in epidemiological studies [66]. Patients homozygous for SULT1A1*2 have 10-fold lower phenol SULT activity than those homozygous for SULT1A1*1 and are found to have an increased risk for lung, breast, and other cancers [67, 68].

GST-conjugation reaction depends on GSH forming less reactive products that are readily excreted. Intracellular level of GSH is high, reaching 8–10 mM in some tissues, which is important to GST activity [69]. The end products of GST conjugation are either excreted by transport proteins into the bile or converted to N-acetyl cysteine conjugates and excreted into the urine. Like most other conjugative enzymes, GSTs have endogenous substrates such as prostaglandins. Most GSTs are cytosolic enzymes, but there are examples of membrane-bound GSTs such as MGST1, MGST2, and MGST3. These membrane-bound GSTs are located in the ER. Like UGTs and SULTs, there are different isoforms of GSTs. There are seven classes of cytosolic GSTs: alpha, mu, pi, sigma, theta, omega, and zeta [70]. Polymorphisms are also found within each class of GSTs; the greatest impact of GST polymorphism on drug metabolism arises in the mu and theta classes. Individuals carrying the incompetent genes have no active enzyme and are associated with an increased risk for certain cancers [71].

Methyltransferases catalyze the transfer of a methyl group from a cosubstrate known as S-adenosylmethionine (SAM) to an oxygen, sulfur, or nitrogen functional group on the substrate. There are over 100 MTs; however, only a few have been found to catalyze the metabolism of drugs, including catechol methyltransferase (COMT), thiol methyltransferase (TMT), thiopurine methyltransferase (TPMT), and nicotinamide N-methyltransferase (NNMT) (see Figure 3.7). COMT catalyzes the O-methylation of catechols and catecholamines such as dopamine. COMT requires Mg^{2+} for its catalytic activity [72]. COMT is found in most tissues including brain, lung, and red blood cells, but the highest levels are expressed in the liver and kidney. The endogenous substrates of COMT include l-dopa, dopamine, norepinephrine, and epinephrine [73]. Drugs metabolized by COMT include rimiterol, dobutamine, dihydroxyphenylserine, isoprenaline, and carbidopa. Methylation differs from sulfonation, glucuronidation, and GSH conjugation, as methyl conjugates are less polar than the parent drug and therefore may not be more easily eliminated. For DDI, COMT is subjected to inhibition by nitrocatechol drugs such as entacapone, nitecapone, and tolcapone [72]. COMT is polymorphic in humans. Genetic polymorphism cause 3- to 4-fold reduced enzyme activity and is linked to an increased risk of breast cancer.

Arylamine N-acetyltransferases catalyze the transfer of an acetyl group from acetyl coenzyme A to aryl amines and N-hydroxyaryl amines such as ρ-aminobenzoic acid. Similar to methylation, N-acetylation differs from other conjugation reactions as it also produces metabolites that are less polar (less water soluble) than their parent drug forms (see Table 3.2). There are two human forms of NAT, NAT1 and NAT2; the kinetics of these enzymes proceed via a ping-pong bi-bi reaction mechanism. Drugs that are NAT substrates include isoniazid, hydralazine, phenelzine, sulphamethazine, endralazine, p-aminosalicylic acid, procainamide, nitrazepam, debrisoquine, and dapsone [74–76]. Despite the similarities between the two isoforms, they are quite different in their substrate specificity and tissue distribution. NAT1 is widely expressed in many tissues, while NAT2 is found primarily in the liver and gut. For DDI, drugs such as ketoprofen, ibuprofen, paclitaxel, and salicylamide are inhibitors of NATs [77]. Both forms are polymorphic; however, the polymorphism of NAT2 has been more extensively studied and has been implicated in the variability of the therapeutic activity and toxicity of isoniazid and other hydrazine drugs. The rate of occurrence of poor/intermediate metabolizers in various ethnic populations has been found to be highly variable but has been reported to be $\sim 10 - 30\%$ in Asians and 50–60% in Caucasians [78]. Poor metabolizers treated with isoniazid are at a higher risk to develop hepatic disorders and peripheral neuropathy [79]. NAT2 polymorphisms have also been attributed. NAT1 polymorphisms have been linked to myeloma, lung, bladder, and other cancers; NAT2 polymorphisms have been linked to liver, colorectal, non-Hodgkin lymphoma, bladder cancer, and other cancers [80].

3.2.3 Phase III Metabolism

The phase III biotransformation is referred to as the process of excretion or elimination of drug metabolites into bile or urine via transporter proteins. This is the only mechanism to translocate polar conjugative metabolites from hepatocytes to bile or to systemic circulation, then elimination by renal proximal tubular cells. Adenosine triphosphate (ATP)-mediated efflux active transport mechanism mostly is needed to get metabolites generated from phase II conjugation pathways to cross biological membranes. Furthermore, transport proteins play a major role in the uptake of phase I and II metabolites of drugs and other xenobiotics (including endogenous biomarkers) from systemic circulation, distributing them to body organs and then excreting them into bile and urine. There are multiple types of transport proteins that modulate the uptake and excretion of drugs and metabolites into and out of cells. Drug transporters are divided into two families, the ATP-binding cassette (ABC) and the solute carrier (SLC) family. These families are further divided into subfamilies such as multidrug resistance transporters (MDRs), MRPs, OATP, and organic anion transporters (OATs).

Phase III, phase II, and phase I metabolism are modulated by the same mechanism of regulation by nuclear receptors (NRs), such as pregnane X receptor (PXR), constitutive androstane receptor (CAR), peroxisome proliferator-activated receptor (PPAR), and the aryl hydrocarbon receptor (AhR), which indicates that these reactions are subject to inductions by similar ligands. Furthermore, the three reactions have overlapping substrate specificity. Numbers of the phase I, II, and III enzymes are implicated in clinical DDIs. Drug transporters and their various members will be discussed in chapter 4 in more detail.

3.2.4 Localization of Drug Metabolism in Organ Cells

Clearly, as previously described, the main site of phase I and phase II drug metabolism reactions is the liver cells, the hepatocytes. There most of the DMEs are expressed and their metabolic capacity is responsible for drug clearance. The smooth endoplasmic reticulum (SER) and the cytosolic compartments are the two subcellular fractions of hepatocytes where DMEs (especially phase I oxidation and the significant conjugation reactions) are localized, though special reactions, such as β-oxidation, occur mainly in the mitochondria and peroxisomes (see Table 3.3). When a need to study metabolism in an in vitro system to simplify the investigation and to minimize the variability occurs in in vivo testing, hepatocytes can be used to evaluate all possible metabolic routes in parallel within the same experiment, for example, in phase I, phase II, or phase III drug metabolism. In case specific questions or pathways need to be addressed, a subcellular fraction like microsomes or S-9 can be used. In this case, hepatocytes are lysed by homogenizing liver tissue, followed by differential centrifugation to segregate the desired subcellular fraction [81], the so-called S-9 fraction (supernatant from $9,000 \times g$ sedimentation), which contains cytosol and liposomal fragments of the SER called microsomes. S-9 preparation is a suitable in vitro tool for elucidating almost all important drug-metabolism pathways, but after supplement with cofactor of each reaction, for example, NADPH for oxidation, UDPGA for UGT. Microsomes can then be separated from cytosol by more forceful sedimentation of the S-9 fraction at $105,000 \times g$, allowing membrane-bound drug-metabolism enzymes (e.g., P450, FMO) to be prepared. A summary of the metabolic reactions and their subcellular localizations is shown in Table 3.3.

3.3 METABOLIC CLEARANCE AS A CRITICAL FACTOR INFLUENCING DRUG ACTION AND SAFETY

Drug metabolism has critical impacts on (1) clearance of drug from the body; (2) pharmacological activity due to exposure; (3) toxicity due to accumulation or

TABLE 3.3 Major Drug Metabolism Reactions and Localizations in Subcellular Compartments.

Reaction	DME	Abbreviation	Location
Oxidation	Cytochrome P450	P450	Microsomes
	Flavin monooxygenase	FMO	Microsomes
	Aldehyde oxidase	AO	Cytosol
	Xanthine oxidase	XO	Cytosol
	Alcohol & aldehyde dehydrogenase	ADH, ALDH	Mitochondria and cytosol
	Monoamine oxidase	(MAO)	Mitochondria
Hydrolysis	Carboxyestrase	CES	Microsomes, plasma, blood, cytosol
	Epoxide hydrolyase	EH	Microsomes and cytosol
Reduction	Aldo-keto reductase	AKR	Cytosol
	Quinone reductase	NQOI	Microsomes and cytosol
	Reductive dehydrogenase		Microsomes
	Azo and nitro reductase		Microsomes
Conjugation	Glucuronide	UGT	Microsomes
	Sulfate	SULT	Cytosol
	Methyltransferase	COMT	Microsomes, cytosol, blood
	N-acetyltransferase	NAT	Cytosol, mitochondria
	Glutathione transferase	GST	Microsomes, cytosol
	Amino acid transferase		Microsomes, mitochondria

metabolite formed; and (4) extent of DDIs. As described in Figure 3.1, the reader can review the significance of the interplay between phase I/II DME-mediated metabolism in intestine and liver and phase III uptake and efflux drug transporters in regard to the efficacy and safety of drugs. The most critical value of the relationships among phase I/II/III is their influence in metabolic clearance, where the metabolites are eliminated in urine or in the bile. In this section, the approaches used to accurately predict the metabolic clearance parameter prior to conducting clinical studies in humans will be listed. The impact of drug metabolism on the pharmacokinetics/pharmacodynamics (PK/PD) and safety will also be discussed.

As mentioned above, drug metabolism influences the % of bioavailability, hence the extent of drug exposure in the blood circulation to be distributed to body organs. In addition, drug metabolism will have a great impact on total clearance, especially the metabolic clearance and biliary clearance. The presence of drug in the body is usually quantitated as the PK half-life ($t_{1/2}$) or the mean residence time (MRT), both of which are typically given in units of hours or days. To quantitate $t_{1/2}$ of a drug, the extent (capacity) of drug metabolism is measured. The extent of metabolism is directly related to the extent of drug clearance from the body, which is the most direct quantitative measurement of an organ's ability to eliminate a drug. Plasma clearance of hepatically cleared compounds is dependent on blood perfusion of the liver, intrinsic clearance ($CL_{intrinsic}$), transport processes, and, in some cases, the extent of protein binding.

For drugs that are stable to metabolism by DMEs, the elimination process of intact drugs can be either renal, pulmonary, biliary, or by combination of two of these pathways or all. Hepatic clearance followed by renal clearance are the two pathways that have been associated with elimination of most drugs that have been developed to date [see Figure 3.1(A)]. Renal clearance of drugs occurs predominantly by filtration, secretion, and reabsorption, the latter two processes requiring transporters (this will be discussed in detail in the next chapter). In contrast to renal clearance, hepatic clearance can occur predominantly by metabolism or by biliary excretion, or both.

As mentioned previously, DMEs and transporters both play important roles in the hepatic clearance of drugs. The rate of metabolism of the drug by the particular enzyme catalyzing the metabolic reaction(s) is determined by the Michaelis–Menten equation of enzyme kinetics:

$$\text{Rate} = \frac{V_{max} \times [\text{Drug}]}{K_m + [\text{Drug}]}. \tag{3.1}$$

To calculate intrinsic clearance for a particular enzyme pathway, Michaelis–Menten constants V_{max} and K_m can be determined experimentally. The V_{max}/K_m ratio can be scaled to whole-body intrinsic clearance according to Equations (3.2) and (3.3):

$$CL_{int} = V_{max} / K_m \ (\text{ml/min/mg}) \times \text{Scaling Factor} \ (\text{mg/kg}), \tag{3.2}$$

$$\text{Scaling Factor} = [\text{MPPG (mg/g)} \times \text{liver wt (g)}] / \text{body wt (kg)}, \tag{3.3}$$

where MPPG is milligram microsomal protein per gram of liver. In vivo clearance of drug can then be calculated based on in vitro data using the well-stirred model of hepatic disposition, as shown in Equation (3.4):

$$\text{Predicted CL} = Q_H \times f_u \times Cl_{int} / (Q_H + f_u \times Cl_{int}). \tag{3.4}$$

3.3.1 Effect of Physiological Factors on Drug Metabolism-Mediated Drug Clearance

As mentioned earlier, systemic clearance of drugs can involve metabolic clearance, renal clearance, and biliary clearance. With metabolic clearance, enzyme abundance and activities are the key determinants in the metabolic clearance of drugs that are moderately or extensively metabolized by liver enzymes. However, many physiological parameters also play a role in influencing the metabolic clearance, such as protein binding, hepatic flow, liver size, milligram microsomal protein per gram (MPPG) liver, age, sex, genetic polymorphism, and disease states. These factors could alter the extent of metabolism and hence drug clearance and disposition among humans.

3.3.1.1 ***Protein Binding*** Plasma protein binding (PPB) is believed to have a significant influence on the rate of drug diffusion between plasma and tissues (influx and efflux) and therefore influence clearance (and V_{dss}) of drugs. In general, it is desirable to avoid highly plasma protein-bound drug. Drugs that have a lower free fraction in tissues tend to be sequestered in the tissue compartment (showing a high V) relative to drugs with a higher free fraction in tissue. The free drug is also the entity that is amenable to metabolism by DMEs, glomerular filtration, active secretion, and active reabsorption. The determinants of the organ metabolic clearance [82–85] are indicated by the following well-stirred model equation:

$$\text{Predicted } CL_{org} = Q_{b,org} \times f_{u,b} \times Cl_{int} / (Q_{b,org} + f_{u,b} \times Cl_{int}), \quad (3.5)$$

where CL_{org} and CL_{int} are organ clearance and intrinsic clearance (volume/time), Q_{org} is the organ perfusion rate or organ blood flow (volume/time), and $f_{u,b}$ is fraction unbound in blood.

For drugs that are efficiently metabolized by DMEs and have high intrinsic clearance values, the above equation collapses to the following:

$$\text{Predicted } CL_{org} = Q_{b\,org} \times f_{u,b} \times Cl_{int} / f_{u,b} \times Cl_{int}. \quad (3.6)$$

Thus,

$$\text{Predicted } CL_{org} = Q_{b\,org}. \quad (3.7)$$

The organ clearance is thus perfusion rate limited and independent of both protein binding and intrinsic clearance.

For drugs that are not good substrates for DMEs (low intrinsic clearance), the equation for organ metabolic clearance then rearranges to the following:

$$\text{Predicted } CL_{org} = Q_{org} \times f_{u,b} \times Cl_{int} / (Q_{b,org}). \quad (3.8)$$

Thus,

$$\text{Predicted } CL_{org} = f_{u,b} \times Cl_{int}. \quad (3.9)$$

The free fraction in blood, or $f_{u,b}$, is now a crucial determinant of hepatic metabolic clearance in addition to the low intrinsic capacity (intrinsic clearance) of DMEs that can biotransform the drug.

The use of the well-stirred model to calculate clearance requires a consideration of drug-unbound fraction (f_u) as shown in Equation (3.4). The protein binding of drug as discussed in chapter 2 indicated that it varies not only due to difference in physiochemical properties (such as lipophilicity) and chemical structures but also due to changes in subject profile; therefore, the values of f_u of a drug may vary as a function of disease state, age, pregnancy, and diet. It is thus recommended that in vivo hepatic metabolic drug clearance might be predicted in vitro by measuring f_u in the specific human populations for which the clinical trial is intended [86]. It has been shown that plasma protein levels are lower in the newborn and gradually increase with age [86], as will be discussed later.

3.3.1.2 Hepatic Blood Flow (Q_H) This important parameter is poorly understood, though Q_H may influence hepatic clearance when hepatic elimination is limited by the rate of hepatic drug transport and not enzyme activity [64]. With drugs that are blood flow rate limited, changes in metabolic enzymes will impact hepatic clearance only when the enzyme activity is exceedingly poor (absent or undeveloped), such as when enzyme activity becomes the rate-limiting factor for overall drug elimination.

3.3.1.3 Liver Size Relative to Body Weight The major organs responsible for drug clearance are the liver and kidneys. Their relative size can affect the extent of metabolism, clearance, and in turn the PK and PD of drugs. The changes in this parameter among patient populations can alter the PK/PD of drugs; the most prominent observation is that the mass of liver (and kidney) relative to age is several-fold greater in preschool-age children than in adults [64]. Based on this observation, the ratio of liver to body mass is not a constant parameter; it is considerably greater in infants and young children than in adults. This may contribute to greater hepatic clearance in children. Because the weight ratio of different organs may vary with age, it is recommended that the actual ratio for each age population be used in calculating the clearance and estimation of doses [87]. Yanni et al. (2010) [87] determined the factors responsible for the higher clearance of the metabolically cleared antifungal drug voriconazole in children age 2–8 compared to adults and concluded that the ratio of liver size to body mass seems not to contribute to the difference in clearance (although the scaling factor that is the product of milligrams of microsomal protein and ratio of gram liver to kg body mass was statistically different, greater in children compared to adults).

3.3.1.4 Milligram Microsomal Protein per Gram of Liver Similarly, like relative liver to body weight, relative microsomal protein to gram liver ratio can also be different among various human populations, the most observed cases in pediatric populations compared to adults. When analyzing various microsomal samples, the results suggested an increase in microsomal protein content from birth (26 mg/g liver) to the average microsomal protein content of a 30-year-old (40 mg/g liver), which then declined to 31 mg/g liver for the average 60-year-old [64]. When correlation analysis was performed between observed and calculated clearance for 21 different drugs in adults at different milligram MPPG of liver (77, 45, 21, and 34 mg/g), the best correlation was found when MPPG is 34 mg/g for young adults. MPPG values determined with samples from pediatric livers indicated that MPPG increases with age. The mean MPPG value for young infants < 1 year of age was as low as 10 mg/g, while the mean value for children < 10 years of age was 23 mg/g, and for adults the mean value was 31 mg/g [64]. It is clear that the change in MPPG with age has an impact on altering the scaling factor value and the hepatic clearance of drugs in human populations, especially between young children and adults.

3.3.2 Role of Drug Transporters

In addition to DMEs, transporter-mediated processes also play critical roles in the metabolic clearance and overall disposition as well as serious DDIs of numerous therapeutic drugs. Besides their roles in xenobiotic transport, drug transporters often mediate important physiologic functions via transport of endogenous substrates such as amino acids, bile acids, and hormones that influence the metabolism and, most importantly, that affect the efficacy in target organs [88–90]. Drug transporters are expressed in all body organs; however, they are most extensively investigated in those organs that are responsible for drug disposition and clearance, such as the liver, the kidney, and the intestine [89]. Drug transporters can be generally separated into two major classes: uptake and efflux transporters. The role of uptake transporters is to facilitate the translocation of drugs into cells, like hepatocytes. Major members of the uptake transporters are OATP (SLCO), sodium-taurocholate cotransporting polypeptide (NTCP, SLC10A1), OAT (SLC22A), and organic cation transporter (OCT, SLC22A) families. Among the uptake transporters, certain transporters have been shown to exhibit both influx and efflux properties, while others act only as uptake transporters. For example, the OATPs are primarily considered uptake transporters that mediate transport via sodium-independent mechanisms, in contrast to sodium-dependent uptake transporters such as NTCP. Efflux transporters export drugs from the intracellular to the extracellular milieu, often against high concentration gradients. The ABC efflux transporters are P-gp (ABCB1), the bile salt export pump (BSEP, ABC11), the MRP (ABCC) family, and the breast cancer resistance protein (BCRP, ABCG2). The efflux transporter, P-gp, has different functions in various tissues. For example, in the liver and renal tubules, P-gp facilitates the excretion of substrate-xenobiotics into bile and urine, respectively [91]. Like in the other organs, drug transporters are either localized in the basolateral membrane or apical (canalicular) membrane to enhance the uptake from blood into the hepatocytes or the efflux into bile, respectively. When uptake transporters facilitate drug uptake to the hepatocytes, this should governs the extent of drug metabolism. Transporters can modulate the extent of metabolism of drug candidates by affecting their exposure to the DMEs [92, 93]. Ability of transporters to modulate metabolism is also dependent on the Biopharmaceutics Classification System (BCS) class of the NCE under investigation [94]. For drugs that fall in Class I of the BCS characterized by high solubility/high permeability, it is unlikely that the uptake or efflux transporters would have any effect on the metabolism/exposure of these drugs. In contrast, both uptake and efflux transporters are expected to modulate metabolism/exposure of drug candidates that are classified as low solubility and high permeability (Class II), high solubility and low permeability (Class III), and low solubility and low permeability (Class IV).

In addition, as indicated in literature, DDIs can occurs due to the modulation of DME, such as the DDI between cisapride and erythromycin, statins and protease inhibitors, terfenadine and ketoconazole, or fluvoxamine and astemizole. There are also examples of DDIs due to the modulation of transporters and DME, such as those between digoxin and quinidine, probenecid and acyclovir, and talinolol and verapmil. As expected, the interplay between DMEs and transporters is found to complicate the cases of DDI. For example, gemfibrozil causes 6- to 8-fold increases in exposures of coadministered cerivastatin [95–100] or repaglinide, not only due to inhibition of uptake inhibitor OATP1B1 and efflux transporter P-gp but also due to inhibition of P450 enzymes CYP2C8 and CYP3A4. Inhibition of CYP3A4 and

P-gp by ketoconazole warranted a better exposure and better therapeutic results in the increase of HIV protease inhibitors ritonavir and squinavir concentrations in plasma and cerebrospinal fluid (CSF). Recently, a report by Tornio et al. (2014) [101] showed that interaction between cerivastatin and repaglinide—both are substrates of cytochrome P450 (CYP2C8, CYP3A4, and OATP1B1—resulted in increased risk of rhabdomyolysis in patients using cerivastatin with clopidogrel. In healthy volunteers, repaglinide area under the concentration-time curve (AUC0–∞) was increased 5.1-fold by a 300-mg loading dose of clopidogrel and 3.9-fold by continued administration of 75 mg clopidogrel daily. In vitro, we identified clopidogrel acyl-β-D-glucuronide as a potent time-dependent inhibitor of CYP2C8. Using a physiologically based pharmacokinetic (PBPK) model indicated that inactivation of CYP2C8 by clopidogrel acyl-β-D-glucuronide leads to uninterrupted 60–85% inhibition of CYP2C8 during daily clopidogrel treatment. Clopidogrel is a strong CYP2C8 inhibitor via its acyl-β-D-glucuronide, causing the DDI to occur, in contrast to what has been previously hypothesized: that the hepatic uptake by OATP is responsible for the observed DDI between cerivastatin with clopidogrel.

3.3.3 Effect of Age on Drug Metabolism and Clearance

The changes in phase I and phase II DMEs as afunction of age is one of the factors affecting drug clearance in pediatric populations that has been extensively investigated in vitro by measuring both the expression and activities of DMEs in tissues from different age groups [102–106]. Furthermore, it has been investigated in the clinical setting (in vivo) by comparing the hepatic metabolic clearance of DME probe substrates in various age groups and in silico using PBPK models [107–109]. These studies showed that age-associated changes in phase I/II DMEs can have a significant impact on PK/PD of drugs whose elimination is predominantly dependent on the metabolic pathway mediated by those enzymes. As far as drug metabolism is concerned, important differences have been found in the pediatric populations compared with adults, such as:

1. Phase I enzymes: oxidative enzymes (P450 CYP1A2, and CYP3A7 versus CYP3A4), reductive enzymes, and hydrolytic enzymes
2. Phase II enzymes: NAT, UGT, SUL
3. Phase III enzymes: uptake and efflux transporters (OATP, P-gp)

Generally, the major enzyme differences observed in comparison with the adult age are in newborn infants, although for some enzymes (e.g., CYP3A7 vs. CYP3A4, UGTs and other phase II enzymes, and uptake drug transporters), significant differences still exist between infants, toddlers, and adults. With these differences that have an impact on drug disposition, PK/PD, and safety, altered dosage regimens, alternative therapy, and/or administrative routes have been recommended to optimize drug exposure and achieve therapeutic efficacy and safety in children [107–109].

For example, in adult liver, CYP3A4 comprises 10–50% of the total P450 enzymes [110], while CYP3A7 is the dominant enzyme in fetal liver [110]. A temporal switch was observed between CYP3A7 and CYP3A4 in terms of their expression and activity in liver as a function of age; while CYP3A7 mRNA was detected in fetal liver, it was not detected in adult livers. For CYP3A4 expression, the opposite was observed: it is not expressed in fetal liver but is expressed in adult. The clinical implications of the temporal switch between CYP3A7 and CYP3A4 are shown with the PK difference of cisapride, which is commonly used in pediatric patients with feeding intolerance, apnea, and bradycardia due to gastroesophageal reflux. Due to a higher than acceptable incidence of cisapride-related QT prolongation in pediatric patients, cisapride was removed from the market and used only through a limited access program for feeding intolerance in neonates. A study by Kearns et al. (2003) [102] in the same patient population demonstrated a significant increase in QT prolongation not associated with adult risk factors [111, 112]. To elucidate the mechanism underlying these observed adverse events, in vitro studies were conducted and results indicated that cisapride is metabolized primarily by CYP3A4 but not by CYP3A5 or CYP3A7 [112]; consequently it was hypothesized that neonates would show deficient metabolic activity toward cisapride [113–116], and in vitro metabolism measured in 50% of the livers from fetuses or neonates less than 7 days old was low. Oral clearance of midazolam [113] is also markedly lower in preterm infants relative to that of younger children; it is higher in children aged 2–12 years than in adolescents aged 12–16 years [115]. Likewise, in several drug therapies like tacrolimus, amlodipine, clonazepam, lopinavir, and saquinavir, the clearance was different in pediatrics than in adults [64]. Due to these differences in clearance, weight-corrected doses of drugs predominantly metabolized by P450 enzymes in children, CYP3A4 in particular, are 2-fold higher than adult doses to achieve equivalent therapeutic concentrations [117, 118].

Regarding other oxidative pathways, neonates are born with high FMO-1 and low FMO-3 expression, but this reverses rapidly and FMO-3 becomes the dominant form that increases toward adulthood [119].

In contrast to CYPs and other oxidative pathways, UGT capacity is very low at birth and only increases very gradually [64, 120]. The major UGT isozyme that metabolizes bilirubin is UGT1A1, and its activity is low around the time of birth. It can affect many neonates born with a degree of mild jaundice, the condition that can be severe in babies that carry UGT1A1 polymorphic gene [121]. Adult levels of the UGTs appear from aged 6 months onward [122], although in general conjugative pathways they may take longer to mature than oxidative ones. Premature neonates have little drug-metabolizing capacity even in comparison with term neonates, so that they present a particularly difficult therapeutic management problem.

At birth, human serum albumin (HSA) concentrations are close to adult levels (75–80%), while α_1-acid glycoprotein (AAG) is initially half the adult concentration. McNamara and Alcorn (2002) [123, 124] and Yanni et al. (2011) [86] have reported that age-dependent changes in f_u as a function of age should be conducted, especially with drugs that are extremely lipophilic. Disease state and some physiological and biomedical condition cause alteration of PPB (as shown in chapter 2) and could alter the metabolic clearance and disposition of drugs in these specific populations.

It is a highly questionable ethical procedure to carry out clinical research on individuals who can't or refuse to give informed consent. Without parental approval, pediatric patients can't be recruited for any clinical studies, with few exceptions. Pediatric patients are also fragile and have low blood volume (a problem in repeated PK sampling) and low renal function as well as relatively poor drug biotransformation capacity. Retrospective investigations were set up to understand factors that alter metabolic clearance in children compared to adults, as seen in investigations of the mechanism underlying the higher clearance of voriconazole in children compared to adults (Yanni et al. (2010) [87]), so that prospective clinical investigation can be well designed for delivering a safer and more effective drug therapy in pediatric populations in any ongoing treatment programs.

3.3.4 Effect of Hormones on Metabolic Clearance and Gender Difference in Drug Metabolism

It might be expected that the age-mediated changes in drug metabolism, especially in children > 8 years, related to changes in levels of hormone. There is considerable evidence in animal models that drug clearances vary between the sexes, partly related to possession or lack of possession of particular CYPs [125]. The effects of gender in humans are not so well characterized, but some CYPs do appear to have higher activities in males, such as CYP1A2, which leads to increased clearance in males with drugs such as clozapine [126, 127]. Overall, females have lower UGT and CYP2E1, higher CYP3A4 and CYP2B6, but the same CYP2C9 and CYP2D6 activities [5, 128]. Interestingly, probably the most dangerous drug side effect in terms of rapidly lethality, TdP, is more common in women due to estrogen effects and so more than 60% of the cases are female [128]. Women are also more at risk from other drug adverse reactions, such as rashes. There are other factors that also govern female drug clearance; these factors include lower total body water, different basal metabolic rates, lower renal glomerular filtration rate, and generally lower body size and volume in comparison with men. The enzymes of drug metabolism are even different in men and women, leading to gender differences in PKs. Recently, Liu et al. (2013) demonstrated that gender, and perhaps more importantly the differences in plasma sex hormone levels in female species, drives the gender-dependent differences in expression profiles of UGT enzymes and efflux transporters. These differences significantly affect the metabolic clearance and the systemic bioavailability of compounds eliminated via this disposition pathway [129].

3.3.5 Effects of Disease on Drug Metabolism

Several biomedical defects, ranging from specific renal impairment to hepatic organ malfunction to metabolic disorders, may have a remarkable dimension of metabolic clearance and metabolites of many drugs. These conditions may affect CYP expression, as seen in diabetes, which induces CYP2E1 activity as indicated by the significant increase in clearance of the prototype substrate chlorzoxazone in obese diabetics [130]. Changes in liver and intestinal DME activities were lowered in patients with chronic kidney disease (CKD) [131]; the changes were manifested in the decrease of clearance of drugs with clearance mechanism dependent on the hepatic metabolism in CKD patients. However, the major focus in terms of disease on the metabolic capability is usually in the liver. Any severe disruption of hepatic ultrastructure or metabolism either through disease (hepatitis or cancer) or abuse (cirrhosis) will strongly affect the ability of the organ to fulfill its main biochemical and physiological functions as well as affect enzyme expression. Hepatitis viral infections, such as the C and B variants, can lead to cirrhosis and hepatocellular carcinoma, and they affect liver cellular oxidative defense systems by reducing GSH maintenance, which increases hepatic oxidative stress [132–136]. The physical loss of functional hepatocytes, combined with poor organ perfusion, is the major factor in the drastic loss of the metabolic capacity of the cirrhotic liver. In addition, CYP expression, such as CYP3A and CYP2E1, is significantly reduced in the surviving hepatocytes. Hepatitis C causes down-regulation in the expression of most CYPs, such as CYP2B6, CYP2C9, CYP2C19, CYP2E1, and CYP3A5, but it increases expression of others, such as CYP3A7. Several autoimmune conditions also lead to antibody formation that is directed at various CYPs, although the mechanisms at work are still not understood. In general, CYP capability is reduced in hepatic disease and will affect drug clearance.

3.3.6 Genetic Polymorphism and Ethnic Variability Effect on Metabolic Clearance

Many of the DMEs and drug transporters show genetic polymorphism, which mostly influences the levels of expression and function [137–143]. Polymorphisms of DMEs such as CYP2C19, CYP2D6, CYP3A, CYP2B6, UGT1A1, NAT, or transporters such as OATPs, OCTs, and P-gp are found to significantly contribute to the variability in human drug PK. Hence, knowing the enzymes and transporters that are involved in the clearance of new drug candidates along with polymorphisms that exist for these enzymes and transporters is a critical step, one that is commonly conducted in the discovery stage prior to initiating clinical investigations. Because the safety of drugs and a patient's response to any medication are indirectly influenced by drug metabolism, any factors may critically affect the metabolism, such as genetic polymorphism. During the research investigation of NCEs, the so-called reaction phenotype that companies strategically set up to ensure a safer and therapeutic effective drug is developed usually addresses the genetic polymorphism involved with the DMEs and drug transporters.

In a recent retrospective investigation to determine the mechanism underlying the adverse events observed in bosentan drug therapy, the investigation revealed that CYP2C9 genetic polymorphism correlated well with the observed elevation of alanine aminotransferase (ALT, a biomarker for liver toxicity) in patients treated with the drug. This finding contradicted with the previous hypothesis that bosentan inhibition of BSEP might be the sole cause of the observed liver injury in humans following treatment with the drug [144]. The authors then recommended that patients should be profiled for CYP2C9*2 to "individualize" the drug therapy by selecting the suitable dose for each patient based on the genetic fingerprint of CYP2C9. This study is one example of hundreds of case studies with drugs where the decrease in their total plasma clearance correlates well with the decrease in activity of metabolizing enzymes and the elevation in adverse drug reactions, as seen in case of bosentan.

As was shown earlier, metabolism is governed primarily by the P450s; many of these enzymes show genetic polymorphisms with frequencies that variy among different ethnicities, causing variations among human populations. For example, the variability in PK of voriconazole is high among Japanese patients compared to Caucasian patients due to the genetic polymorphism of CYP2C19, the primary enzyme responsible for voriconazole metabolism and clearance [145]. Thirty percent of Japanese were characterized as poor metabolizers of CYP2C19, while only 5% in the Caucasian population. When individuals carry the genetic polymorphism of enzyme responsible for the clearance of tested drug, the drug dose is adjusted. Optimization of drug responses by identifying the genetic fingerprint is a crucial step toward "individualized medicine."

In general, human populations can be genetically characterized as (1) ultrametabolizers (Ums), which metabolize substrate so quickly because of gene duplication and drugs effectively maintaining high clearance, thus the therapeutic levels often cannot be reached in clinical practice in these individuals; (2) poor metabolizers (PMs), individuals who carry incompetent (homozygous) enzyme function that results in a lack of metabolism of CYP substrates; and (3) intermediate metabolizers (IMs), heterozygotes with partial expression of competent CYPs, which result in intermediate drug clearance. The most severe side effects are often noticed with IMs and PMs, and substrates that require metabolic activation may demonstrate suboptimal clinical efficacy.

The three genetic polymorphism categories of human populations are found to vary the disposition of drugs that depend on CYP2D6 for clearance. The variability in the metabolism and the clearance is a major concern for many of the antipsychotic drugs: (1) they are primarily metabolized by CYP2D6, and (2) they possess narrow therapeutic index [146, 147]. Approximately 7–10% of Caucasians, 0–5% Africans, and 0–1% Asians lack CYP2D6 activity because of the presence of one or several mutant alleles at the CYP2D6 gene locus, and these individuals are known as PMs. Drug therapy with codeine that metabolizes by CYP2D6 to form the active metabolite morphine was not effective because of the low formation of the active metabolite in PMs (e.g., Chinese population) [148].

For other CYP2D6 substrates or other genetic polymorphic enzymes, PMs demonstrate markedly greater AUC values for parent drugs that are metabolized by these enzymes compared to IMs and UMs, and therefore require lower doses to achieve therapeutic effects.

In addition, Caucasian patients require higher warfarin (metabolized by CYP2C9) doses than Asians to attain a comparable anticoagulant effect where ~ 20% of Asians and 3% Caucasians are PMs of CYP2C9 and CYP2C19 [148]. Also, in treatment with the anticancer drug indisulam [149], the patients that carry heterozygous CYP2C9*3, heterozygous CYP2C19*2, and heterozygous CYP2C19*3 mutations showed a reduced elimination rate compared to patients that have competent CYP2C9 and CYP2C19 activities. The authors of this report recommended that screening for these CYP2C polymorphisms may guide dose adaptation and aid in the selection of an optimized initial indisulam dosage [149].

3.4 SPECIES DIFFERENCES IN DRUG METABOLISM

The species differences in drug metabolism by DMEs phase I, II, and III between animal species and humans are common and frequently complicate preclinical drug development programs, especially within pharmacology and toxicology investigations. Because the animal DMEs may have different activities and specificities from those of human, the drug metabolism data from animal studies are usually compromised, though the species differences in drug metabolism where active or toxic metabolites are involved can have dramatic implications with human investigations [150]. Generally, the differences in drug metabolism between animals and humans can be related to

1. Lack of morphological similarity or complete absence of specific enzyme among animal species or between animal model compared to human.
2. Differences in enzyme (isoform) composition, gene regulation, protein expression, or catalytic activities.
3. Differences in enzyme kinetics that can cause difference in metabolite profile presence in systemic circulation or those trapped in tissues. For example, GDC-0834 was investigated as a potential treatment for rheumatoid arthritis using in vitro methodologies that revealed the predominant formation of an inactive metabolite (M1) via amide hydrolysis in humans that appeared to be NADPH-independent in human liver microsomes. The maximum rate of M1 formation (V_{max}) was substantially higher in humans compared with that in other species. In contrast, the Michaelis–Menten constant (K_m) was comparable among species. Intrinsic clearance (V_{max}/K_m) of GDC-0834 from M1 formation in humans was 23- to 169-fold higher than observed in rats, dogs, and monkeys [151].
4. Differences in regulation by nuclear receptors that affect the number of drugs that have the ability to induce various sets of DMEs in the liver and extrahepatic tissues via transcriptional activation through nuclear hormone receptors or transcription factors. That has significant impact on the observed species differences of ligand binding and ultimately induction of DMEs. Consequently, enzyme induction in any single species does not correlate with potential induction of the same enzyme in other species, in particular, human [152, 153].
5. Differences in substrate specificity, expression profile of isozymes, and tissue distribution of various phase I/II/III metabolic enzymes [154]. Because of species differences in expression levels and substrate specificity of transporters, extrapolation of data obtained in preclinical species to humans is not feasible, both with respect to predicting PK and DDI potential. Current research is focusing on using in silico approaches, such as building whole-body PBPK models

for drug transporters, as well as improving in vitro–in vivo extrapolation to determine the absolute protein levels of transporters in human tissues.

As described, understanding species differences in drug metabolism is helpful for scientists to select the animal models from which the metabolism data can be most accurately extrapolated to humans [155]. Unlike certain physiologic parameters, such as QH and liver wt/kg body weight, that can be allometrically scaled across species in order to predict human PK from preclinical data, no allometric relationship can be applied to the biochemical parameters that regulate the differences in DMEs across species. Most significant investigation using animal models in various stages of drug discovery and development is in assessment of DDI potentials and their underlying mechanisms in relation to drug metabolism. In vivo data in well-selected animal models can be significantly valuable, especially when these data are assessed against others data generated from in vitro models, which can substantiate the extrapolation of in vitro human data to clinical outcomes.

For species similarities and differences in DMEs, particularly P450 enzymes, no single animal species is identical to humans at the functional level for any DME, but more similarities are found in higher species like monkeys and dogs [156]. Recent advancements in molecular biology have provided valuable insights into these observed similarities and differences at the molecular level. When P450 isoforms are extensively involved in drug metabolism, higher species generally exhibit a higher degree of amino acid sequence identity to humans, as reported by Tang and Prueksaritanont (2010) [156]. Consistently, a greater similarity in substrate specificity was found; for example, human CYP3A4, rhesus monkey CYP3A64, and dog CYP3A12 are more selective than other CYP3A isoforms of the respective species for testosterone 6β-hydroxylation, midazolam 1′-hydroxylation, and nifedipine oxidation, but the selectivity weakens between rat CYP3A1 and 3A2. On the other hand, slight differences in protein sequence homology can lead to enormous changes in catalytic efficiency as seen in diclofenic hydroxylation mediated by CYP2C9 in humans versus monkeys, which share $\geq$ 92% homology. In case of DME inhibition, potent reversible human CYP3A inhibitor ketoconazole appears to have reasonable cross-activity in many species, as shown by the comparable K_i or IC_{50} values determined in multiple species for CYP3A as in cynomolgus monkeys and in rats. Thus, DDIs caused by CYP3A inhibition can be reasonably predicted by multiple animal models. For DDIs caused by CYP induction, it may be assessed from data generated using rhesus monkeys and cynomolgus monkeys for CYP3A. Induction data generated from dogs potentially estimates CYP1A, CYP2B, and CYP3A that exhibits characteristics intermediate to those of rodents and humans. Kyokawa et al. [157] also reported comparable induction of beagle dog intestinal CYP3A12 by rifampin at a dose (10 mg/kg/day) to the clinical dose for humans (600 mg/man/day). However, because of the known different response to CYP inducers, rodents are not comparable to humans [158]. To explain these data, rifampicin is known to be a potent PXR ligand in humans, dogs, and rabbits, but in the case of mice ad rats, the prototypical PXR agonists are dexamethasone and pregnenolone 16α-carbonitrile (PCN) [159]. Similar to rifampin, troglitazone is an activator of human PXR and inducer of human CYP3A4 but does not induce CYP3A in rodent models or activate rodent PXR [160].

In addition to DDI, species difference was detected in metabolite profile. Using midazolam 1′-hydroxylase (1′-OH-MDZ) as a measure of CYP3A activity [161], mouse, rat, rabbit, dog, and monkey microsomal levels of activity were found to be approximately 103%, 9.2%, 20%, 137%, and 128%, respectively, of the average human activity. A remarkable and significant interspecies variability in metabolism in midazolam metabolism, with higher rates of midazolam 1′-hydroxylation, were observed in dogs and monkeys in comparison with humans; however, no significant differences were observed between cynomolgus and rhesus monkeys [162]. Also, as it was known from midazolam metabolism in vitro and in vivo, the midazolam 1′-hydroxylation activity was the predominant reaction in humans, while the 4-hydroxylation was observed in rodents.

In contrast to the species difference in CYP3A prototype substrate midazolam, no species differences are shown in CYP3A4 prototype substrate testosterone. Testosterone 6β-hydroxylation is a commonly used marker activity for CYP3A enzymes in all species. There is no significant difference in the reaction among the species, with the exception of a higher K_m value observed in dogs and female rats; the latter has lower V_{max} values. Also, rats not only show higher rates of testosterone 6β-hydroxylation than humans but also generates 2β− and 16β−hydroxyltestosterone. In regard to testosterone 6β− hydroxylation kinetics across species liver microsomes, mouse and male rat appear to be the most similar species to human.

As described, the selection of the best animal model to be used in development of a new drug in humans is difficult. Caution should be taken in extrapolation of animal data to humans; however, when in vitro metabolism studies with liver microsomes, hepatocytes, liver slices, and recombinant enzymes are available, the information can be incorporated into the decision-making process to aid in the selection of the appropriate in vivo animal model. It is highly unlikely that any animal model can be the appropriate model for all aspects of drug metabolism for a given new drug candidate. A common practice among pharmaceutical companies is employing different species to assess each parameter associated with drug biotransformation, as will be addressed later.

3.5 TRANSLATIONAL TECHNOLOGIES AND METHODOLOGIES AND REGULATORY RECOMMENDATION FOR DRUG METABOLISM

Because of the key role metabolism plays in unraveling the bioavailability, clearance, half-life, exposure, and potential dose selection for optimum efficacy and minimum toxicity of new drug candidates, understanding the metabolism is a critical aspect of drug development ability. It is always important to understand the full set of

1. Metabolic reactions
2. Kinetics of these reaction in relation to potential DDI

TABLE 3.4 Strategic Applications of in vitro Models in Drug Metabolism and its Mechanisms.

Specific DM Testing	In Vitro Tools					
	Single Enzyme	Subcellular Fraction			Hepatocytes	
		S-9	Cytosol	Microsomes	Suspension	Cultured
Metabolic stability		XX		XX	XX	
Metabolic kinetics & clearance	X	XX		XXX	XXX	X
Species difference		XX		XX	XX	
Metabolite ID	XX		XX	XX	XX	
Inhibition	XXX			XXX		
Inhibition mechanism				XXX		
Induction						XXX
Reaction phenotype	XXX		XX	XXX	XX	
Biosynthesis /bioreactor	XXX		XX	XXX	XXX	

Notes: X = uncommonly used; XX = used in selected studies; XXX = used in almost all preclinical investigations.

3. Enzymes involved in the metabolism of a compound in both preclinical models as well as in humans, to avoid any unnecessary complications or failures in drug development re the pharmacokinetic, pharmacological, and toxicological aspects

There are various in vitro, in vivo, and in silico models that are often employed throughout drug discovery and development. The purpose of each model system in providing necessary information to elucidate the metabolic disposition of new and existing drug molecules will be discussed next, although the use of in silico models will be discussed in later chapters of this book.

3.5.1 In Vitro Models of Drug Metabolism

In vitro models of drug metabolism can be single enzymes, intact cells such as hepatocytes, or whole organ infusion models. Each model system has its own application in the study of drug metabolism in drug discovery and development (see Table 3.4). The most commonly employed in vitro model systems are S-9, microsomes, and hepatocytes [163–167]. However, recombinant expressed individual enzymes are also frequently used.

3.5.1.1 Single-cDNA Expressed Enzymes The single enzyme system is a useful tool to assess the molecular mechanism of drug metabolism and clearance mediated by phase I–III DMEs. Also, the expressed enzyme can be an optimum tool to determine the reaction phenotype, that is, identification of enzymes involved in metabolism of a drug candidate and qualitatively determine the % contribution of those enzymes that might be responsible for generation of certain metabolite(s). It is also used to determine DDI potential (inhibition). The assay is conducted in controlled incubations of enzyme content, buffer, and cofactors of all tested reactions, which can be easily manipulated. In the incubation, abundant enzyme protein is added at a concentration the same as its physiological content in the specific organ under investigation in vivo, which is extremely low; hence it is considered as an advantage, as it diminishes the potential of nonspecific binding. For the preparation, enzymes are individually expressed as recombinant enzymes through cDNA-directed expression systems in a variety of host cell lines (mammalian, bacterial, insect, and yeast) that are commercially available [166, 168]. The most commonly employed single-enzyme family is the P450 isozymes; however, other oxidative enzyme reactions and drug transporters are now available, such as FMOs, UGTs, GSTs, SULT, OATPs, MRPs, and P-gp [43, 169]. In general, the single-enzyme system will tend to overemphasize the importance or potency of metabolic reactions; nonetheless, they are extremely useful in breaking down complex metabolic pathways into individual pathways or enzymes [43].

3.5.1.2 Subcellular Fractions Subcellular fractions such as S9, cytosol, and microsomes prepared from whole tissue homogenate that process in buffer [26, 170] are the most used tools to study drug metabolism, DDI, and metabolite profile in every body organ and in animal species. Depending on the tissue, subcellular isolations can be prepared from homogenization of the whole organ or from homogenization of specific segment. For example, when preparing liver subcellular fractions, no specific segment is used but usually the entire liver is homogenized to form the desired subcellular fractions, especially in rats or small animals. However, for intestinal or kidney, the enterocytes or kidney cortex are used in preparation of subcellular fractions, respectively. Either fresh or frozen tissues can be used in preparing these subcellular fractions, the fractions can be stored frozen at −80°C for years, and they are widely available commercially for drug metabolism and toxicity-related studies [167, 168, 171–173]. The most predominant subcellular fraction for evaluating drug metabolism is the microsomal fraction. The microsomal fraction is the most commonly employed in vitro model because most drugs are metabolized by the P450s, UGTs, and FMOs that are found in the microsomal fraction, and it is the tool used in nearly every study that reveals the metabolism and toxicity-related studies except enzyme induction [see Table 3.4]. The cytosolic fraction is used less frequently; however, when the major enzyme involved in metabolism or inhibition of SLT or aldehyde oxidase (AO) of a particular drug is located in the cytosol, then this model becomes extremely useful. The S-9 fraction also has utility with multiple different metabolic pathways that involve microsomal enzymes (e.g., P450s and UGTs) as well as cytosolic enzymes (e.g., SULT, NAT, GST). Strategically, when the metabolic pathways become too complex, the hepatocyte model is used.

TABLE 3.5 Example for the Prediction of Drug X from in vitro Rate of Metabolism in Liver Microsomes (nmol/min/mg) and in Hepatocytes (pmol/min/million cells) to in vivo Clearance (mL/min/Kg) in Animal Model and Human (data listed for [177]).

Species	Liver Microsomes		Hepatocytes		Observed In
	Rate nmol/ min/mg	Predicted CL mL/min/kg	Rate Pmol/ min/106 cells	Predicted CL mL/min/Kg	Vivo Blood CL mL/min/Kg
Mouse	0.06	42	30	49	46
Rat	0.13	32	49	32	21
Dog	0.23	24	24	15	16
Monkey	0.13	26	20	16	23
Human	0.22	16	26	11	–

3.5.1.3 Cellular Systems Primary cells isolated from tissues and immortalized cell lines are both types of cells that have been utilized in assessing a variety of drug-metabolism functions; however, the primary cells tend to be employed more often in routinely determining induction of DMEs, metabolite profile, and assessing the role of drug transporters in drug disposition and DDI. For high throughput application in early drug discovery, immortalized cell lines are often used [174, 175]. Like the microsomes, hepatocytes can also be cryopreserved and stored for long periods of time, which makes them highly useful for routine or even high throughput assays to assess metabolic stability, metabolite identification, induction, or cytotoxicity [176]. Several publications documented the prediction of in vivo metabolic clearance using hepatocytes as well as liver microsomes from animal models and humans and compared to in vivo blood clearance. Data from one publication is depicted in Table 3.5 as an example. This data indicate a good agreement between predicted CL using liver microsomes and hepatocytes as well as between in vitro systems and in vivo observed data in all species; the least of those was observed in rats [177].

A hepatocytes cell model is prepared first by liver perfusion using a calcium-free buffer followed by a buffer containing collagenase to prepare a suspension of single cells [163]. Hepatocytes can be used as a cell suspension within 2–3 h of isolation, or after culture of the cells over collagen-plated wells (for several days) or in sandwich format [178]. Hepatocytes in suspension represent a fully functional unit of drug metabolism from the liver, containing normal concentrations of enzymes and cofactors with the ability to perform complex and multistep metabolic reactions [179]. Hepatocyte suspension for this reason is commonly used to determine metabolic stability, predict human metabolic clearance, and compare metabolic pathways or metabolite identification across multiple species to assess species differences in drug metabolism [179]. Also, it is used in bioreactors for the biosynthesis of metabolites that can't be prepared from chemical synthesis. Unfortunately, when in suspension, the viability of the cells will begin to decline somewhere around 2–3 h, limiting the length of incubation time and, therefore, the degree or extent of metabolism. Hepatocytes in culture will survive for several days or even weeks because of the artificial environment created in culture with an extracellular matrix and cell-to-cell contact.

Hepatocytes in culture are predominantly used to assess enzyme induction (transcriptional activation) and the potential for DDIs or cytotoxicity [180]. Unfortunately, even in culture, the hepatocytes will begin to dedifferentiate and the phenotypic properties of the cells may start to change within 24–48 h in culture. Most important of these changes are those of the CYP enzymes, which in general will decrease with time in culture; however, some CYPs have been shown to increase and others to decrease and then increase with culture time. Modern advances in cell culturing techniques such as micropatterned hepatocyte coculture systems and different three-dimensional cultures such as spheroids have made significant advances in improving the stability and longevity of DMEs in culture [181].

3.5.2 In Vivo Models of Drug Metabolism

In vitro models of drug metabolism are extremely useful to study specific reactions, mechanisms, or structure activity relationships; however, in vivo systems are required to understand additional factors that may contribute to the type of metabolism, rate of metabolism, or other elimination pathways in the overall disposition/exposure of a drug or metabolites. Although these issues can be determined within in vitro studies, an in vivo study will provide formation such as the contribution of extrahepatic metabolism, protein-binding effects, first-pass metabolism, and circulating metabolite profile under treatment regimen [182]. Several approaches can be established in vivo to determine the drug metabolism studies. These can be ex vivo/in situ using perfused organs, preclinical studies in intact or cannulated animal species (mass balance studies [MBSs]), or phase I PK clinical studies in humans [183].

3.5.2.1 Preclinical Animal Studies The most common in vivo drug-metabolism studies are based on animal biological matrices including plasma, urine, and bile, as well as feces in some situations. Plasma and urine are sampled from intact animals; however, bile collection requires surgical bile duct–cannulated animals [183]. Plasma, urine, bile, and feces can be collected and examined for parent drug and metabolites, usually after dosing animals with radiolabeled drugs [184] as part of MBS.

The MBS is considered as an essential in vivo animal study is conducted in rodent and nonrodent species. In this study, animals are dosed either orally or intravenously (or both) with a radiolabeled drug and radioactivity is measured from bile, urine, feces, exhaled $^{14}CO_2$, animal carcasses, and cage washings [183]. The purpose of this study is to (1) determine the major elimination pathways of drug and drug-related material; (2) determine the total recovery of radiolabeled material (what goes in should come out); and (3) conduct metabolic profiling from the various matrices collected to verify that the human metabolites generated are the same as those formed in sufficient quantities in the toxicology models. Sometimes, specific tissues may be collected at various time points for determination of total radioactivity or metabolites to determine tissue distribution (e.g., brain, liver, adrenal). Additional studies related to drug metabolism in preclinical species can include regulation of DMEs, understanding the mechanism of low systemic exposure, and determination of hepatic extraction.

Drug-metabolism-mediated DDI can also be conducted in vivo preclinical and in humans. For example, enzyme induction in animals can be assessed after dosing animals for multiple days (generally 1–2 weeks). Liver is then removed and the expression and enzyme activity of DMEs are measured (primary choice is CYP3A). In some situations, the administration of specific or global enzyme inhibitors can help elucidate the enzymes involved in metabolism or determine the mechanism of low exposure. For example, the global CYP450 inhibitor 1-aminobenzotriazole (ABT) can be administered to rats to address gut versus liver contributions to first-pass metabolism [185] or to assess the cause of low systemic exposure (poor absorption vs. high first-pass metabolism).

By using in vivo study, hepatic extraction can be determined by infusing the drug via the portal vein and measuring systemic parent drug and/or metabolites [182]. This approach is commonly conducted in rats; however, it is also conducted in dogs or monkeys but after implanting vascular access ports that lead to the portal vein [186].

3.5.2.2 Genetically Modified Animal/Chimeric Mouse Model/Ex Vivo/In Situ Organ Perfusion Developing improved animal models for predicting systemic drug and metabolite exposure in man as early as possible in the drug discovery process is seen as critical in avoiding project delays, in contributing toward the early safety evaluation of drug candidates, and in estimating the starting dose first in clinical trials. Recent advancements in technology have evolved and expanded toward the use of a variety of genetically modified animal (GEMA) models to gain insight into the role that DMEs, nuclear transcriptional factors, and transporters play in the disposition of drugs and in the regulation of endogenous substrates, and hence to assess the mechanism(s) associated with drug-induced toxicities. There are numerous GEMA models generated for drug metabolism and transporter proteins that have provided greater characterization and mechanistic understanding of proteins that contribute to the disposition of drugs [187, 188], and the authors have focused on GEMAs that have been widely studied and listed the GEMA models that are available to study DMEs and transcriptional factors involved in the up-regulation of genes and drug transporters.

While in vitro systems, such as human liver microsomes or hepatocytes, provide variable data when used to predict actual human in vivo metabolic profiles of drug candidates, these sometimes provide poor prediction of biotransformation of drugs undergoing complex metabolic pathways with large numbers of secondary metabolites, which are often implicated in serious drug-induced toxicity in vivo. The Food and Drug Administration (FDA), according to metabolites in safety testing (MIST) regulation, as discussed in more detail in chapter 11, mandates that the appropriate testing of significant metabolites in the preclinical and clinical stages of drug discovery is fulfilled and included in the investigational new drug (IND)/new drug application (NDA) filling. One of the concerns is that no adequate test, of not only the parent but also the major human metabolites arising from administration of the drug, is possible since in vitro systems can sometimes fail. Where the laboratory species used for safety studies did not generate the particular human metabolite, then it was mandated that the metabolite itself needed to be synthesized and appropriate safety studies needed to be conducted in the preclinical animal species on the metabolite itself. Clearly, a preclinical animal species that reciprocates human metabolism and human-like metabolic profile would have significant advantages for toxicity studies in humans. A move toward "humanizing" laboratory animal species is the technological breakthrough of being able to delete the gene encoding for the animal homologue of a particular gene and transfecting the human homologue into the mouse genome. After injection of human hepatocytes into the mouse spleen, over time the mouse liver will become repopulated with human hepatocytes replacing $> 70 - 90\%$ of the native mouse hepatocytes.

The chimeric mouse model has been shown to be useful in evaluating the metabolism, disposition, PK, pharmacology, and toxicity of many xenobiotics [189, 190]. Using chimeric mice, several authors have been able to demonstrate metabolic profiles that were specific to humans and different from wildtype mice, for example, dexamethasone (CYP3A4), diclofenac (CYP2C9), paclitaxel (CYP2C8), and debrisoquine (CYP2D6) [189]. The model has also been shown to be a predictor of circulating human metabolites, especially metabolites derived from multiple metabolic reactions [190]. In addition to P450, the chimeric mouse model expresses human phase II enzymes and transporters [189–193]. The chimeric mouse model was also able to predict the human disposition of the drug cefmetazole. In mice and rats, cefmatazole is eliminated in feces (bile), whereas in humans, urinary elimination of cefmatazole is found. When dosed to chimeric mice, cefmatazole was eliminated in urine, similar to the human elimination pathway and different than the wildtype mouse [189]. DDIs have also been evaluated in the humanized mouse liver model. For example, the human-specific CYP3A4 inducer rifampicin was shown to induce human CYP3A4 in the chimeric mice but not in wildtype mice. Also, inhibition of human CYP2D6 was demonstrated by using quinidine as an inhibitor and debrisoquine as the probe substrate. The AUC and C_{max} of 4-hydroxy debrisoquine decreased significantly in the humanized mice when codosed with debrisoquine and quinidine but did not change in the control mice [189]. In PK, biotransformation and hepatic transporter effects of troglitazone were investigated in chimeric mice with humanized livers. The clinical chemistry revealed no consistent pattern of changes associated with troglitazone treatment in the PXB mouse. Human MRP2 but not mouse mrp2 is down-regulated following troglitazone treatment. Oxidative and conjugative metabolic pathways were identified, with the sulfate being the predominant metabolite in humanized mice compared to control mice, where the glucuronide conjugate was predominate. There was no evidence of GSH conjugation. The primary oxidative pathways were mono- and di-oxidations, which may also be attributed to quinone or hydroquinone derivatives. Several metabolites were observed in chimeric mice only. The metabolic profiles in the chimeric mouse were similar to reported human data, indicating that the chimeric mouse model can provide a useful first insight into circulating human metabolites of xenobiotics metabolized in the liver [194].

For perfused organs, liver, heart, lung, kidney, brain, intestine, and intestine-liver preparations were used to study metabolism, liver and intestine were reported. Organ perfusion is generally limited to smaller animals such as rats and, to a lesser extent, rabbits. Organ perfusions, such as the liver or intestine, offer several advantages: (1) the concentration of drug, buffer, and rate of drug delivery can all be easily manipulated during the experiment and tailored to the needs of the investigator; (2) the conditions can resemble the physiological conditions; (3) the perfusions for liver or the combination of intestine-liver perfusions can provide valuable information not only on the first-pass metabolism but also the interplay between DMEs and transporter-mediated disposition [195]. However, one of the most significant disadvantages of the isolated perfused organ is that the study is conducted for only a single organ/experiment/per animal, unlike the intact in vivo animal (or in vitro experiments), where several conditions with several organs can be tested during the same experiment.

REFERENCES

1. Gonzalez EJ, Nebert DW. Evolution of the P450 gene superfamily: animal-plant "warfare," molecular drive and human genetic differences in drug oxidation. Trends Genet 1990;6:182–186.
2. Guengerich EP. Cytochromes p450, drugs, and diseases. MoL Intervent 2003;3:194–204.
3. Meyer UA. Overview of enzymes of drug metabolism. J. Pharmacokinet. Biopharm 1996;24:449–459.
4. Nebert DW, Nelson DR, Coon MJ, Estabrook RW, Feyereisen R, Fujii-Kuriyama Y, Gonzalez FJ, Guengerich FP, Gunsalus IC, Johnson EF, et al. The P450 superfamily: update on new sequences, gene mapping, and recommended nomenclature. DNA Cell Biol 1991;10:1–14.
5. Nelson DR, Koymans L, Kamataki T, Stegeman JJ, Feyereisen R, Waxman DJ, Waterman MR, Gotoh O, Coon MJ, Estabrook RW, Gunsalus IC, Nebert DW. P450 superfamily: update on new sequences, gene mapping, accession numbers and nomenclature. Pharmacogenetics 1996;6:1–42.
6. Banoglu E. Current status of the cytosolic sulfotransferases in the metabolic activation of promutagens and procarcinogens. Curr Drug Metab 2000;1:1–30.
7. King CD, Rios GR, Green MD, Tephly TR. UDP-glucuronosyltransferases. Curr Drug Metab 2000;1:143–161.
8. Chen C, Yu R, Owuor ED, Kong AN. Activation of antioxidant-response element (ARE), mitogen-activated protein kinases (MAPKs) and caspases by major green tea polyphenol components during cell survival and death. Arch Pharm Res 2000;23:605–612.
9. Sugatani J, Kojima H, Ueda A, Kakizaki S, Yoshinari K, Gong QH, Owens IS, Negishi M, Sueyoshi T. The phenobarbital response enhancer module in the human bilirubin UDP-glucuronosyltransferase UGT1A1 gene and regulation by the nuclear receptor CAR. Hepatology 2001;33:1232–1238.
10. Brinkmann U, Eichelbaum M. Polymorphisms in the ABC drug transporter gene MDRI. Pharmacogenomics J 2001;1:59–64.

11. Kerb R, Hoffmeyer S, Brinkmann U. ABC drug transporters: hereditary polymorphisms and pharmacological impact in MDR1, MRP1 and MRP2. Pharmacogenomics 2001;2:51–64.
12. Tirona RG, Kim RB. Pharmacogenomics of organic anion-transporting polypeptides (OATP). Adv Drug Deriv Rev 2002;54:1343–1352.
13. Kim RB. Organic anion-transporting polypeptide (OATP) transporter family and drug disposition. Eur J Clin Invest 2003;33 Suppl 2:1–5.
14. Mizuno N, Niwa, T, Yotsumoto Y, Sugiyama Y. Impact of drug transporter studies on drug discovery and development. Pharmacol Rev 2003;55:425–461.
15. Staudinger JL, Madan A, Carol KM, Parkinson A. Regulation of drug transporter gene expression by nuclear receptors. Drug Metab Dispos 2003;31:523–527.
16. Wang H, LeCluyse EL. Role of orphan nuclear receptors in the regulation of drug-metabolising enzymes. Clin Pharmacokinet 2003;42:1331–1357.
17. Hinson JA, Forkert PG. Phase II enzymes and bioactivation. Can J Physiol Pharmacol 1995;73:1407–1413.
18. Penner N, Klunk LJ, Prakash C. Human Radiolabeled Mass Balance Studies: Objectives, Utilities and Limitations. Biopharm. Drug Dispos 2009;30:185–203.
19. Schilter B, Turesky RJ, Juillerat M, Honegger P, Guigoz Y. Phase I and phase II xenobiotic reactions and metabolism of the food-borne carcinogen 2-amino-3,8-dimethylimidazo[4,5-f]quinoxaline in aggregating liver cell cultures. Biochem Pharmacol 1993;45:1087–1096.
20. Rushmore TH, Kong AN. Pharmacogenomics, regulation and signaling pathways of Phase I and II drug metabolizing enzymes. Curr Drug Metab 2002;3:481–490.
21. Sinz MA. In vitro and in vivo models of drug metabolism. Part I. Basics overview of drug metabolism. In: Encyclopedia of drug metabolism and interactions, ed. Lyubimov AV. Hoboken, NJ: Wiley, 2012, 1:1–31.
22. Clarke SE, Jones BC. Human cytochromes P450 and their role in metabolism-based drug–drug interactions. In: Drug–drug interactions, 2nd ed., ed. Rodrigues AD. New York: Informa Healthcare, 2008, 53–85.
23. Xu L, Das B, Prakash C. CYP450 enzymes in drug discovery and development: an overview. In: Encyclopedia of drug metabolism and interactions, ed. Lyubimov AV. Hoboken, NJ: Wiley, 2012, 10: 1–35.
24. Brown CM, Reisfeld B, Mayeno AN. Cytochrome P450: a structure-based summary of biotransformations using representative substrates. Drug Metab Rev 2008;40:1–100.
25. Guengerich FP. Cytochrome p450 and chemical toxicology. Chem Res Toxicol 2008;21:70–83.
26. Omura T, Sato R. A new cytochrome in liver microsomes. J Biol Chem 1962;237:1375–1376.
27. Klingenberg M. Pigments of rat liver microsomes. Arch Biochem Biophys 1958;75:376–386.
28. Guengerich FP. Cytochrome p450 and chemical toxicology. Chem Res Toxicol 2008;21(1):70–83.
29. Nelson DR. Gene nomenclature by default, or BLASTing to Babel. Hum Genomics 2005;2:196–201.
30. Ortiz de Montellano PR, ed. Cytochrome P450: structure, mechanism, and biochemistry, 3rd ed. New York: Kluwer Academic/Plenum Publishers, 2005.

31. Paine MF, Khalighi M, Fisher JM, et al. Characterization of interintestinal and intra-intestinal variations in human CYP3A-dependent metabolism. J Pharmacol Exp Ther 1997;283:1552–1562.
32. Williams PA, Cosme J, Vinkovic DM, et al. Crystal structures of human cytochrome P450 3A4 bound to metyrapone and progesterone. Science 2004;305:683–686.
33. Kuehl P, Zhang J, Lin Y, et al. Sequence diversity in CYP3A promoters and characterization of the genetic basis of polymorphic CYP3A5 expression. Nat Genet 2001;27:383–391.
34. Huang W, Lin YS, McConn DJ, et al. Evidence of significant contribution from CYP3A5 to hepatic drug metabolism. Drug Metab Dispos 2004;32:1434–1445.
35. Ekins S, Stresser DM, Williams JA. In vitro and pharmacophore insights into CYP3A enzymes. Trends Pharmacol Sci 2003;24:161–166.
36. Gellner K, Eiselt R, Hustert E, et al. Genomic organization of the human CYP3A locus: identification of a new, inducible CYP3A gene. Pharmacogenetics 2001;11:111–121.
37. Parkinson A, Ogilvie BW. Biotransformation of xenobiotics, 7th ed. New York: McGraw-Hill, 2008.
38. Fura A, Shu YZ, Zhu M, et al. Discovering drugs through biological transformation: role of pharmacologically active metabolites in drug discovery. J Med Chem 2004;47:4339–4351.
39. Zhou S, Chan E, Duan W, et al. Drug bioactivation, covalent binding to target proteins and toxicity relevance. Drug Metab Rev 2005;37:41–213.
40. Penner N, Woodward C, Prakash C. Drug metabolizing enzymes and biotransformation reactions. New York: Wiley, 2010.
41. Guengerich FP. Common and uncommon cytochrome P450 reactions related to metabolism and chemical toxicity. Chem Res Toxicol 2001;14:611–650.
42. Ling J, Johnson KA, Miao Z, et al. Metabolism and excretion of erlotinib, a small molecule inhibitor of epidermal growth factor receptor tyrosine kinase, in healthy male volunteers. Drug Metab Dispos 2006;34:420–426.
43. Yanni SB, Annaert PP, Augustijns P, Bridges A, Gao Y, Benjamin DK Jr, Thakker DR. Role of flavin-containing monooxygenase in oxidative metabolism of voriconazole by human liver microsomes. Drug Metab Dispos 2008;36(6):1119–1125.
44. Miners JO, Mackenzie PI. Drug glucuronidation in humans. Pharmacol Ther 1991;51:347–369.
45. Otake Y, Hsieh F, Walle T. Glucuronidation versus oxidation of the flavonoid galangin by human liver microsomes and hepatocytes. Drug Metab Dispos 2002;30(5):576–581.
46. Cappiello M, Giuliani L, Pacifici GM. Distribution of UDP glucuronosyltransferase and its endogenous substrate uridine 5′-diphosphoglucuronic acid in human tissues. Eur J Clin Pharmacol 1991;41:345–350.
47. Strassburg CP, Nguyen N, Manns MP, et al. UDP glucuronosyltransferase activity in human liver and colon. Gastroenterology 1999;116:149–160, 259–265.
48. Kurkela M, Garcia-Horsman A, Luukkanen L, et al. Expression and characterization of recombinant human UDP-glucuronosyltransferases (UGTs). J Biol Chem 2003;278(6):3536–3544.
49. Ouzzine M, Barre L, Netter P, et al. The human UDP-glucuronosyltransferases: structural aspects and drug glucuronidation. Drug Metab Rev 2003;35(4):287–303.

50. Csala M, Margittai E, Banhegyi G. Redox control of endoplasmic reticulum function. Antioxid Redox Signal 2007;13(1):77–108.

51. Burchell B, Lockley DJ, Staines A, et al. Substrate specificity of human hepatic UDP-glucuronosyltransferases. Methods Enzymol 2005;400:3–57.

52. Kiang TKL, Ensom MHH, Chang TKH. UDP-glucuronosyltransferases and clinical drug–drug interactions. Pharmacol Ther 2005;106:97–132.

53. Williams JA, Hyland R, Jones BC, et al. Drug–drug interactions for UDP-glucurmonosyltransferase substrates: a pharmacokinetic explanation for typically observed low exposure (AUC_I/AUC) ratios. Drug Metab Dispos 2004;32(11): 1201–1208.

54. Garnett WR. Lamotrigine-interaction with other drugs. In: Anti-epileptic drugs, 5th ed., ed. Levy RH, Mattson RH, Meldrum BS, et al. Philadelphia: Lippincott Williams & Wilkins, 2002, 300–331.

55. Miners JO, Mackenzie PI. Drug glucuronidation in humans. Pharmacol Ther 1991;51:347–369.

56. Guillemette C. Pharmacogenomics of human UDP-glucuronosyltransferase enzymes. Pharmacogenomics J 2003;3:136–158.

57. Bosma PJ, Chowdhury JR, Bakker C, et al. The genetic basis for the reduced expres sion of bilirubin UDP-glucuronosyltransferase 1 in Gilbert's syndrome. N Engl J Med 1995;333:1171–1175.

58. Al-Shurafa HA, Bassas AF, Broering DC, et al. Management of Crigler-Najjar Syndrome type I. Saudi Med J 2001;22(6):486–489.

59. Van Der Veere CN, Sinaasappel M, McDonagh AF, et al. Current therapy for Crigler-Najjar syndrome type I: report of a world registry. Hepatology 1996;24(2):311–315.

60. Evans WE, Reilling M. Pharmacogenomics: translating functional genomics into rational therapeutics. Science 1999;286:487–491.

61. Cappiello M, Franchi M, Giuliani L, et al. Distribution of 2-naphthol sulphotransferase and its endogenous substrate adenosine 3′-phosphate 5′-phosphosulfate in human tissues. Eur J Clin Pharmacol 1989;37:317–320.

62. Pacifici GM. Sulfation of drugs. In: Human cytosolic sulfotransferases, ed. Pacifici GM, Coughtrie MW. Boca Raton, FL: CRC Press, 2005, 157–177.

63. Gamage N, Barnett A, Hempel N, et al. Human sulfotransferases and their role in chemical metabolism. Toxicol Sci 2006;90(1):5–22.

64. Yanni S, Allegaert K. Drug metabolism and interactions in pediatric patients. In: Encyclopedia of drug metabolism and interactions, ed. Lyubimov AV. Hoboken, NJ; Wiley, 2012, 11:1–44.

65. Strott CA. Sulfonation and molecular action. Endocr Rev 2002;23:703–732.

66. Hempel N, Barnet A, Gamage N, et al. Human SULT1A SULTs. In: Human cytosolic sulfotransferases, ed. Pacifici GM, Coughtrie MW. Boca Raton, FL: CRC Press, 2005, 179–230.

67. Raftogianis RB, Wood TC, Otterness DM, et al. Phenol SULT pharmacogenetics in humans: association of common SULT1A1 alleles with TS PST phenotype. Biochem Biophys Res Commun 1997;239:298–304.

68. Wang Y, Spitz MR, Tsou AM, et al. Sulfotransferase (SULT) 1A1 polymorphism as a predisposition factor for lung cancer: a case control analysis. Lung Cancer 2002;35:137–142.
69. Meister A. Glutathione metabolism. Methods Enzymol 1995;251:3–7.
70. Hayes JD, Flanagan JU, Jowsey IR. Glutathione transferases. Annu Rev Pharmacol Toxicol 2005;45:51–88.
71. Townsend DM, Tew KD, Tapiero H. The importance of glutathione in human disease. Biomed Pharmacother 2003;57:145–155.
72. Mannisto PT, Kaakkola S. Catechol-O-methyltransferase (COMT): biochemistry, molecular biology, pharmacology, and clinical efficacy of the new selective COMT inhibitors. Pharmacol Rev 1999;51(4):593–628.
73. Guldberg HC, Marsden CM. Catechol-O-methyltransferases, pharmacological aspects and physiological role. Pharmacol Rev 1975;27:135–206.
74. Wu H, Dombrovsky L, Tempel W, et al. Structural basis of substrate-binding specificity of human arylamine N-acetyltransferases. J Biol Chem 2007;282(41):30189–30197.
75. Butcher NJ, Boukouvala S, Sim E, et al. Pharmacogenetics of the arylamine N-acetyltransferases. Pharmacogenetics J 2002;2:30–42.
76. Graf T, Broly T, Hoffman F, et al. Prediction of phenotype for acetylation and for debrisoquine hydroxylation by DNA-test in healthy human volunteers. Eur J Clin Pharmacol 1992;43:399–403.
77. Makarova SI. Human N-acetyltransferases and drug-induced hepatotoxicity. Curr Drug Metab 2008;9(6):538–545.
78. Levy GN, Weber WW. Interindividual variability of N-acetyltransferases. In: Interindividual variability in human drug metabolism, ed. Pacifici GM, Pelkonen O. London: Taylor and Francis, 2001, 333–357.
79. Yamamoto M, Sobue G, Mukoyama M, et al. Demonstration of slow acetylator genotype of IV-acetyltransferase in isoniazid neuropathy using an archival hematoxylin and eosin section of a sural nerve biopsy specimen. J Neurol Sci 1996;135:51–54.
80. Agundez JAG. Polymorphisms of human N-acetyltransferases and cancer risk. Curr Drug Metab 2008;9:520–531.
81. Ozols J. Preparation of membrane fractions. Methods Enzymol 1990;182:225–235.
82. Rowland M, Tozer TN. Clinical pharmacokinetics and pharmacodynamics. New Delhi: Wolters Kluwer India Private Limited, 2011.
83. Jambhekar SS, Breen PJ. Basic pharmacokinetics. London: Pharmaceutical Press; 2009.
84. Barton P, Austin RP, Fessey RE. In vitro models for plasma binding and tissue storage. In: Comprehensive medicinal chemistry II, ed. Testa B, van De Waterbeemd H. Oxford: Elsevier, 2007, 5:321–340.
85. Venkateshwarlu V. Biopharmaceutics and pharmacokinetics. Hyderabad, India: Pharma Book Syndicate, 2005.
86. Yanni SB, Smith PB, Benjamin DK Jr, Augustijns PF, Thakker DR, Annaert PP. Higher clearance of micafungin in neonates compared with adults: role of age-dependent micafungin serum binding. Biopharm Drug Dispos 2011;32(4):222–232.
87. Yanni SB, Annaert PP, Augustijns P, Ibrahim JG, Benjamin DK Jr, Thakker DR. In vitro hepatic metabolism explains higher clearance of voriconazole in children versus adults: role of CYP2C19 and flavin-containing monooxygenase 3. Drug Metab Dispos 2010;38:25–31.

88. Giacomini KM, Huang SM, Tweedie DJ, et al. Membrane transporters in drug development. Nat Rev Drug Discov 2010;9:215–236.
89. Ho RH, Kim RB. Transporters and drug therapy: implications for drug disposition and disease. Clin Pharmacol Ther 2005;78:260–277.
90. Chandra P, Brouwer KL. The complexities of hepatic drug transport: current knowledge and emerging concepts. Pharm Res 2004;21:719–735.
91. Sharom FJ. The P-glycoprotein multidrug transporter. Essays Biochem 2011;50(1):161–178.
92. Benet LZ. The drug transporter-metabolism alliance: uncovering and defining the interplay. Mol Pharm 2009;6(6):1631–1643.
93. Benet LZ, Cummins CL, Wu CY. Unmasking the dynamic interplay between efflux transporters and metabolic enzymes. Int J Pharm 2004;277(1–2):3–9.
94. Benet LZ. Predicting drug disposition via application of a biopharmaceutics drug disposition classification system. Basic Clin Pharmacol Toxicol 2009;106(3):162–167.
95. Watanabe T, Maeda K, Kondo T, et al. Prediction of the hepatic and renal clearance of transporter substrates in rats using in vitro uptake experiments. Drug Metab Dispos 2009;37:1471–1479.
96. Watanabe T, Kusuhara H, Watanabe T, et al. Prediction of the overall renal tubular secretion and hepatic clearance of anionic drugs and a renal drug–drug interaction involving organic anion transporter 3 in humans by in vitro uptake experiments. Drug Metab Dispos 2011;39:1031–1038.
97. Watanabe T, Kusuhara H, Maeda K, et al. Investigation of the rate-determining process in the hepatic elimination of HMG-CoA reductase inhibitors in rats and humans. Drug Metab Dispos 2010;38:215–222.
98. Soars MG, Webborn PJ, Riley RJ. Impact of hepatic uptake transporters on pharmacokinetics and drug–drug interactions: use of assays and models for decision making in the pharmaceutical industry. Mol Pharm 2009;6:1662–1677.
99. Nakanishi T, Tamai I. Genetic polymorphisms of OATP transporters and their impact on intestinal absorption and hepatic disposition of drugs. Drug Metab Pharmacokinet 2012;27:106–121.
100. Yoshida K, Maeda K, Sugiyama Y. Transporter-mediated drug–drug interactions involving OATP substrates: predictions based on in vitro inhibition studies. Clin Pharmacol Ther 2012;91(6):1053–1064.
101. Tornio A, Filppula AM, Kailari O, Neuvonen M, Nyrönen TH, Tapaninen T, Neuvonen PJ, Niemi M, Backman JT. Glucuronidation converts clopidogrel to a strong time-dependent inhibitor of CYP2C8: a phase II metabolite as a perpetrator of drug-drug interactions. Clin Pharmacol Ther 2014;96(4):498–507.
102. Kearns GL, Abdel-Rahman SM, Alander SW, et al. Developmental pharmacology: drug disposition, action, and therapy in infants and children. N Engl J Med 2003;349:1157–1167.
103. Alcorn J, McNamara PJ. Using ontogeny information to build predictive models for drug elimination. Drug Discov Today 2008;13:507–512.
104. Johnson TN. The development of drug metabolizing enzymes and their influence on the susceptibility to adverse drug reactions in children. Toxicology 2003;192:37–48.
105. Lacroix D, Sonnier M, Moncion A, et al. Expression of CYP3A in the human liver. Evidence that the shift between CYP3A7 and CYP3A4 occurs immediately after birth. Eur J Biochem 1997;247:625–634.

106. Barter ZE, Bayliss MK, Beaune PH, et al. Scaling factors for the extrapolation of in vivo metabolic drug clearance from in vitro data: reaching a consensus on values of human microsomalprotein and hepatocellularity per gram of liver. Curr Drug Metab 2007;8:33–45.
107. Rostami-Hodjegan A, Tucker GT. Simulation and prediction of in vivo drug metabolism in human populations from in vitro data. Nat Rev Drug Discov 2007;6:140–148.
108. Ginsberg G, Hattis D, Sonawanec B. Incorporating pharmacokinetic differences between children and adults in assessing children's risks to environmental toxicants. Toxicol Appl Pharmacol 2004;198:164–183.
109. Johnson TN, Rostami-Hodjegan A, Tucker GT. Prediction of the clearance of eleven drugs and associated variability in neonates, infants and children. Clin Pharmacokinet 2006;45(9):931–956.
110. Shimada T, Yamazaki H, Mimura M, et al. Characterization of microsomal cytochrome P450 enzymes involved in the oxidation of xenobiotic chemicals in human fetal liver and adult lungs. Drug Metab Dispos 1996;24:515–522.
111. de Wildt SN, Ito S, Koren G. Challenges for drug studies in children: CYP3A phenotyping as example. Drug Discov Today 2009;14:6–15.
112. Stevens JC, Hines RN, Gu C, et al. Developmental expression of the major human hepatic CYP3A enzymes. J Pharmacol Exp Ther 2003;307:573–582.
113. Bernardini S, Semama DS, Huet F, et al. Effects of cisapride on QTc interval in neonates. Arch Dis Child Fetal Neonatal Ed 1997;77:F241–F243.
114. Treluyer JM, Bowers G, Cazali N, et al. Oxidative metabolism of amprenavir in the human liver, effect of the CYP3A maturation. Drug Metab Dispos 2003;31:275–281.
115. Pearce RE, Gotschall RR, Kearns GL, et al. Cytochrome P450 involvement in the biotransformation of cisapride and racemic norcisapride in vitro: differential activity of individual CYP3A isoforms. Drug Metab Dispos 2001;29:1548–1554.
116. Treluyer JM, Rey E, Sonnier M, et al. Evidence of impaired cisapride metabolism in neonates. Br J Clin Pharmacol 2001;52:419–425.
117. Anderson GD, Lynn AM. Optimizing pediatric dosing: a developmental pharmacologic approach. Pharmacotherapy 2009;29:680–690.
118. Mahmood I, Tammara V. A comparison of two sparse sampling population pharmacokinetic approaches for the estimation of pharmacokinetic parameters in children. Int J Clin Pharmacol Ther 2004;42:240–245.
119. Hines RN, Koukouritaki SB, Poch MT, et al. Identification and functional analysis of human FMO3 genetic variants. FASEB J 2006;20:A264–A264.
120. Allegaert K, Vanhole C, Vermeersch S, et al. Both postnatal and postmenstrual age contribute to the interindividual variability in tramadol glucuronidation in neonates. Early Hum Dev 2008;84:325–330.
121. Kilic I, Koseler A, Cakaloz I, et al. Screening for G71R mutation of the UDP-glucuronosyltransferase 1 (UGT1A1) gene in neonates with pathologic and prolonged hyperbilirubinemia in Turkey. Int J Clin Pharmacol Ther 2010;48:504–508.
122. McCarver DG, Hines RN. The ontogeny of human drug-metabolizing enzymes: phase II conjugation enzymes and regulatory mechanisms. J Pharmacol Exp Ther 2002;300(2):361–366.
123. McNamara PJ, Alcorn J. Protein binding predictions in infants. Pharm Sci 2002;4:E4.

124. Alcorn J, McNamara PJ. Using ontogeny information to build predictive models for drug elimination. Drug Discov Today 2008;13:507–512.

125. Coleman MD, Tingle MD, Winn MJ, et al. Gonadal influence on the metabolism and hematological toxicity of dapsone in the rat. J Pharm Pharmacol 1990;42:698–703.

126. Ng W, Uchida H, Ismail Z, et al. Clozapine exposure and the impact of smoking and gender: a population pharmacokinetic study. Ther Drug Monit 2009;31:360–366.

127. Gunes A, Ozbey G, Vural EH, et al. Influence of genetic polymorphisms, smoking, gender and age on CYP1A2 activity in a Turkish population. Pharmacogenomics 2009;10:769–778.

128. Anderson GD. Gender differences in pharmacological response: epilepsy in women: the scientific basis for clinical management. Int Rev Neurobiol 2008;83:1–10.

129. Liu W, Kulkarni K, Hu M. Gender-dependent differences in uridine 5′-diphospho-glucuronosyltransferase have implications in metabolism and clearance of xenobiotics. Expert Opin Drug Metab Toxicol 2013;9(12):1555–1569.

130. Wang Z, Hall SD, Maya JF, et al. Diabetes mellitus increases the in vivo activity of cytochrome P450 2E1 in humans. Br J Clin Pharmacol 2003;55:77–85.

131. Nolin TD, Naud J, Leblond FA, Pichette V. Emerging evidence of the impact of kidney disease on drug metabolism and transport. Clin Pharmacol Ther 2008;83(6):898–903.

132. Tsunedomi R, Iizuka N, Hamamoto Y, et al. Patterns of expression of cytochrome P450 genes in progression of hepatitis C virus-associated hepatocellular carcinoma. Int J Oncol 2005;27:661–667.

133. Miyakawa H, Kitazawa E, Kikuchi K, et al. Immunoreactivity to various human cytochrome P450 proteins of sera from patients with autoimmune hepatitis, chronic hepatitis B, and chronic hepatitis C. Autoimmunity 2001;33:23–32.

134. Bogdanos DP, Dalekos GN. Enzymes as target antigens of liver-specific autoimmunity: the case of cytochromes P450s. Curr Med Chem 2008;15:2285–2292.

135. Villeneuve JP, Pichette V. Cytochrome P450 and liver diseases. Curr Drug Metab 2004;5:273–282.

136. Lu Y, Arthur I, Cederbaum AI. CYP2E1 and oxidative liver injury by alcohol. Free Radic Biol Med 2008;44:723–738.

137. Dietrich CG, Geier A. Effect of drug transporter pharmacogenetics on cholestasis. Expert Opin Drug Metab Toxicol 2014;10(11):1533–1551.

138. Taur SR, Kulkarni NB, Gandhe PP, Thelma BK, Ravat SH, Gogtay NJ, Thatte UM. Association of polymorphisms of CYP2C9, CYP2C19, and ABCB1, and activity of P-glycoprotein with response to anti-epileptic drugs. J Postgrad Med 2014;60(3):265–269.

139. He XJ, Jian LY, He XL, Tang M, Wu Y, Xu YY, Sun XJ, Zhao LM. Association of ABCB1, CYP3A4, EPHX1, FAS, SCN1A, MICA, and BAG6 polymorphisms with the risk of carbamazepine-induced Stevens–Johnson syndrome/toxic epidermal necrolysis in Chinese Han patients with epilepsy. Epilepsia 2014;55(8):1301–1306.

140. Goswami S, Yee SW, Stocker S, Mosley JD, Kubo M, Castro R, Mefford JA, Wen C, Liang X, Witte J, Brett C, Maeda S, Simpson MD, Hedderson MM, Davis RL, Roden DM, Giacomini KM, Savic RM. Genetic variants in transcription factors are associated with the pharmacokinetics and pharmacodynamics of metformin. Clin Pharmacol Ther 2014;96(3):370–379.

141. Dhoro M, Ngara B, Kadzirange G, Nhachi C, Masimirembwa C. Genetic variants of drug metabolizing enzymes and drug transporter (ABCB1) as possible biomarkers for adverse drug reactions in an HIV/AIDS cohort in Zimbabwe. Curr HIV Res 2013;11(6):481–490.

142. Dennis BB, Bawor M, Thabane L, Sohani Z, Samaan Z. Impact of ABCB1 and CYP2B6 genetic polymorphisms on methadone metabolism, dose and treatment response in patients with opioid addiction: a systematic review and meta-analysis. PLoS One 2014 29;9(1):e86114.

143. Kim KA, Joo HJ, Lee HM, Park JY. Influence of ABCB1 and CYP3A5 genetic polymorphisms on the pharmacokinetics of quetiapine in healthy volunteers. Pharmacogenet Genomics 2014;24(1):35–42.

144. Markova SM, De Marco T, Bendjilali N, Kobashigawa EA, Mefford J, Sodhi J, Le H, Zhang C, Halladay J, Rettie AE, Khojasteh C, McGlothlin D, Wu AH, Hsueh WC, Witte JS, Schwartz JB, Kroetz DL.Association of CYP2C9*2 with bosentan-induced liver injury. Clin Pharmacol Ther 2013;94(6):678–686.

145. Kimura M, Yamagishi Y, Kawasumi N, Hagihara M, Hasegawa T, Mikamo H Clinical implication of therapeutic drug monitoring on voriconazole from the aspect of the analysis for CYP2C19 gene. Jpn J Antibiot 2010;63(3):255–264.

146. Gassó P, Mas S, Papagianni K, Ferrando E, de Bobadilla RF, Arnaiz JA, Bioque M, Bernardo M, Lafuente A. Effect of CYP2D6 on risperidone pharmacokinetics and extrapyramidal symptoms in healthy volunteers: results from a pharmacogenetic clinical trial. Pharmacogenomics 2014;15(1):17–28.

147. Hendset M, Molden E, Knape M, Hermann M. Serum concentrations of risperidone and aripiprazole in subgroups encoding CYP2D6 intermediate metabolizer phenotype. Ther Drug Monit 2014;36(1):80–85.

148. Xie HG, Kim RB, Wood AJ, et al. Molecular basis of ethnic differences in drug disposition and response. Annu Rev Pharmacol Toxicol 2001;41:815–850.

149. Zandvliet AS, Huitema AD, Copalu W, Yamada Y, Tamura T, Beijnen JH, Schellens JH. CYP2C9 and CYP2C19 polymorphic forms are related to increased indisulam exposure and higher risk of severe hematologic toxicity. Clin Cancer Res 2007;13(10):2970–2976.

150. Baillie TA, Rettie AE. Role of biotransformation in drug-induced toxicity: influence of intra- and inter-species differences in drug metabolism. Drug Metab Pharmacokinet 2010;26(1):15–29.

151. Liu L, Halladay JS, Shin Y, Wong S, Coraggio M, La H, Baumgardner M, Le H, Gopaul S, Boggs J, Kuebler P, Davis JC Jr, Liao XC, Lubach JW, Deese A, Sowell CG, Currie KS, Young WB, Khojasteh SC, Hop CE, Wong H. Significant species difference in amide hydrolysis of GDC-0834, a novel potent and selective Bruton's tyrosine kinase inhibitor. Drug Metab Dispos 2011;39(10):1840–1849.

152. Martignoni M, Groothuis GM, de Kanter R. Species differences between mouse, rat, dog, monkey and human CYP-mediated drug metabolism, inhibition and induction. Expert Opin Drug Metab Toxicol 2006;2:875–894.

153. Su T, Ding X. Regulation of the cytochrome P450 2A genes. Toxicol Appl Pharmacol 2004;199:285–294.

154. Chu X, Bleasby K, Evers R. Species differences in drug transporters and implications for translating preclinical findings to humans. Expert Opin Drug Metab Toxicol 2013;9(3):237–252.

155. DeKeyser JG, Shou M. Species differences of drug-metabolizing enzymes. Part I. Basics overview of drug metabolism. In: Encyclopedia of drug metabolism and interactions, ed. Lyubimov AV. Hoboken, NJ: Wiley, 2012, 1:1–38.
156. Tang C, Prueksaritanont T. Use of in vivo animal models to assess pharmacokinetic drug–drug interactions. Pharm Res 2010;27(9):1772–1787.
157. Kyokawa Y, Nishibe Y, Wakabayashi M, Harauchi T, Maruyama T, Baba T, et al. Induction of intestinal cytochrome P450 (CYP3A) by rifampicin in beagle dogs. Chem Biol Interact 2001;134:291–305.
158. Gibson GG, Plant NJ, Swales KE, Ayrton A, El-Sankary W. Receptor-dependent transcriptional activation of cytochrome P450 3A genes: induction, mechanisms, species differences and interindividual variations in man. Xenobiotica 2002;32:165–206.
159. Lam JL, Shugarts SB, Okochi H, Benet LZ. Elucidating the effect of final-day dosing of rifampin in induction studies on hepatic drug disposition and metabolism. J Pharmacol Exp Ther 2006;319:864–870.
160. Martignoni M, Groothuis GMM, de Kanter R. Species differences between mouse, rat, dog, monkey and human CYP-mediated drug metabolism, inhibition and induction. Expert Op Drug Metab Toxico 2006;2:875–894.
161. Kronbach T, Mathys D, Umeno M, et al. Oxidation of midazolam and triazolam by human liver cytochrome P450IIia4. Mol Pharmacol 1989;36(1):89–96.
162. Sharer JE, Shipley LA, Vandenbranden MR, et al. Comparisons of phase I and phase II in vitro hepatic enzyme activities of human, dog, rhesus monkey, and cynomolgus monkey. Drug Metab Dispos 1995;23(11):1231–1241.
163. LeCluyse EL, Alexandre E, Hamilton GA, et al. Isolation and culture of primary human hepatocytes. Methods Mol Biol 2005;290:207–229.
164. Guengerich FP. Analysis and characterization of enzymes. In: Principles and methods of toxicology, ed. Wallace HA. New York: Raven Press, 1989, 777–814.
165. Waterman M, Johnson E. Cytochrome P450. New York: Elsevier, 1991.
166. Tang W, Wang RW, Lu AY. Utility of recombinant cytochrome P450 enzymes: a drug metabolism perspective. Curr Drug Metab 2005;6(5):503–517.
167. Hayes R, Pool W, Sinz M, et al. Recent developments in drug metabolism methodology. In: Pharmacokinetics: regulatory, industrial, academic perspectives, ed. Welling P, Tse F. New York: Marcel Dekker, 1995, 201–222.
168. Carlson TJ, Fisher MB. Recent advances in high throughput screening for ADME properties. Comb Chem High Throughput Screen 2008;11(3):258–264.
169. Bourcier K, Hyland R, Kempshall S, et al. Investigation into UDP-glucuronosyltransferase (UGT) enzyme kinetics of imidazole- and triazole-containing antifungal drugs in human liver microsomes and recombinant UGT enzymes. Drug Metab Dispos 2010;38(6):923–929.
170. Yanni SB, Chism JP, Rickert DE, Serabjit-Singh CJ.Diminished drug metabolism due to hormonal effects of cobalt mesoporphyrin on flavoproteins. Chem Biol Interact 1993;89(1):73–87.
171. Obach RS. Predicting clearance in humans from in vitro data. Curr Top Med Chem 2011;11(4):334–339.
172. Mohutsky M, Wrighton S, Ring B. In vitro metabolism: subcellular fractions. In: Handbook of drug metabolism, ed. Pearson P, Wienkers L. New York: Informa Healthcare, 2009, 445–464.

173. Gomez-Lechon MJ, Lahoz A, Gombau L, et al. In vitro evaluation of potential hepatotoxicity induced by drugs. Curr Pharm Des 2010;16(17):1963–1977.
174. Sinz M, Kim S. Stem cells immortalized cells, and primary cells in ADMET assays. Drug Discov Today Technol 2006;3(1):79–85.
175. Guo L, Dial S, Shi L, et al. Similarities and differences in the expression of drug-metabolizing enzymes between human hepatic cell lines and primary human hepatocytes. Drug Metab Dispos 2011;39(3):528–538.
176. Mao J, Mohutsky MA, Harrelson JP, et al. Prediction of CYP3A mediated drug–drug interactions using human hepatocytes suspended in human plasma. Drug Metab Dispos 2011;39(4):591–602.
177. Kamath AV, Wang J, Lee FY, Marathe PH. Preclinical pharmacokinetics and in vitro metabolism of dasatinib (BMS-354825): a potent oral multi-targeted kinase inhibitor against SRC and BCR-ABL. Cancer Chemother Pharmacol 2008;61(3):365–376.
178. Yanni SB, Augustijns PF, Benjamin DK Jr, Brouwer KL, Thakker DR, Annaert PP. In vitro investigation of the hepatobiliary disposition mechanisms of the antifungal agent micafungin in humans and rats. Drug Metab Dispos 2010;38(10):1848–1856.
179. Foster JA, Houston JB, Hallifax D. Comparison of intrinsic clearances in human liver microsomes and suspended hepatocytes from the same donor livers: clearance-dependent relationship and implications for prediction of in vivo clearance. Xenobiotica 2011;41(2):124–136.
180. Chu V, Einolf HJ, Evers R, et al. In vitro and in vivo induction of cytochrome P450: a survey of the current practices and recommendations: a pharmaceutical research and manufacturers of America perspective. Drug Metab Dispos 2009;37(7):1339–1354.
181. Wang WW, Khetani SR, Krzyzewski S, et al. Assessment of a micropatterned hepatocyte coculture system to generate major human excretory and circulating drug metabolites. Drug Metab Dispos 2010;38(10):1900–1905.
182. Landskroner KA, Hess P, Treiber A. Surgical and pharmacological animal models used in drug metabolism and pharmacokinetics. Xenobiotica 2011;41(8):687–700.
183. Penner N, Klunk LJ, Prakash C. Human radiolabeled mass balance studies: objectives, utilities and limitations. Biopharm Drug Dispos 2009;30(4):185–203.
184. U.S. Department of Health and Human Services. Food and Drug Administration Guidance for Industry. Safety testing of drug metabolites. FDA, 2008.
185. Strelevitz TJ, Foti RS, Fisher MB. In vivo use of the P450 inactivator 1-aminobenzotriazole in the rat: varied dosing route to elucidate gut and liver contributions to first-pass and systemic clearance. J Pharm Sci 2006;95(6):1334–1341.
186. Salyers K. Preclinical pharmacokinetic models for drug discovery and development. In: Handbook of drug metabolism, ed. Pearson P, Wienkers L. New York: Informa Healthcare, 2009, 659–674.
187. Bessire A, Youdim K, Hurst S, Lee AC. Utility of genetically modified animal models for drug metabolism and drug transporters Part VI. Methods and protocols for prediction and evaluation of drug metabolism and drug interaction studies. In: Encyclopedia of drug metabolism and interactions, ed. Lyubimov AV. Hoboken, NJ: Wiley, 2012, 6:1–88.
188. Chen C, Stock JL, Liu X, Shi J, Van Deusen JW, DiMattia DA, Dullea RG, de Morais SM. Uility of a novel Oatp1b2 knockout mouse model for evaluating the role of Oatp1b2 in the hepatic uptake of model compounds. Drug Metab Dispos 2008;36(9):1840–1845.

189. Katoh M, Tateno C, Yoshizato K, et al. Chimeric mice with humanized liver. Toxicology 2008;246(1):9–17.

190. Kamimura H, Nakada N, Suzuki K, et al. Assessment of chimeric mice with humanized liver as a tool for predicting circulating human metabolites. Drug Metab Pharmacokinet 2010;25(3):223–235.

191. Masakazu Kakuni M, Yamasaki C, Tachibana A, Yoshizane Y, Ishida Y, Tateno C. Chimeric mice with humanized livers: a unique tool for in vivo and in vitro enzyme induction studies. Int J Mol Sci 2014;15(1):58–74.

192. Katoh M, Matsui T, Okumura H, Nakajima M, Nishimura M, Naito S, Tateno C, Yoshizato K, Yokoi T. Expression of human phase II enzymes in chimeric mice with humanized liver. Drug Metab Dispos 2005;33:1333–1340.

193. Okumura H, Katoh M, Sawada T, Nakajima M, Soeno Y, Yabuuchi H, Ikeda T, Tateno C, Yoshizato K, Yokoi T. Humanization of excretory pathway in chimeric mice with humanized liver. Toxicol Sci 2007;97:533–538.

194. Schulz-Utermoehl T, Sarda S, Foster JR, Jacobsen M, Kenna GJ, Morikawa Y, Salmu J, Gross G, Wilson ID. Evaluation of the pharmacokinetics, biotransformation and hepatic transporter effects of troglitazone in mice with humanized livers. Xenobiotica 2012;42(6):503–517.

195. Xu X, Hirayama H, Pang KS. First-pass metabolism of salicylamide. Studies in the once-through vascularly perfused rat intestine-liver preparation. Drug Metab Dispos 1989;17(5):556–563.

4

EXCRETION: PRINCIPLE, METHODS, AND APPLICATIONS FOR BETTER THERAPY

4.1 OUTLINE OF DRUG EXCRETION AND MECHANISMS

Drug uptake, followed by metabolism and excretion mediated by the function of key body organs such as liver and kidney, determines the systemic clearance and exposure of many drugs. In this regard, both liver hepatocytes and kidney proximal tubule epithelia can play a critical role in drug safety and efficacy. For example, statins and angiotensin-converting enzyme (ACE) inhibitor drugs transport from the blood to the bile or urine via an uptake, and efflux transporters are a significant determinant of drug exposure [1]. The interplay of enzymes and transporters was characterized to understand the factors that determine the intrinsic unbound drug clearance in the intestine, liver, and kidney [2,3]. According to the Biopharmaceutics Classification System (BCS) discussed in chapter 1, highly permeable drugs are those with neither their active uptake nor efflux a rate-limiting step [4,5]. In prediction of organ clearance, physiologically based pharmacokinetic (PBPK) models that incorporate blood flow rate, extent of protein binding, and intrinsic metabolic and excretion clearance may fit. In contrast, most anionic drugs and some hydrophilic organic cationic and zwitterionic drugs exhibit poor membrane permeability [4–6], thus, these types of drugs require transporters for their efficient penetration into or out of the liver or kidney. In prediction of intrinsic organ clearance, a combination of intrinsic clearances for cellular uptake and efflux into the systemic circulation may work.

Organ intrinsic clearance means the relationship between metabolism and biliary/renal excretion. The uptake intrinsic clearance, metabolic intrinsic clearance, or efflux

Translational ADMET for Drug Therapy: Principles, Methods, and Pharmaceutical Applications, First Edition. Souzan B. Yanni.

intrinsic clearance can be rate determining for drugs with their disposition mediated by drug-metabolizing enzymes (DMEs) and drug transporters [7]. In this case, any changes in metabolism and transport activity due to potential interactions with coadministered drugs or genetic polymorphisms with both players should predict potential changes in the systemic exposure of drugs, safety, and efficacy [8]. These factors were described in detail in chapter 3 and are shown in Figure 3.1. As has been reported for most of the drugs on the market, the expression and function of DMEs and drug transporters are important factors in dominating the absorption, disposition, metabolism, and excretion (ADME) and safety in human patients. Each metabolic enzyme and transporter consists of a wide variety of isoforms, and the substrate specificity of each isoform is very broad. In that way it provides an evolutionary ability to protect the body against numerous kinds of toxic effects mediated by xenobiotics exposure. It is not only the broad specificity that is broad but also the broad localizations at several tissues in the body. The efficiency of DMEs and drug transporters is found to be at the highest with tissues responsible for detoxification and elimination of xenobiotics, such as intestine, liver, and kidney. For example, in the liver, efficient detoxification can be achieved by the sequential processing of compounds, such as in the cellular uptake from the blood circulation to hepatocytes via influx transporters, phase I and II metabolism, and biliary excretion via efflux transporters. To enhance the efficiency of detoxification, basal-to-apical directional transcellular transport, called vectorial transport, of compounds is achieved with the aid of uptake and efflux transporters [9]. For example, in the liver and kidney, several uptake transporters are expressed on the basal membrane facing the blood circulation, and efflux transporters are expressed on the apical membrane facing the bile or urine. This allows the compounds to be concentrated into the bile and urine from the systemic blood.

The mechanisms of drug elimination by these processes will be discussed, and choice of tools that are commonly available and used to determine the elimination kinetics and mechanisms will be listed.

4.2 EXCRETION OF DRUGS IN HUMANS AS FUNCTION OF DRUG TRANSPORTERS

4.2.1 Biliary and Renal Excretion

There are certain factors that determine the drug elimination pathway to biliary or renal excretion, such as molecular weight (MW) of drug, its polarity, and its metabolism [10]. It has been found that the MW cutoff of drugs eliminated in humans by biliary excretion is in the 500–600 Da range and strongly polar; however, in rats, guinea pigs, and rabbits, the MW cutoffs are around 325, 400, and 475 Da, respectively [11]. Biliary excretion can be the route for most conjugated forms, but nonconjugated drugs can predominantly be excreted in bile. Although these factors are involved in determining the route of elimination, a definitive observation indicates that the uptake and excretion of drugs in the liver and kidney is largely determined by transporters based on the selectivity of drug and metabolites. Furthermore, it has been observed that biliary excretion of β-lactam antibiotics by multidrug resistance protein MRP2 increased with increase in the MW of these drugs [12]. In a general sense, drugs excreted into bile are strongly polar. Therefore, it can be concluded that most drugs excreted into bile are conjugated forms, though biliary secretion is not a selective pathway for conjugated metabolites only. Now it becomes clear that drug MW, lipophilicity, and metabolism are not the only determinates for the elimination pathways. It is, rather, the uptake and excretion of drugs by transporters in the liver and kidney that largely mediate the elimination pathways and mechanisms.

As discussed in previous chapters, transport proteins are key players in oral drug absorption, distribution, and metabolism. Due to their roles in these processes and their impact on drug disposition, transporter proteins are most highly expressed in the intestine, liver, and kidney, and to a lesser extent in the brain (see Figure 5.4). The localizations, substrate specificity, and functions vary among these body organs. For the excretion, which is the focus of this chapter, the transport of drugs and metabolites is accomplished by the adenosine triphosphate (ATP)-binding cassette (ABC) or the solute carrier (SLC) superfamilies. In the human ABC family, they are divided into seven subfamilies (A to G) [13]. The MW of transporters in the ABC family ranges from 150 to 200 kDa, and each has two transmembrane domains and two ATP-binding domains [14]. The binding and hydrolysis of ATP is required to mediate transport. For the SLC family, it consists of 12 putative transmembrane domains and ranges in MW from 50 to 100 kDa [15,16]. The uptake and excretion transporters are both responsible for the biliary and renal elimination. The uptake transporters are organic cation transporter (OCT, SLC), organic cation/ergothioneine transporter (OCTN, SLC), organic anion transporter (OAT, SLC), organic anion transporting polypeptide (OATP, SLC), urate transporter (URAT, SLC), sodium-taurocholate cotransporting polypeptide (NTCP, SLC), peptide transporter (PEPT, SLC), and multidrug and toxin extrusion protein (MATE, SLC). For efflux transporters they are P-glycoprotein (P-gp, MDR1, ABC), multidrug resistance proteins (MRPs, ABC), bile salt export pump (BSEP, ABC), and breast cancer resistance protein (BCRP, ABC) [17]. Several of the uptake SLC transporters, for example, the OATs, are bidirectional and transport substrates in both directions as shown in Figure 4.1 and according to the article by Kusuhara and Sugiyama (2009) [3].

4.2.1.1 Biliary Excretion One of the key functions of liver is production of bile that forms in the hepatocytes [18]. Bile produced in hepatocytes (~1 liter/day) is collected in the bile canaliculi, where it flows (range from 1.5 to 2 mL/min/kg) into the bile ducts where bile is either directed into the duodenum to aid in digestion or flows into the gallbladder for storage [19]. Most of the bile salts in the intestine are reabsorbed in the terminal ileum and returned to the liver via the portal vein, where they are reused by the hepatocytes [20]. Transporters that are specific for transporting the bile acids (BAs), conjugates, and other endogenous biomarkers such as bilirubin are all expressed at the sinusoidal membrane of the hepatocytes and apical bile canaliculi (as shown in Figure 4.2), according to an article by Kosters and Karpen (2008) [18]. These transporters are capable of taking and secreting drugs and their metabolites [21].

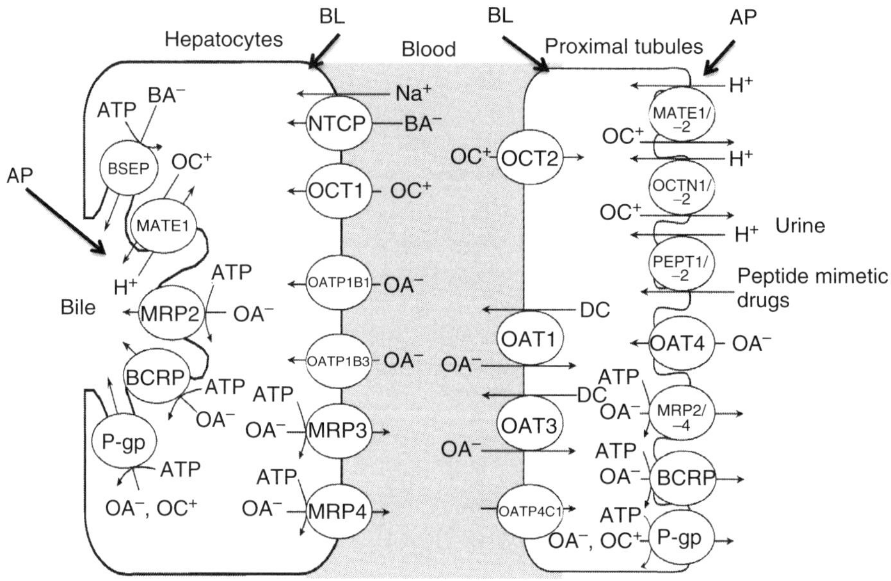

Figure 4.1 Drug transporters at the AP and BL side of hepatocytes and proximal tubular cells. From Ref [3] with permission.

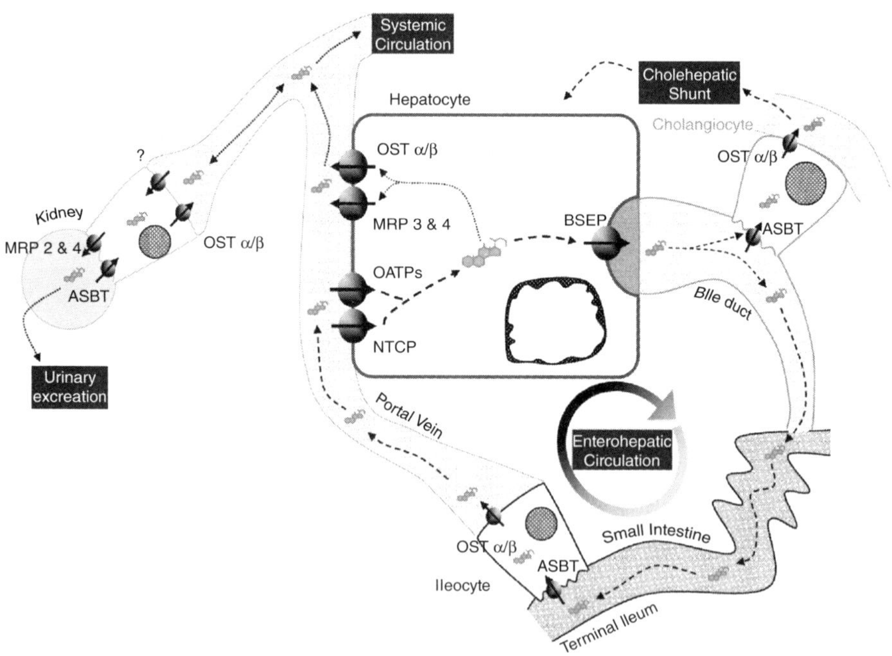

Figure 4.2 Roles for bile acid transporters in enterohepatic and systemic circulations. Figure presented by Ref [18].

Once a drug or metabolites (generated at the gastrointestinal [GI] tract) are taken up by the sinusoidal (BL) membrane transporters into the hepatocyte, which is delivered via portal vein circulation, there are two paths it can follow. It can either be secreted back into systemic circulation via sinusoidal efflux transporters or excreted as unchanged drug and/or as metabolite into the bile via apical canalicular efflux transporters. Drugs and metabolites excreted into the sinusoidal blood via the basolateral (BL) sinusoidal membrane typically end up being eliminated in the urine, while those excreted in bile via the canicular membrane end up in the intestine where they are eliminated in the feces or can potentially undergo reabsorption via a process known as enterohepatic recirculation, as indicated in Figure 4.2. In excretion via the hepatobiliary route, parent drug and/or metabolites must first be taken up into the hepatocyte by transporter proteins such as OAT, OATP, OCT, and NTCP [16]. Transporters involved in the biliary excretion out of the hepatocyte into the bile or sinusoidal blood include BCRP, BSEP, P-gp, several of the MRPs, and MATE1. While the role of BSEP in the transport of drugs and conjugated metabolites remains unclear, several drugs have been shown to inhibit BSEP (e.g., cyclosporine A, bosentan, fluvastatin, glibenclamide, troglitazone, and rifampicin), potentially leading to cholestasis [22,23]. Metabolites, such as the hydrophilic conjugated metabolites, cannot passively diffuse out of hepatocytes and must therefore be actively transported. The main transport proteins involved in the biliary and BL excretion of glucuronide, sulfate, and glutathione (GSH) metabolites are the efflux canalicular transporters, MRP2 and BCRP, or the sinusoidal membrane efflux transporters, MRP3 and MRP4 [17,22]. Most glucuronide conjugates that are excreted into the bile are MRP2 substrates (e.g., conjugates of ceftriazone, ampicillin) [24], whereas most sulfate conjugates are BCRP substrates [17,22]. ATP-mediated transporters responsible for BL excretion into the sinusoidal blood are MRP3, MRP4, MRP5, and MRP6 [22]. Of these, MRP3 is efficient in transporting both glucuronide and sulfate conjugates [17,22]. MRP3 is responsible for transport of metabolites such as morphine-3-glucuronide and morphine-6-glucuronide, while MRP4 has been shown to have high affinity to sulfate conjugates but is also capable of transporting glucuronide and GSH conjugates. GSH conjugates are also transported by MRP5 and MRP6. However, little is known of MRP5 and MRP6 in their ability to transport conjugated organic anions [17,22]. MRPs can also transport unconjugated metabolites and drug molecules, such as micafungin [25].

4.2.1.2 Renal Excretion The major route of drug excretion is by renal elimination via the kidney functions. There are three major renal excretion mechanisms that are specifically responsible for excretion of drugs based on the drug physiochemical properties and structures. These mechanisms are (1) glomerular filtration, (2) tubular secretion, and (3) tubular reabsorption. These mechanisms depend on the kidney functional unit, the nephrons. The nephron consists of several components, including (1) Bowman's capsule, (2) proximal tubule, (3) the loop of Henle, (4) distal tubule, and (5) collecting duct. The renal excretion mechanism that depends on these compartments' renal excretion mechanisms is based on the properties of the drug; for example, (1) when low MW molecules are renally eliminated, they filter in

Bowman's capsule; (2) when weak electrolyte drugs such as acids are actively secreted and reabsorbed, this occurs in the proximal tubules; and (3) reabsorption of water occurs in the loop of Henle and passive reabsorption of water and lipid soluble drugs occurs in the distal tubule. The following paragraphs describe the renal elimination pathways in further detail.

1. *Glomerular filtration* (GF): The kidneys receive about 20% of the cardiac output, and the excretion by drugs by GF is carried out in the glomerular compartment by filtering molecules from arterial blood and only low molecules <5 kDa in MW can pass through the membranes of the glomeruli [26]. Consequently, protein-bound drugs, or those that incorporate into red blood cells, do not pass through the glomeruli. It is known that 10% of the blood that enters the glomerulus is filtered. This filtration rate is often measured by determining the renal clearance of inulin. Inulin is readily filtered in the glomerulus and is not subject to tubular secretion or reabsorption. Thus inulin clearance is equal to the GF rate, which varies from individual to individual depending on age, body weight, sex, disease state, and ethnicity. In healthy individuals the normal range is 110 to 130 ml/min (≈180 L of filtered blood/day). Drugs or compounds that are extensively secreted, such as *p*-aminohippuric acid (PAH), may have clearance values approaching the renal plasma flow rate of 425 to 650 ml/min and are used clinically to measure this physiological parameter.The rate of renal clearance of drugs (CL_r) is estimated by this equation:

$$CL_r = R_r/C_p, \tag{4.1}$$

where R_r is the renal excretion rate and C_p is the plasma concentration. When no other factors are involved, such as protein binding or secretion/reabsorption, renal clearance of the drug will be equivalent to the glomerular filtration rate (GFR). Although most drugs are filtered from blood in the glomerulus, the overall renal excretion is controlled by what happens in the tubules. More than 90% of the filtrate is reabsorbed. Normal urine output is much less, $\sim 1 - 2$ liters per day. In a simpler term, renal clearance is the hypothetical volume of plasma from which the substance is eliminated by the kidney appearing in the urine per unit time. A physiologic definition of renal clearance (CL_R) is a ratio of the instantaneous urinary excretion rate (dA_e/dt) and the plasma concentration of drug at the time when the urinary sample is collected (C_p):

$$CL_R = \frac{(dA_e/dt)}{C_p}. \tag{4.2}$$

This equation assumes that urine is collected over a finite period during which the plasma concentration of drug is changing following first-order kinetics. In a more practical sense, renal clearance is calculated as the ratio of the amount of drug excreted unchanged $A_{e(\Delta t)}$ during a urine collection interval (Δt) to the plasma concentration of drug at the midpoint of the interval (C_{mid}):

$$CL_R = \frac{A_{e(\Delta t)}/\Delta t}{C_{mid}}. \tag{4.3}$$

By rearranging the above equation, the slope of a linear regression of multiple urine collections ($A_{e(\Delta t)}/\Delta t$) versus C_{mid} is equal to average renal clearance. When taking integration time 0 to infinity, the numerator dA_e becomes $A_{e(\infty)}$ and the denominator $C_p dt$ is the corresponding area under the plasma concentration–time curve of drug ($AUC_{0-\infty}$):

$$CL_R = \frac{A_{e(\infty)}}{(AUC_{0-\infty})}. \tag{4.4}$$

In a clinical setting, collection of plasma and urine at the same time might be difficult, especially for drugs with a long terminal elimination half-life, as typically a sample collection period of at least five terminal half-lives can be considered equivalent to infinite term. Equation (4.4) is practically modified as follows:

$$CL_R = \frac{A_e\ (t_2 - t_1)}{AUC\ (t_2 - t_1)}. \tag{4.5}$$

2. *Tubular secretion*: In the proximal tubule, an active secretion of molecules occurs, and it requires a carrier and a supply of energy. This may be a significant pathway for some compounds such as penicillin, antiviral, metformin, and H-2 antagonist cimetidine [27]. Because tubular secretion is an active process (mediated by drug transporters to be discussed in the following sections), there may be competitive inhibition of the secretion of one compound by another as observed in the inhibition of penicillin excretion by competition with probenecid, which reduces the excretion of the penicillin, thereby prolonging penicillin plasma concentrations (PDR). Since then it has been shown that probenecid also alters the distribution of penicillins to various tissues, causing more drug to distribute out of plasma, and hence causing even more exposure and less clearance. This could also be used to reduce the excretion of cephalosporins.
3. *Tubular reabsorption*: Drugs present in the glomerular filtrate can be reabsorbed in the tubules. In the distal tubule there is passive excretion and extensive reabsorption of lipid soluble drugs that readily permeate the cell membrane. Water in the filtrate is reabsorbed and therefore the concentration gradient is now in the direction of reabsorption. Thus if a drug is nonionized or in the un-ionized form it may be readily reabsorbed. The pH of filtrate can greatly influence the extent of tubular reabsorption for many drugs that are either weak bases or acids. When urine is acidic, weak acid drugs tend to be reabsorbed, whereas when urine is more alkaline, weak bases are more extensively reabsorbed. Thus,

acidic urine can cause less reabsorption of weak bases or enhanced excretion and vice versa for basic urine. These changes can be quite significant, as urine pH can vary from 4.5 to 8.0 by variations in diet (e.g., meat lowers urine pH) or as an effect of specific drugs. In the case of a drug overdose, it is possible to increase the excretion of some drugs by suitable adjustment of urine pH. For example, with an overdose of pentobarbital (a weak acid), it may be possible to increase its excretion by lessening its reabsorption with sodium bicarbonate injection, which makes the urine more alkaline.The process of reabsorption of glucose and lipid-soluble filtrate components occurs along the proximal, distal, and collecting tubules, ending up back in the plasma with only $1-2$ mL/min of the total filtrate volume (filtered by the glomeruli) eliminated in the urine. As explained above, the pH of the tubular fluid (which usually ranges from 4.5 to 8) can determine whether certain drugs are excreted into the urine or reabsorbed. Uncharged nonionic molecules can diffuse passively through the cellular membrane. However, ionic drugs secrete against a concentration gradient, thus the process is an active transport that mediates drug transporters, such as β-lactam antibiotic dicloxacillin, that are reabsorbed by the proximal tubules via active transport [28]. A second mechanism of reabsorption in the kidney is endocytosis. Endocytosis in the kidney occurs by the formation of vesicles along the brush border membrane, which can take up fluids and compounds [29].

Taken all together, renal excretion of drugs is accomplished by three processes: (1) GF, (2) renal tubule secretion (via BL uptake and apical efflux transporters) (Figure 4.1), and (3) renal tubular readsorption (via apical transporters). In the following section, transporter proteins that mediate uptake and excretion as well as secretion processes will be discussed in more detail.

4.2.2 Drug Transporter Function in Renal Excretion

As in liver hepatocytes, the primary site of transport of drugs and metabolites from blood into urine is the proximal tubule cells (PTCs) [29]. The efflux transporters in kidney, as in the liver, are located on the apical (AP, luminal) brush border membrane and can transport an unchanged drug or its metabolites from the proximal tubular cells to urine (Figure 4.1). The uptake of drugs is mediated by several transporter families such as OAT and OATP, as well as OCT at the BL membrane of PTCs [30,31], which transport drug from systemic circulation to the PTC. There are also bidirectional transporters OAT4, URAT1, and PEPT1/2, which transport endogenous substrates and potentially drugs from the urine back into the PTC. The PEPT2 for example, is involved in the reabsorption of peptide-like drugs, while the bidirectional transporters URAT1 and OAT4 thus far have only been shown to be involved in the reabsorption of endogenous substrates such as urate [16,32]. The renal excretion of drugs and metabolites out of the kidney PTCs into the urine involves the transporters with bidirectional uptake function OAT4 and OCTN1 and those with efflux function such as of MRP2, MRP4, MDR1/P-gp, and MATE1 [16,17].

The organic anion transport system in the kidney comprises the two well-investigated transporters OAT1 and OAT3 expressed in the basolateral membranes of the proximal tubules, as well as OAT2 and OATP4C1, which are functionally involved with renal drug excretion in humans. There are other transporters in the OAT family such as OAT4 that also express in human kidney, but its significant function in drug dispositions and elimination in humans is still unknown. Both OAT1 and OAT3 mediate active transport and benzylpenicillin is a relatively specific inhibitor of OAT1 and OAT3. By comparing the inhibition potency, the results suggest that both OAT1 and OAT3 contribute to the renal uptake of organic anions. OAT1 contributes to the uptake of low MW organic anions, such as p-aminohippurate and antiviral drugs (adefovir, cidofovir, and tenofovir), whereas OAT3 makes a greater contribution to the renal uptake of bulky amphipathic organic anions, such as statins, sartans, and benzylpenicillin, which are substrates of hepatic organic anion transporters as well [3]. Furthermore, OAT3 accepts some cationic drugs, such as H2 receptor antagonists, and zwitterions such as fexofenadine [3].

For the organic cation transport system, OCT2/SLC22A2 is the predominant organic cation transporter in the BL membrane of the PTC. OCT2 mediates the facilitated diffusion of various cationic drugs, including metformin and cimetidine. The luminal (AP) efflux mechanism for hydrophilic or oganic cations is implied to be active based on the external concentration gradient of H^+, which may enhance the uptake of typical cationic compounds into brush-border membrane vesicles. OCTN1/SLC22A4 and OCTN2/ SLC22A5 mediated the antiport of tetraethylammonium (TEA), a typical substrate of renal organic cation transporters with H^+ considered as a significant transporter in this regard, but more recently, MATE proteins have also been identified, comprising MATE1/SLC47A1 and MATE2K/ SLC47A2 in the human kidney. The MATE proteins accept organic cations, which are excreted into the urine without metabolism and considered to be responsible for the antiport of organic cations with protons in the kidney.

ABC transporters mediating primary active transport, such as BCRP (ABCG2) and MRP2 (ABCC2) and MRP4 (ABCC4), have been shown to constitute a third efflux mechanism in the kidney. The lack of that transporter will not affect tubular clearance with regard to the plasma concentrations, but it will instead significantly increase the kidney concentration of drugs such as trichlorothiazide, cephalosporins (ceftizoxime and cefazolin), and antiviral drugs (adefovir and tenofovir) when they are given intravenously by constant infusion. In addition to MRP4, subjects with a single-nucleotide polymorphism (SNP) or mutation in MRP2 exhibited delayed elimination of methotrexate via systemic elimination, suggesting the involvement of MRP2 in the luminal efflux [33,34]. As in hepatobiliary excretion, MRP2 is primarily involved in the renal excretion of glucuronide, sulfate, and GSH conjugates [35].

4.3 TRANSLATIONAL TOOLS TO DETERMINE THE BILIARY AND RENAL CLEARANCE

4.3.1 In Vitro Methods in Determination of Biliary Clearance

The prominent role of the liver in the biotransformation and elimination of xenobiotics in animals and humans is well recognized. Many in vitro and in situ tools,

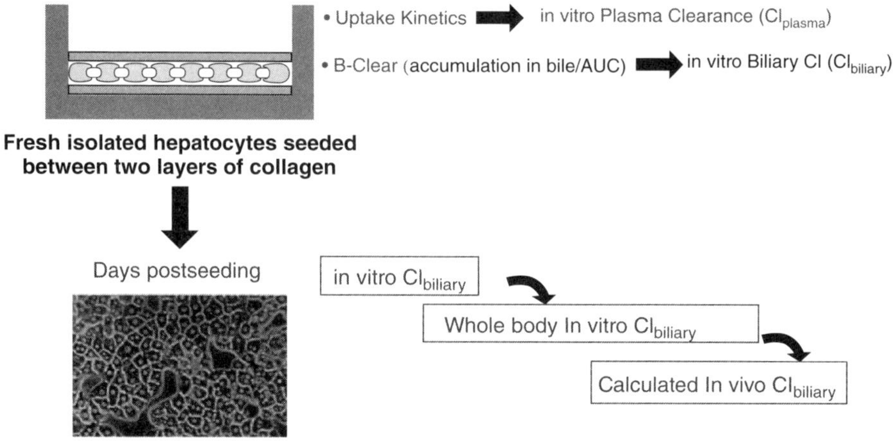

Figure 4.3 Sandwich-cultured hepatocytes, an in vitro model for hepatobiliary disposition.

as well as animal models, have been established to study processes involved in the metabolism and hepatobiliary disposition of drugs [36,37]. Despite the availability of in vitro models to estimate the biliary clearance ($Cl_{biliary}$) of compounds in vivo, little information exists on the predictive accuracy of these methods in humans owing to the absence of systematic clinical studies measuring the $Cl_{biliary}$ of drugs. Accurate $Cl_{biliary}$ values are needed to compare in vitro and in vivo systems [38,39]. However, due to the complexity of the procedures required to measure hepatobiliary clearance of compounds in humans as well as the species-dependent differences in hepatic metabolism and transport, it remains challenging to predict the human $Cl_{biliary}$ of drugs and the potential for interactions with endogenous and exogenous substances cleared by the biliary route [40,41]. This often translates into costly failures of drugs at late stages in development due to unexpected liver toxicity or drug interactions associated with hepatic enzymes or transport proteins [42–44].

Biliary excretion of drugs can be investigated using in vitro systems with varying degrees of complexity, from a simple model such as the primary isolated fresh hepatocyte to as complex as canalicular or sinusoidal membrane vesicles prepared from a cell line transfected with specific transporter proteins [40]. Human hepatocytes express the full array of metabolizing enzymes and transport proteins, and sandwich-cultured human hepatocytes (SCHH) repolarize, form canalicular networks, express and properly localize relevant and functional drug transport proteins, and can be used to characterize uptake and efflux of xenobiotics by the liver [40,45] (Figure 4.3). By using this system, it is possible to determine the biliary excretion of drugs by disrupting the tight junctions via depletion of calcium from the medium, as seen in Figure 4.4. This approach has been used by several investigators quantitatively and qualitatively:

1. Hepatobiliary disposition and mechanism
2. Transporters responsible for the biliary clearance

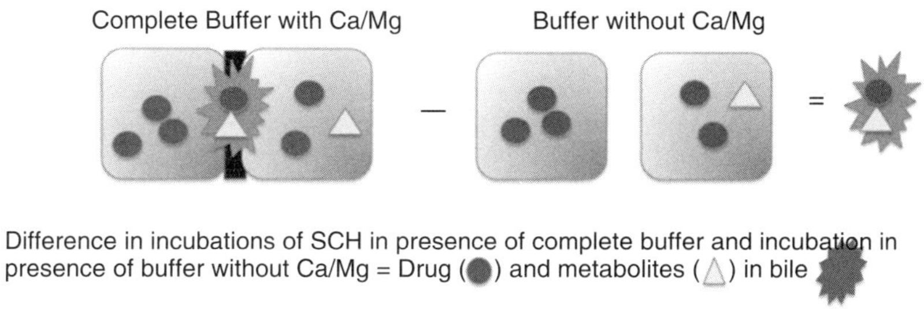

Figure 4.4 In vitro extrapolation of hepatobiliary disposition.

3. Rate limiting step
4. Potential drug–drug interactions (DDIs)
5. Hepatotoxicity and drug-induced cholestasis
6. Variability among human populations

In these studies, the $Cl_{biliary}$ values of three compounds were estimated from data generated in SCHH. These values were compared with in vivo $Cl_{biliary}$ values measured in humans in order to evaluate the utility of this in vitro model in the prediction of hepatobiliary drug clearance in vivo. In most of the investigations a good correlation between in vitro and in vivo data was identified [40,45–48].

The intrinsic biliary clearance in SCHH using the B-CLEAR technology (Qualyst Transporter Solutions, Raleigh, NC) is calculated as follows:

$$\text{In vitro } Cl_{biliary} = \left[\frac{\text{Accumulation (cells + bile)} - \text{Accumulation (cells)}}{AUC_{medium}}\right], \quad (4.6)$$

where Accumulation (cells + bile) is accumulated drug in the presence of complete buffer (with calcium and magnesium), while Accumulation (cells) is the accumulation in calcium and magnesium-free buffer. AUC medium is equal to the product of incubation time and initial unbound substrate concentration.

To ensure the utility of SCHH as a predictive model for the biliary clearance in humans, it is essential that the in vitro system maintain the expression and function of hepatic drug transport proteins similar to the in vivo situation. The data provide the first evidence that SCHH are capable of accurately predicting the $Cl_{biliary}$ of compounds in humans and enable the comparison of drugs with different chemical structures and diverse mechanisms of hepatic uptake and biliary excretion. In investigation by Ghibellini et al. (2007) [49], the biliary clearance of three drugs was tested and found to be accurately ranked, but the $Cl_{biliary}$ predicted from SCHH was consistently lower than the in vivo $Cl_{biliary}$ for the three drugs evaluated in the study. These findings could be related to less extensive canalicular network formation in culture compared to liver tissue in vivo. In the liver, adjacent hepatocytes form bile canaliculi at each cell-cell contact, and the interconnected canalicular spaces drain bile away from the hepatocytes and into small bile ducts. When cultured in a sandwich configuration, primary human hepatocytes repolarize over time and form bile canaliculi; however, the extent of bile canalicular network formation varies from donor to donor, and is not as extensive as that observed in the native tissue. If less surface area per hepatocyte is occupied by bile canaliculi, it is probable that less drug would be excreted into bile in sandwich-cultured hepatocytes compared to the in vivo situation. Despite these limitations, this preliminary work suggests that the SCHH system may be used to distinguish between drugs with different extents of hepatic clearance, and it is capable of predicting whether a compound would be extensively, moderately, or negligibly extracted by the liver and excreted into bile in vivo in humans. In a recent report by Chatterjee et al. (2014) [50], the authors demonstrate the development and validation of a new approach using sandwich-cultured hepatocytes (SCH) to identify compounds that potentially cause cholestasis by altering BA disposition. The approach is based on exposing human and rat SCH (for (24–48 h) to known cholestatic and/or hepatotoxic compounds, in the presence or in the absence of a concentrated mixture of BAs.

Urea assay was used to assess (compromised) hepatocyte functionality at the end of the incubations. The cholestatic potential of the compounds was expressed by calculating a drug-induced cholestasis index (DICI), reflecting the relative residual urea formation by hepatocytes coincubated with BAs and test compound as compared to hepatocytes treated with test compound alone. Compounds with clinical reports of cholestasis, including cyclosporin A, troglitazone, chlorpromazine, bosentan, ticlopidine, ritonavir, and midecamycin, showed enhanced toxicity in the presence of BAs (DICI ≤ 0.8) for at least one of the tested concentrations. In contrast, the in vitro toxicity of compounds causing hepatotoxicity by other mechanisms (including diclofenac, valproic acid, amiodarone, and acetaminophen) remained unchanged in the presence of BAs. A safety margin (SM) for drug-induced cholestasis was calculated as the ratio of the lowest in vitro concentration for which DICI ≤ 0.8, to the reported mean peak therapeutic plasma concentration. SM values obtained in human SCH correlated well with reported % incidence of clinical drug-induced cholestasis, while no correlation was observed in rat SCH. According to the results listed in this report, the authors concluded that this in vitro model may enable early identification of drug candidates causing cholestasis by disturbed BA handling.

4.3.2 In Vitro Methods in Determination of Renal Clearance

The proximal tubule plays a central role in the tubular secretion of xenobiotics. Tubular secretion can be considered a three-step process consisting of uptake across the BL membrane, intracellular accumulation, and efflux across the apical membrane. The uptake and efflux steps are mediated by a range of transport proteins located at the BL and apical membranes of PTCs [51]. To understand the role of drug transporters in PTCs is important so that the prevention of any potential DDI or drug-transporter-mediated nephrotoxicity can be achieved. However, because of the unavailability of a suitable experimental model that addresses all issues regarding renal elimination and mechanisms as well as toxicity, minor advancements on xenobiotic renal secretion were made. In Table 4.1, few models that were

TABLE 4.1 List of Established Models Used in Vitro to Determine Kidney Excretion, Mechanisms, and Nephrotoxicity.

Model	Application	Advantages and Disadvantages
Isolated perfuse kidney	Monitor renal function; examine drug transport process and metabolism	Identical morphology as in vivo, but for short term use; difficult surgery
Kidney slices	Active transport of organic anions and metabolic functions used as in vitro and ex vivo models	Easy to prepare; maintain architecture and cellular structure of intact kidney for short time <2 h that limits its use
Isolated cells suspension	Proximal tubule enriched suspension used in nephrotoxicity, uptake, and metabolism studies	2 h preparations, large amount of cells within selective section, enough O_2, short viability
Cultured proximal tubular cells + distal cells mixture in TW formal	Polarized cells, TJ, mechanistic transport and renal disposition and function studies	Polarized cells derived from intact tissue, subcultured/ frozen for several studies
Pure cultured proximal tubular cells in TW format	Discrete active/efflux transport mechanism and metabolism	Further purification using Percoll gradient or antibodies/flow cytometry, low yield

prepared primarily from human kidney and used to test specific applications are listed, including the advantages and disadvantages of using these models. For example, to understand the mechanism of drug secretion in human PT, a mixed and pure primary culture of human PTCs and distal tubule/collecting duct cells (DTCs) was grown as a polarized monolayer on filter supports to use the model for transport studies [52–56]. The scheme of the preparation of the PTC model is shown in Figure 6.2. The validation of the model as a tool to investigate the renal clearance and mechanism was discussed in a series of studies [52–56] and selected results are depicted in Figures 4.5 and 4.6. By measuring the mRNA, protein levels, and localization of transporters such as OCT2, OCTN2, OAT1, OAT3, OAT4, MDR1, MRP2, and BCRP, the utility of the model as a tool to assess the transporter-mediated drug elimination was achieved. Analysis of the expression of the ABC efflux pumps MDR1, MRP2, and BCRP confirmed their apical membrane localization. At the functional level, tubule cell monolayers retain the necessary machinery to mediate the net secretion of the prototypic substrates, PAH, and creatinine. PAH secretion across the monolayer consists of the uptake of PAH across the BL membrane by OAT1 and OAT3 and the apical exit of PAH by a probenecid and MK571-sensitive route consistent with actions of MRP2 or MRP4. Creatinine secretion was by OCT2-mediated uptake at the BL membrane and via MDR1 at the apical membrane. Functional expression of MDR1 and BCRP at the apical membrane was also demonstrated. Similarly, measurement of calcein efflux demonstrated the functional expression of MRP2 at the apical membrane of

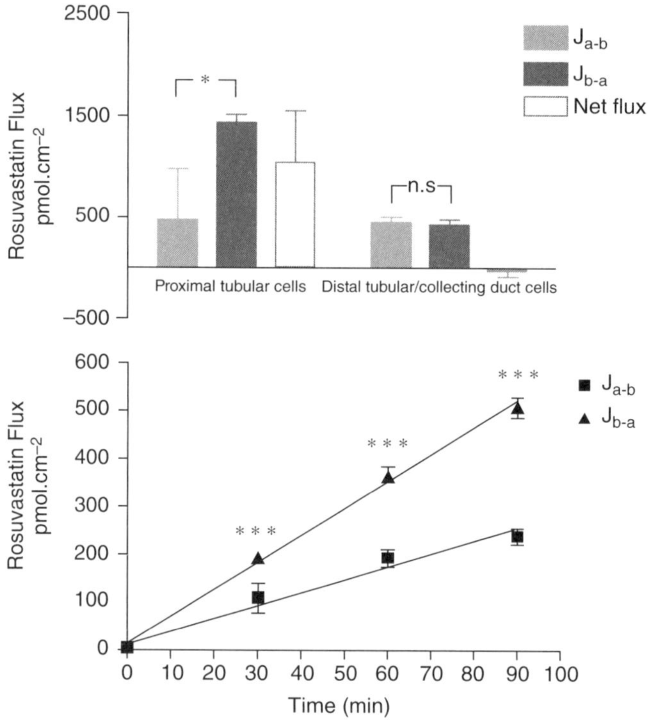

Figure 4.5 The comparison between PTC and DTC in efflux of statin drug rosuvastatin. PTC or DTC model grown in Transwell format. Efflux of MRP2-mediated rosuvastatin vs. passively diffused marker mannitol. Ja-b is the value for flux from apical side to basolateral side across the cell monolayer grown on semipermeable membrane in Transwell format, while Jb-a is the flux from basolateral to apical side. Ref [52] with permission.

cell monolayers. Thus the investigation concluded that human PTC may provide an insight to investigate renal xenobiotic disposition. In a study by the same group, Verhulst et al. (2008) [53] investigated clearance of rosuvastatin (known to be cleared from the body by both biliary and renal clearance). The renal clearance of rosuvastatin is believed to be due to active tubular secretion. In this study, the in vitro human PTC model (see Figures 4.5 and 4.6) was used to elucidate the mechanism of renal elimination. The authors focused on measuring renal clearance using freshly isolated PTC from humans that polarized by growing the cells on Transwell format, as discussed earlier. Because rosuvastatin (and other statins) may alter proximal tubular function, the authors in this study aimed to characterize the mechanisms of tubular rosuvastatin secretion to define the factors that could influence the presence/concentration of rosuvastatin in PTC. It has been found that rosuvastatin net secretion across PTCs was saturable and the BL uptake step was rate limiting mediated by OAT3. In addition, results indicated that rosuvastatin efflux at the apical membrane was mediated by MRP2/4 and BCRP together with a

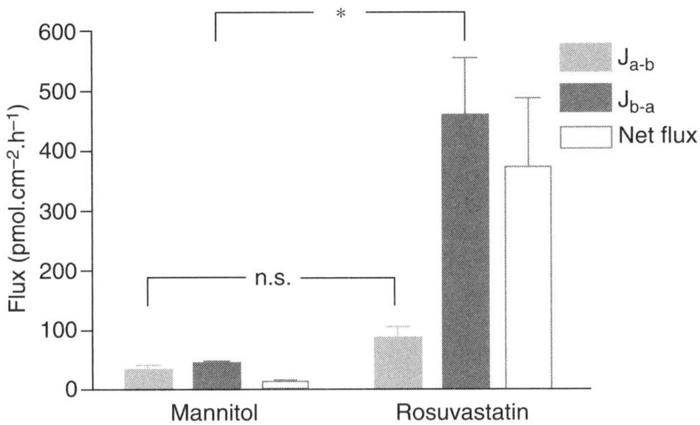

Figure 4.6 The validation of drug transporter efflux function of PTC model grown in Transwell format. Efflux of MRP2-mediated rosuvastatin vs. passively diffused marker mannitol. Ja-b is the value for flux from apical side to basolateral side across the cell monolayer grown on semipermeable membrane in Transwell format, while Jb-a is the flux from basolateral to apical side. Ref [52] with permission.

small contribution from P-gp. The study concluded that the polarized PTC grown in Transwell format can provide a detailed insight into rosuvastatin's renal disposition and the possible factors influencing it. These researchers have used the PTC model to predict the renal clearance of other drugs.

4.3.3 In Vivo Methods in Determination of Biliary and Renal Clearances

The most common and well-performed study to assess renal elimination, similar to the case of biliary elimination, is by using radiolabeled drug in a mass balance study (MBS), which is usually conducted in an in vivo animal model at the late stage discovery or preclinical stage of drug development. MBS is also conducted in Phase 1 clinical investigation in humans to monitor the pathway of elimination for parent drug and metabolites.

4.3.3.1 MBSs in Humans Radiolabeled excretion studies in preclinical animal species used for long-term safety assessment and human volunteers with the aid of the radiolabeled drug provide the total fate of drug-related material. One of the primary objectives of these studies is to demonstrate that the administered dose is readily eliminated from the body, ideally after exerting the intended therapeutic effect. By quantitative collection of the excreta for sufficient time, it not only provides valuable information on mass balance, route, and extent of excretion of the administered radioactive dose but also provides critical information on the clearance mechanism of the drugs.

The most comprehensive way to quantify drug and metabolite elimination in feces and urine is through a human MBS. In MBSs, radiolabeled (usually ^{14}C) drug is dosed via the therapeutic administration route (oral or IV). The position of the radiolabel in the compound is chosen such that it is metabolically stable to potential biotransformation pathways.

Samples, such as urine and feces (as well as plasma), are collected, and the amount of radioactivity in the samples is quantitated using a variety of analytical techniques. The recovery of samples in excreta is measured to assess the route of elimination or samples that remained bound to tissues and raises safety issues. In preclinical species, bile duct–cannulated animals as well as intact animals are frequently used to assess the mechanism of drug excretion, renally versus biliary or both.

To measure the renal and biliary clearance in preclinical species, renal clearance is determined from the following equation:

$$CL_r = \frac{\text{Amount of parent drug excreted in urine up to time t}}{\text{AUC (0 - t) iv}}, \tag{4.7}$$

while the biliary clearance can be determined from the following equation:

$$CL_b = \frac{\text{Amount of parent drug excreted in bile up to time t}}{\text{AUC (0 - t) iv}} \tag{4.8}$$

Several studies describing the procedure for determining the excretion of drugs using [^{14}C]– and [^{3}H] – labeled compounds have been published [57–59]. Mass balance and excretion study of torcetrapib in rats, monkeys, and mice by oral administration of [^{14}C] torcetrapib, has been reported [57]. The results showed that the administered radioactive dose was quantitatively recovered in all species. Excretion of the radioactivity was rapid and nearly complete within 48 h after dosing. In rats and monkeys, the majority of the dose was excreted in feces, while in the mice, the dose was recovered equally in urine and feces. In separate studies using bile duct–cannulated rats and monkeys, only $<10\%$ of the administered radioactivity was recovered in bile, revealing that a major portion of the dose excreted in the feces of rats and monkeys was mainly due to unabsorbed dose.

In humans, the Food and Drug Administration (FDA) requires that MBS should be included in each new drug application (NDA). Complete mass balance (100% recovery of radioactive material) is rarely achieved in humans (and sometimes in preclinical species) because of sampling or analytical issues. However, recoveries of greater than 80% are normally achievable in human MBSs, and no less than 90% recovery has been recommended in animal studies as a minimum cutoff for acceptability [60]. Recovery in human MBSs is generally lower than that obtained from rat or dog studies, most likely due to the difficulty in managing or controlling the collection of human excreta samples from human subjects. For compounds and metabolites with a very long half-life, it is more difficult to achieve mass balance, and sample collection times and study objectives should be adjusted accordingly. Low recovery (e.g., lower than 80% in human studies) does not necessarily invalidate the study if the objectives of the MBS have been largely met. However, significantly lower recovery could raise safety concerns regarding the drug due to the possibility of covalent binding of drug to macromolecules (or noncovalent tissue sequestration).

It has been shown that quantitative information of unchanged drug and its metabolites that are eliminated via the excreta that generated from MBSs can provide important data regarding the following [61–65]:

1. Clearance mechanism of the drug and metabolites, especially the toxic ones
2. Levels of circulating metabolites versus levels of circulating parent drug
3. Ratio of blood/plasma levels of the drug and metabolites
4. Suitability of the species used in the preclinical safety studies (compared the route of elimination of parent and metabolites in human vs. toxic animal model)
5. Proportion of unchanged drug eliminated
6. The toxicity effect of human metabolite in animal species
7. The pharmacokinetic (PK) profile of parent versus metabolite to determine the drug species that drive the observed overall PK of the drug in humans

In a human MBS, the fraction of the amount of drug entering the systemic circulation that is excreted unchanged in urine (f_e) is an important PK parameter that indicates the major pathway of drug elimination through either the excretion of unchanged drug via the kidney or metabolism in the liver followed by the excretion of the metabolite(s) in urine. As expected, the value of f_e ranges between 0.0 and 1.0 depending on the extent to which a drug is renally excreted compared to the overall drug elimination (theoretically 1.0). When the value is low, excretion is a minor pathway of drug elimination, suggesting that the most amount of the drug administered is metabolized. If renal excretion is the major route of elimination, the value of f_e will be higher with an upper limit of 1.0, which means that renal excretion is the only route of drug elimination. The value of f_e for a particular drug will remain constant independent of the dose of the drug administered via IV injection and is estimated by

$$f_e = \frac{U^\infty}{\text{Dose}}, \tag{4.9}$$

where U^∞ is total drug excreted unchanged. If the drug is administered orally, f_e is defined by

$$f_e = \frac{U^\infty}{F_x \ \text{Dose}}, \tag{4.10}$$

where F represents the absolute bioavailability of a drug. When drug in plasma and urine is determined, the amount of unchanged drug excreted in urine is calculated by multiplying urine drug concentration and the urine volume collected within a time interval and summing the amount from all intervals, as shown earlier. In practice, collecting urine should be taken care of for at least five half-lives. In those situations in which total urine collection is impossible, f_e may be estimated by the ratio of renal and total clearances,

$$f_e = \frac{K_r}{\lambda_z} = \frac{CL_R \times C}{CL \times C} = \frac{CL_R}{CL}, \tag{4.11}$$

where k_r is the urinary excretion rate constant, λ_Z is the first-order elimination rate constant, C is the plasma drug concentration, CL_R the renal clearance, and CL the total clearance. The CL_R can be estimated from the urine data described in the previous section, while CL can be determined from the plasma data,

$$CL = \frac{\text{Dose}}{\text{AUC}}, \tag{4.12}$$

where AUC is the total area under the plasma concentration–time curve. This calculation of clearance is independent of the shape of the concentration–time profile. For a drug with linear PK, the value of clearance is also independent of dose.

4.3.4 In Vivo Model to Study Excretion and Toxicity: Chimeric Mice with Humanized Liver

A chimeric mouse model, as described in chapter 3, is a model that is genetically altered by replacing the mouse native hepatocytes with hepatocytes from a human donor. The new generated liver of the mouse has hepatocytes with functions that resemble those of the human donor. These include metabolism and formation of human metabolites and response to human drug inducer and inhibitors, and it has similar drug transporter function as in human liver [66,67]. Immunodeficient mice that overexpress urokinase-type plasminogen activator (uPA) and with a knockout of the fumarylacetoacetate hydrolase (Fah) genes are the two most common mouse models for these studies. These mice are called chimeric or "humanized" because the liver is 70–90% repopulated with human cells. Several investigations are currently taking place to assess the strengths and weaknesses of the humanized mouse model toward its full usage as an ADME and toxicity model that can mimic human drug disposition and safety [68]. However, the results from these investigations indicate the value of including the model in early drug development programs prior to first injection in humans (FIH) study. That will support identity issues with circulating metabolites, the excretion pathway, DDIs, and short- and long-term investigation of potential toxicity of drugs or metabolites with significant human exposure [66–68].

4.4 IMPAIRMENT OF DRUG ELIMINATION

4.4.1 Hepatic Impartment: Cholestasis

When liver function is impaired, the hepatic clearance of most drugs cleared via the metabolic pathway will be compromised, leading to drug accumulation and toxicity. Dose adjustment of drugs excreted in human bile is considered when used to treat impaired hepatic function, and a dose reduction is typically required. Drugs such as antibiotics like ampicillin, doxycycline, erythromycin, gentamycin, tetracycline, troglitazone, and cephazolin; other therapeutic agents such as anticancer and immune suppression agents like indomethacin, irinotecan, lorazepam, and methotrexate; or

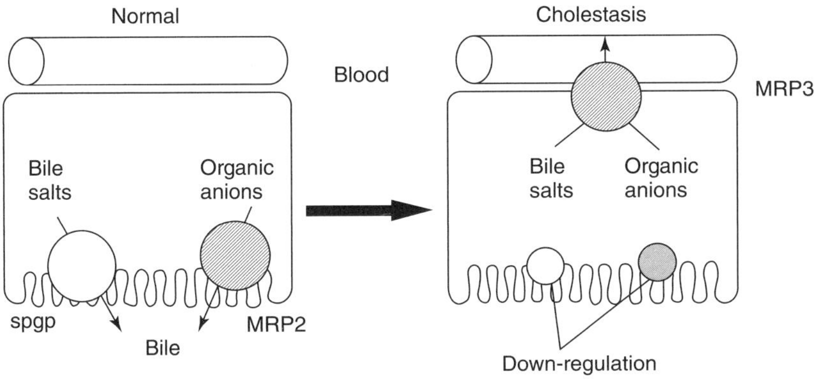

Figure 4.7 Schematic presentation of potential synergy between MRP2 and MRP3 during cholestasis events. Ref [3] with permission.

other drugs such as morphine phenytoin, sulfamethoxazole testosterone, diazepam, digoxin, and valproic acid need to be monitored. The P450 enzymes are generally those that are implicated to a greater degree in the liver impairment than the phase II conjugation enzymes [69,70].

In addition to liver impairment due to incompetent DME activities as mentioned above, liver impairment and potential cholestasis can occur due to impaired function of drug transporters that are responsible for excreting xenobiotics and endogenous molecules as well as their metabolites. As shown in Figure 4.7, potential synergy between MRP2 and MRP3 during cholestasis events has been reported by Kusuhara and Sugiyama (2009) [3]. As discussed earlier, normal hepatobiliary secretion and enterohepatic circulation are required for the elimination of endogenous compounds such as cholesterol, bilirubin, and their metabolites from the body. They are also required to maintain lipid and BA homeostasis. Furthermore, it is the primary process that the body uses to eliminate drugs, carcinogens, and potentially toxic by-products generated from the breakdown of xenobiotics and endobiotics, including BAs. The disposition of BA is as follows: it is taken up by the liver from portal blood and secreted into the bile by distinct transport proteins that have specific affinity and capacity to their substrates located at the sinusoidal and canalicular membranes of hepatocytes. In addition, BA secretion into the bile occurs against a concentration gradient. With very high concentration of BAs in bile, the transport of BA is by an active transport mechanism. Consequently, the regulation of bile and its formation and secretion rely upon the functional competency of membrane transporters.

Over the past 5–10 years, extensive efforts have been invested in understanding the mechanisms of BA transport and homeostasis in order to fully uncover the means of regulation of their expression in health and disease states as well as in development [18]. Many BA transporters, including MRP (ABCC) and OATP (SLCO) families, have been discovered that are principally involved in maintaining the flow of the enterohepatic circulation. Transporters such as NTCP and BSEP were also involved in uptake and excretion of BA and other drugs; mutation or inhibition of these transporters can cause severe adverse drug reactions, as reported by International Transporter Consortium (ITC) team efforts (2013) [71]. For example, mutations in the BSEP gene lead to the inherited cholestatic disorder progressive familial intrahepatic cholestasis type 2, or PFIC2 [72,73]. This liver disease is characterized by severe jaundice, hepatomegaly, failure to thrive, and pruritus from infancy onset, whereas liver histology shows portal inflammation and giant-cell hepatitis. In addition, high BAs and aminotransferases (ALT) are detected in serum; however, a clinical hallmark of this disease is presenting with normal γ-glutamyl transferase (GGT) levels. The disease has a rapid progressive course and generally leads to cirrhosis and liver failure with a need for liver transplantation within the first decade of life. A number of studies have identified polymorphisms and sequence variability in the human ABCB11 gene [74–78]. Expression variability among the white population is due to a specific SNP at C-allele position 1457 in the ABCB11 gene associated with low BSEP protein expression levels. Tens of other polymorphisms were found in diversified human populations (Caucasians, Koreans, and African Americans) where the polymorphisms were found in exons and, introns as well as in 5′ flanking regions. Most variants were population specific, with the largest variation found in the African American group, but no functionality of the protein was determined for any of the variants [79]. In addition, ~80 polymorphisms were identified in a Japanese population [80], but again this was not associated with functionality studies of the variants. Polymorphisms in BSEP have only been associated with intrahepatic cholestasis of pregnancy [75,81] and in case of drug-induced cholestasis [82].

Furthermore, a recent overview on possible susceptibility factors in liver injury induced by antithyroid agents has been reported [83]. Age, gender, metabolism characteristics, alcohol consumption, underlying diseases, immunologic mechanisms, and drug interactions are involved in enhancing antithyroid drug-induced hepatic damage. The clinically used treatments for antithyroid drug-induced hepatotoxicity and the potential therapeutic strategies found to be effective against this complication are also discussed.

4.4.2 Renal Impartment: Chronic Kidney Disease (CKD)

As in hepatic impairment, renal impairment has led to modifying dosing regimens for elderly patients with compromised kidney function and has resulted in FDA guidelines for when and how to conduct renal impairment studies. Several investigations [84–87] suggested that kidney disease can impact transporters and metabolic enzymes in the liver and gastrointestinal tract, potentially because of the accumulation of uremic toxins. Currently, it is postulated that uremic toxins such as urea, parathyroid hormone, indoxyl sulfate, and cytokines might cause transcriptional/translational modifications or inhibition of DMEs and transporters and data collected thus far seem to support this theory [86]. Assessment of the impact of renal impairment on the systemic exposure of new molecular entities from recently submitted investigational new drug (IND) applications provided evidence that renal impairment can affect the PKs of drugs that are predominately eliminated by nonrenal processes such as metabolism (CYP and non-CYP) and/or active transport [87].

TABLE 4.2 Alteration of PK of Drugs Influenced by DMEs and Transporters in Patients with CKD Compared to Normal adapted data (from Table 2 of [86]).

Drug	Renal Condition	CL Total	CL R	CL NR	% Change in CLR	% Change in CL NR	AUC	$T_{1/2}$	DMEs and Transporters involved
Cerivastatin	Normal	22.2 L/h	NR	NR			13.5 μg.h/1	2.3	CYP2C8, OATP, BCRP, MRP, CYP3A4, P-gp
	CKD	13.3	NR	NR			22.5	3.4	
Lidocaine	Normal	11.9 ml/min/1.73 m2	NR	NR			NR	2.2	CYP1A2, CYP3A4
	CKD	6.0	NR	NR			NR	4.6	
Raloxifene	Normal	48.8 L/h/kg	NR	NR			20.5 ng.h/mL	19.5	CYP3A4, UGT, P-gp
	CKD	28.5	NR	NR			35.1	29.9	
Roxithromycin	Normal	48.8 mL/min	5.5	43.3	−93	−42	115 μg.h/ml	7.9	CYP3A4, P-gp
	CKD	25.3	0.4	25.0			242	15.5	
Reboxetine	Normal	37 ml/min	3	34	−66	−68	2,106 ng.h/mL	12.5	CYP3A4
	CKD	12	1	11			5.923	25.9	
Lomeflaxacin	Normal	229.2 mL/min/1.73 m2	140.4	88.7	−80	−45	27.9 μg.h/ml	8.09	P-gp?
	CKD	77.1	28.4	48.7			93.9		
Felbamate	Normal	0.5 mL/min/kg	0.15	0.35	−87	−34	526 μg.h/mL	19.2	CYP2E1, CYP3A4
	CKD	0.25	0.02	0.23			1,119	33.9	
	CKD	47	2.4	44.6			+77%	7.3	

TABLE 4.3 Epidemiological Studies Linking Uric Acid to Chronic Kidney Disease (CKD). Adapted from Ref [89] with permission.

Reference	Major Findings
[28]	CKD 3–4 and uric acid correlate with death but not with ESRD
[30]	Uric acid >8 mg/dL increased CKD risk 3-fold in men and 10-fold in women
[31]	Uric acid >7 mg/dL increased risk of CKD 1.74-fold in men and 3.12-fold in women
[32]	Higher uric acid quartile conferred 2.14-fold increased risk of ESRD over 25 years
[33]	Elevated uric acid associated with 2.63-fold increased risk of CKD in hypertensive women
[34]	Uric acid associated with prevalent CKD in elderly
[36]	Each 1 mg/dL increase in uric acid increased risk of CKD 7–11%
[38]	Each 1 mg increase in uric acid associated with 1.28 odds ratio of reduced e-GFR at 5 years
[39]	Uric acid > 6.5 mg/dL in men and > 5.3 mg/dL in women associated with hazard ratios of 1.36 for all-cause mortality and 2.14 for incident CKD

Kidney disease is known to alter the renal clearance (i.e., GF) and therefore the PK disposition of drugs and has been incorporated into dosing recommendations for more than 40 years. Although less intuitive, the observation is that kidney disease somehow leads to alterations in nonrenal clearance (CL_{NR}) as well, and the usual dosages of metabolized drugs may need to be altered to individualize therapy for patients with renal failure [88].

Significant advances in molecular biology and clinical pharmacology have been made during the past 30 years, and these have enabled investigators to identify individual DMEs and transporters comprising the CL_{NR} pathways, to characterize alterations in their functional expression and, importantly, to elucidate the interplay between them.

A better understanding of the effects of kidney disease on drug disposition, particularly the enzymes and transporters that predominantly determine nonrenal drug clearance, as well as the mechanism and the clinical relevance of the alterations, may help guide dosing and thereby optimize pharmacotherapy in these patients, who compose a considerable proportion of the U.S. population. Table 4.2 shows the changes in PK of drugs between normal and CKD and potential mechanisms of drug disposition and key factors involved.

In addition, a recent investigation and review article by Bellomo (2013) [89] showed the correlation between uric acid and CKD that can be implicated in deficient excretion function. The relationship between serum uric acid and renal function and disease was discussed based on experimental evidence derived from animal and human studies, evidence gathered from a number of epidemiological studies, and from the few (up to now) studies of uric-acid-lowering therapy. Table 4.3 summarizes the most relevant studies investigating the relationship between serum uric acid (UA) levels and renal function, incidence, and/or progression of renal disease [90–101], which generally support a significant role of UA in CKD.

REFERENCES

1. Kitamura S, Maeda K, Wang Y, Sugiyama Y. Involvement of multiple transporters in the hepatobiliary transport of rosuvastatin. Drug Metab Dispos 2008;36:2014–2023.
2. Shitara Y, Horie T, Sugiyama Y. Transporters as a determinant of drug clearance and tissue distribution. Eur J Pharm Sci 2006;27:425–446.
3. Kusuhara H, Sugiyama Y. In vitro–in vivo extrapolation of transporter-mediated clearance in the liver and kidney. Drug Metab Pharmacokinet 2009;24:37–52.
4. Amidon GL, Lennernas H, Shah VP, Crison JR. A theoretical basis for a biopharmaceutic drug classification: the correlation of in vitro drug product dissolution and in vivo bioavailability. Pharm Res 1995;12:413–420.
5. Wu CY, Benet LZ. Predicting drug disposition via application of BCS: transport/ absorption/ elimination interplay and development of a biopharmaceutics drug disposition classification system. Pharm Res 2005;22:11–23.
6. Waiver of In Vivo Bioavailability and Bioequivalence Studies for Immediate-Release Solid Oral Dosage Forms Based on a Biopharmaceutics Classification System, 2000. [Online]. Available: http://www.fda.gov/cder/guidance/index.htm.
7. Watanabe T, et al. Prediction of the hepatic and renal clearance of transporter substrates in rats using in vitro uptake experiments. Drug Metab Dispos 2009;37:1471–1479.
8. Shugarts S, Benet LZ. The role of transporters in the pharmacokinetics of orally administered drugs. Pharm Res;2009:26:2039–2054.
9. Kusuhara H, Sugiyama Y. Vectorial Transport II. In: Molecular Pharmacokinetics, ed. Kusuhara H, Sugiyama Y. Tokyo: Nanzando, 2008, 364–376. In Japanese.
10. Levine WG. Biliary excretion of drugs and other xenobiotics. Annu Rev Pharmacol Toxicol 1978;18:81–96.
11. Hirom PC, Millbum P, Smith RI, et al. Species variations in the threshold molecular weight factor for the biliary excretion of organic anions. Biochem J 1972;129:1071–1077.
12. Kato Y, Takahara S, Kato S, et al. Involvement of multidrug resistance-associated protein 2 (Abcc2) in molecular weight-dependent biliary excretion of β-lactam antibiotics. Drug Metab Dispos 2008;36(6):1088–1096.
13. Masereeuw R, Russel FGM. Therapeutic implications of renal anionic drug transporters. Pharmacol Ther 2010;126(2):200–216.
14. Locher KP. Structure and mechanism of ABC transporters. Curr Opin Struct Biol 2004;14:426–431.
15. Fredriksson R, Nordstrom KJ, Stephansson O, et al. The solute carrier (SLC) complement of the human genome: phylogenetic classification reveals four major families. FEBS Lett 2008;582:3811–3816.
16. Russel FGM. Transporters in drug absorption, distribution, and removal. In: Enzyme-and transporter-based drug–drug interactions, ed. Pang KS, Rodrigues AD, Peter RM. New York: Springer, 2010, 27–50.
17. Giacomini KM, Huang S-W, Tweedie DJ, et al. The International Transport Consortium. Membrane transporters in drug development. Nat Rev 2010;9:215–236.

18. Kosters A, Karpen SJ. Bile acid transporters in health and disease Xenobiotica 2008;38(7–8):1043–1071.
19. Kumar V, Abbas AK, Fausto N. In: Robbins and Cotran pathologic basis of disease, 8th ed., ed. Schmitt W, Gruliow R. Philadelphia: Elsevier Saunders, 2010, 834.
20. Erlinger S. Bile flow. In: Liver biology and pathobiology, ed. Arias IM, Boyer JL, Fausto N, et al. New York: Raven Press, 1994, 769–786.
21. Chandra P, Brouwer KL. The complexities of hepatic drug transport: current knowledge and emerging concepts. Pharm Res 2004;21(5):719–735.
22. Zamek-Gliszczynski MJ, Hoffmaster KA, Nezasa K-I, et al. Integration of hepatic drug transporters and phase II metabolizing enzymes: mechanisms of hepatic excretion of sulfate, glucuronide, and glutathione metabolites. Eur J Pharm Sci 2006;27:447–486.
23. Stieger B, Meier Y, Meier PJ. The bile salt export pump. Eur J Physiol 2007;453:611–620.
24. Keppler D, Konig J. Expression and localization of the conjugate export pump encoded MRP2 (cMRP/cMOAT) gene in liver. FASEB J 1997;11:509–516.
25. Yanni SB, Smith PB, Benjamin DK Jr, Augustijns PF, Thakker DR, Annaert PP. Higher clearance of micafungin in neonates compared with adults: role of age-dependent micafungin serum binding. Biopharm Drug Dispos 2011;32(4):222–232.
26. Perri D, Ito S, Rowsell V, et al. The kidney—the body's playground for drugs: an overview of renal drug handling with selected clinical correlates. Can J Clin Pharmacol 2003;10(1): 17–23.
27. El-Sheikh AA, Masereeuw R, Russel FG. Mechanisms of renal anionic drug transport. Eur J Pharmacol 2008;585:245–255.
28. Inui K-I, Masuda S, Saito H. Cellular and molecular aspects of drug transport in the kidney. Kidney Int 2000;58:944–958.
29. Masereeuw R, Russel GM. Mechanisms and clinical implications of renal drug excretion. Drug Metab Rev 2001;33(3&4):299–351.
30. Tamai I, Nezu J, Uchino H, et al. Molecular identification and characterization of novel members of the human organic anion transporter (OATP) family. Biochem Biophys Res Commun 2000;273:251–260.
31. Wright SH, Dantzler WH. Molecular and cellular physiology of renal cation and anion transport. Physiol Rev 2004;84:987–1049.
32. Kamal MA, Keep RF, Smith DE. Role and relevance of PEP2 in drug disposition, dynamics, and toxicity. Drug Metab Pharmacokinet 2008;23:236–242.
33. Ranganathan P, Culverhouse R, Marsh S, et al. Methotrexate (MTX) pathway gene polymorphisms and their effects on MTX toxicity in Caucasian and African American patients with rheumatoid arthritis. J Rheumatol 2008;35:572–579.
34. Rau T, Erney B, Gores R, et al. High-dose methotrexate in pediatric acute lymphoblastic leukemia: impact of ABCC2 polymorphisms on plasma concentrations. Clin Pharmacol Ther 2006;80:468–476.
35. Dekant W, Vamvakas S. Biotransformation and membrane transport in nephrotoxicity. Crit Rev Toxicol 1996;26:309–334.
36. Obach RS, et al. The prediction of human pharmacokinetic parameters from preclinical and in vitro metabolism data. J Pharmacol Exp Ther 1997;283:46–58.
37. Obach RS, et al. In vitro cytochrome P450 inhibition data and the prediction of drug–drug interactions: qualitative relationships, quantitative predictions, and the rank-order approach. Clin Pharmacol Ther 2005;78:582–592.

38. Sasaki M, et al. Prediction of in vivo biliary clearance from the in vitro transcellular transport of organic anions across a double-transfected Madin–Darby canine kidney II monolayer expressing both rat organic anion transporting polypeptide 4 and multidrug resistance associated protein 2. Mol Pharmacol 2004;66:450–459.
39. Endres CJ, Hsiao, P, Chung FS, Unadkat JD. The role of transporters in drug interactions. Eur J Pharm Sci 2006;27:501–517.
40. Ghibellini G, Leslie EM, Brouwer, KLR. Methods to evaluate biliary excretion of drugs in humans: an updated review. Mol Pharm 2006;3:198–211.
41. Lin JH. Species similarities and differences in pharmacokinetics. Drug Metab Dispos 1995;23:1008–1021.
42. Lai Y, Tse CM, Unadkat JD. Mitochondrial expression of the human equilibrative nucleoside transporter 1 (Hent1) results in enhanced mitochondrial toxicity of antiviral drugs. J Biol Chem 2004;279:4490–4497.
43. Watkins PB, Seeff LB. Drug-induced liver injury: summary of a single topic clinical research conference. Hepatology 2006;43:618–631.
44. Scheen AJ. Hepatotoxicity with thiazolidinediones: is it a class effect? Drug Saf 2001;24:873–888.
45. Yanni SB, Augustijns PF, Benjamin DK Jr, Brouwer KL, Thakker DR, Annaert PP. In vitro investigation of the hepatobiliary disposition mechanisms of the antifungal agent micafungin in humans and rats. Drug Metab Dispos 2010;38(10):1848–1856.
46. Liu X, LeCluyse EL, Brouwer KR, et al. Use of Ca2+ modulation to evaluate biliary excretion in sandwich-cultured rat hepatocytes. J Pharmacol Exp Ther 1999;289: 1592–1599.
47. Abe K, Bridges AS, Yue W, et al. In vitro biliary clearance of angiotensin II receptor blockers and 3-hydroxy-3-methyl-glutaryl-coenzyme A reductase inhibitors in sandwich-cultured rat hepatocytes: comparison with in vivo biliary clearance. J Pharmacol Exp Ther 2008;326:983–990.
48. Fukuda H, Ohashi R, Tsuda-Tsukimoto M, et al. Effect of plasma protein binding on in vitro–in vivo correlation of biliary excretion of drugs evaluated by sandwich-cultured rat hepatocytes. Drug Metab Dispos 2008;36:1275–1282.
49. Ghibellini G, Vasist LS, Leslie EM, Heizer WD, Kowalsky RJ, Calvo BF, Brouwer KL. In vitro–in vivo correlation of hepatobiliary drug clearance in humans. Clin Pharmacol Ther 2007;81(3):406–413.
50. Chatterjee S, Richert L, Augustijns P, Annaert P. Hepatocyte-based in vitro model for assessment of drug-induced cholestasis. Toxicol Appl Pharmacol 2014;274(1):124–136.
51. Wright SH, Dantzler WH. Molecular and cellular physiology of renal organic cation and anion transport. Physiol Rev 2004;84(3):987–1049.
52. Brown CD, Sayer R, Windass AS, Haslam IS, De Broe ME, D'Haese PC, Verhulst A. Characterisation of human tubular cell monolayers as a model of proximal tubular xenobiotic handling. Toxicol Appl Pharmacol 2008;233(3):428–438.
53. Verhulst A, Sayer R, De Broe ME, D'Haese PC, Brown CD. Human proximal tubular epithelium actively secretes but does not retain rosuvastatin. Mol Pharmacol 2008;74(4):1084–1091.
54. Verhulst A, D'Haese PC, De Broe ME. Inhibitors of HMG-CoA reductase reduce receptor-mediated endocytosis in human kidney proximal tubular cells. J Am Soc Nephrol 2004;15(9):2249–2257.

55. Vervaet BA, Verhulst A, De Broe ME, D'Haese PC. The tubular epithelium in the initiation and course of intratubular nephrocalcinosis. Urol Res 2010;38(4):249–256.
56. Vervaet BA, Verhulst A, D'Haese PC, De Broe ME. Nephrocalcinosis: new insights into mechanisms and consequences. Nephrol Dial Transplant 2009;24(7):2030–2035.
57. Prakash C, Chen W, Rossulek M, et al. Metabolism, pharmacokinetics, and excretion of a cholesteryl ester transfer protein inhibitor, torcetrapib, in rats, monkeys, and mice: characterization of unusual and novel metabolites by high resolution liquid chromatography-tandem mass spectrometry and 1H nuclear magnetic resonance. Drug Metab Dispos 2008;36(10):2064–2079.
58. Prakash C, Kamel A, Anderson W, et al. Metabolism and excretion of the novel antipsychotic drug ziprasidone in rats after oral administration of a mixture of $^{14}C-$ and ^{3}CH – labeled ziprasidone. Drug Metab Dispos 1997;25(2):206–218.
59. Prakash C, Kamel A, Gummerus J, et al. Metabolism and excretion of a new antipsychotic drug, ziprasidone, in humans. Drug Metab Dispos 1997;25(7):863–872.
60. Roffey SJ, Obach RS, Gedge JI, et al. What is the objective of the mass balance study? A retrospective analysis of data I animal and excretion studies employing radiolabeled drugs. Drug Metab Rev 2007;39:17–43.
61. Monro AM. Interspecies comparisons in toxicology: the utility and futility of plasma concentrations of the test substance. Reg Toxicol Pharmacol 1990;12:137–160.
62. Penner N, Klunk LJ, Prakash C. Human radiolabeled mass balance studies: objectives, utilities and limitations. Biopharm Drug Dispos 2009;30:185–203.
63. Beumer JH, Beijnen JH, Schellens JHM. Mass balance studies, with a focus on anticancer drugs. Clin Pharmacokinet 2006;45(26):33–58.
64. Baillie TA, Cayen MN, Fouda H, et al. Drug metabolites in safety testing. Toxicol Appl Pharmacol 2002;182:188–196.
65. Smith DA, Obach RS. Seeing through the MIST: abundance versus percentage. Commentary on metabolites in safety testing. Drug Metab Dispos 2005;33:1409–1417.
66. Strom SC, Davila J, Grompe M. Chimeric mice with humanized liver: tools for the study of drug metabolism, excretion and toxicity. Methods Mol Biol 2010;640:491–509.
67. Schulz-Utermoehl T, Sarda S, Foster JR, Jacobsen M, Kenna GJ, Morikawa Y, Salmu J, Gross G, Wilson ID. Evaluation of the pharmacokinetics, biotransformation and hepatic transporter effects of troglitazone in mice with humanized livers. Xenobiotica 2012;42(6):503–517.
68. Masakazu Kakuni M, Yamasaki C, Tachibana A, Yoshizane Y, Ishida Y, Tateno C. Chimeric mice with humanized livers: a unique tool for in vivo and in vitro enzyme induction studies. Int J Mol Sci 2014;15(1):58–74.
69. Klotz U. Antiarrhythmics: elimination and dosage considerations in hepatic impairment. Clin Pharmacokinet 2007;46(12):985–986.
70. Stangier J, Stahle H, Rathgen K, et al. Pharmacokinetics and pharmacodynamics of dabigatran etexilate, an oral thrombin inhibitor, are not affected by moderate hepatic impairment. J Clin Pharmacol 2008;48:1411–1419.
71. Zamek-Gliszczynski MJ, Lee CA, Poirier A, Bentz J, Chu X, Ellens H, Ishikawa T, Jamei M, Kalvass JC, Nagar S, Pang KS, Korzekwa K, Swaan PW, Taub ME, Zhao P, Galetin A; International Transporter Consortium. ITC recommendations for transporter kinetic parameter estimation and translational modeling of transport-mediated PK and DDIs in humans. Clin Pharmacol Ther 2013;94(1):64–79.

72. Jansen PL, Strautnieks SS, Jacquemin E, Hadchouel M, Sokal EM, Hooiveld GJ, Koning JH, De Jager-Krikken A, Kuipers F, Stellaard F, et al. Hepatocanalicular bile salt export pump deficiency in patients with progressive familial intrahepatic cholestasis. Gastroenterology 1999;117(6):1370–1379.
73. Oude Elferink RP, Paulusma CC, Groen AK. Hepatocanalicular transport defects: pathophysiologic mechanisms of rare diseases. Gastroenterology 2006;130(3):908–925.
74. Pauli-Magnus C, Kerb R, Fattinger K, Lang T, Anwald B, Kullak-Ublick GA, Beuers U, Meier PJ. BSEP and MDR3 haplotype structure in healthy Caucasians, primary biliary cirrhosis and primary sclerosing cholangitis. Hepatology 2004a;39(3):779–791.
75. Pauli-Magnus C, Lang T, Meier Y, Zodan-Marin T, Jung D, Breymann C, Zimmermann R, Kenngott S, Beuers U, Reichel C, et al. Sequence analysis of bile salt export pump (ABCB11) and multidrug resistance p-glycoprotein 3 (ABCB4, MDR3) in patients with intrahepatic cholestasis of pregnancy. Pharmacogenetics 2004b;14(2):91–102.
76. Pauli-Magnus C, Meier PJ. Hepatobiliary transporters and drug-induced cholestasis. Hepatology 2006;44(4):778–787.
77. Pauli-Magnus C, Stieger B, Meier Y, Kullak-Ublick GA, Meier PJ. Enterohepatic transport of bile salts and genetics of cholestasis. J Hepatology 2005;43(2):342–357.
78. Stieger B, Meier Y, Meier PJ. The bile salt export pump. Pflugers Archives 2007;453(5): 611–620.
79. Lang T, Haberl M, Jung D, Drescher A, Schlagenhaufer R, Keil A, Mornhinweg E, Stieger B, Kullak-Ublick GA, Kerb R. Genetic variability, haplotype structures, and ethnic diversity of hepatic transporters MDR3 (ABCB4) and bile salt export pump (ABCB11). Drug Metabolism and Disposition 2006;34(9):1582–1599.
80. Saito S, Iida A, Sekine A, Miura Y, Ogawa C, Kawauchi S, Higuchi S, Nakamura Y. Three hundred twenty-six genetic variations in genes encoding nine members of ATP-binding cassette, subfamily B (ABCB/MDR/TAP), in the Japanese population. J Human Genetics 2002;47(1):38–50.
81. Keitel V, Vogt C, Haussinger D, Kubitz R. Combined mutations of canalicular transporter proteins cause severe intrahepatic cholestasis of pregnancy. Gastroenterology 2006;131(2):624–629.
82. Lang C, Meier Y, Stieger B, Beuers U, Lang T, Kerb R, Kullak-Ublick GA, Meier PJ, Pauli-Magnus C. Mutations and polymorphisms in the bile salt export pump and the multidrug resistance protein 3 associated with drug-induced liver injury. Pharmacogenetics Genomics 2007;17(1):47–60.
83. Heidari R, Niknahad H, Jamshidzadeh A, Abdoli N. Factors affecting drug-induced liver injury: antithyroid drugs as instances. Clin Mol Hepatol 2014;20(3):237–248.
84. Olyaei AJ, Bennett WM. Drug dosing in the elderly patients with chronic kidney disease. Clin Geriatr Med 2009;25:459–527.
85. Huang SM, Temple R, Xiao S, et al. When to conduct a renal impairment study during drug development: US food and drug administration perspective. Clin Pharmacol Ther 2009;86(5):475–479.
86. Nolin TD, Naud J, Leblond FA, et al. Emerging evidence of the impact of kidney disease on drug metabolism and transport. Clin Pharmacol Ther 2008;83(6):898–903.
87. Zhang Y, Zhang L, Abraham S, et al. Assessment of the impact of renal impairment on systemic exposure of new molecular entities: evaluation of recent new drug applications. Clin Pharmacol Ther 2009;85(3):305–311.

88. Hong Sun, Lynda Frassetto, Leslie Z. Benet Effects of renal failure on drug transport and metabolism. Pharmacol Ther 2006;109(1–2):1–11.
89. Bellomo G. Uric acid and chronic kidney disease: A time to act? World J Nephrol 2013;2(2):17–25.
90. Madero M, Sarnak MJ, Wang X, Greene T, Beck GJ, Kusek JW, Collins AJ, Levey AS, Menon V. Uric acid and long-term outcomes in CKD. Am J Kidney Dis 2009;53:796–803.
91. Domrongkitchaiporn S, Sritara P, Kitiyakara C, Stitchantrakul W, Krittaphol V, Lolekha P, Cheepudomwit S, Yipintsoi T. Risk factors for development of decreased kidney function in a southeast Asian population: a 12-year cohort study. J Am Soc Nephrol 2005;16:791–799.
92. Iseki K, Ikemiya Y, Inoue T, Iseki C, Kinjo K, Takishita S. Significance of hyperuricemia as a risk factor for developing ESRD in a screened cohort. Am J Kidney Dis 2004;44:642–650.
93. Obermayr RP, Temml C, Gutjahr G, Knechtelsdorfer M, Oberbauer R, Klauser-Braun R. Elevated uric acid increases the risk for kidney disease. J Am Soc Nephrol 2008;19:2407–2413.
94. Hsu CY, Iribarren C, McCulloch CE, Darbinian J, Go AS. Risk factors for end-stage renal disease: 25-year follow-up. Arch Intern Med 2009;169:342–350.
95. Borges RL, Hirota AH, Quinto BM, Ribeiro AB, Zanella MT, Batista MC. Uric acid as a marker for renal dysfunction in hypertensive women on diuretic and nondiuretic therapy. J Clin Hypertens (Greenwich) 2009;11:253–259.
96. Chen N, Wang W, Huang Y, Shen P, Pei D, Yu H, Shi H, Zhang Q, Xu J, Lv Y. Community-based study on CKD subjects and the associated risk factors. Nephrol Dial Transplant 2009;24:2117–2123.
97. Sturm G, Kollerits B, Neyer U, Ritz E, Kronenberg F. Uric acid as a risk factor for progression of non-diabetic chronic kidney disease? The Mild to Moderate Kidney Disease (MMKD) Study. Exp Gerontol 2008;43:347–352.
98. Weiner DE, Tighiouart H, Elsayed EF, Griffith JL, Salem DN, Levey AS. Uric acid and incident kidney disease in the community. J Am Soc Nephrol 2008;19:1204–1211.
99. Chonchol M, Shlipak MG, Katz R, Sarnak MJ, Newman AB, Siscovick DS, Kestenbaum B, Carney JK, Fried LF. Relationship of uric acid with progression of kidney disease. Am J Kidney Dis 2007;50:239–247.
100. Bellomo G, Venanzi S, Verdura C, Saronio P, Esposito A, Timio M. Association of uric acid with change in kidney function in healthy normotensive individuals. Am J Kidney Dis 2010;56:264–272.
101. Ben-Dov IZ, Kark JD. Serum uric acid is a GFR-independent long-term predictor of acute and chronic renal insufficiency: the Jerusalem Lipid Research Clinic cohort study. Nephrol Dial Transplant 2011;26:2558–2566.

5

DRUG–DRUG INTERACTION: FROM BENCH TO DRUG LABEL

5.1 INTRODUCTION: THE IMPACT OF DRUG–DRUG INTERACTION ON DRUG DISPOSITION AND DRUG SAFETY

Drug–drug interaction (DDI) is one of the major obstacles for the pharmaceutical drug development process; uncovering its potential on any adverse clinical outcomes becomes increasingly important in drug discovery and development. Both in vitro and in vivo preclinical investigations to assess the any clinical outcomes prior to drug launch are taken into account to reveal the potential of DDI and its mechanism of any drug under development. Evaluation of the possible interactions of a drug candidate with other drugs as soon as possible—not only as an inhibitor or inducer (perpetrator) but also as a substrate (victim)—could avoid detrimental DDIs in humans. As will be discussed later, DDIs represent a major mechanism of adverse drug reactions, and consequently their evaluation is critical to studies within all the drug development stages, drug discovery, and regulation of new drug candidates to avoid any serious toxicity that leads to drug withdrawal postmarket. Finally, the ultimate goal of non-clinical and clinical DDI studies is to permit integration of DDI knowledge acquired in the development phase into prescribing guidance in a manner that enables optimal postmarketing risk management following marketing authorization.

Not all the preclinically determined DDIs can be considered as clinically significant (poor correlation between in vitro and in vivo observation). The systemic

---

Translational ADMET for Drug Therapy: Principles, Methods, and Pharmaceutical Applications, First Edition. Souzan B. Yanni.

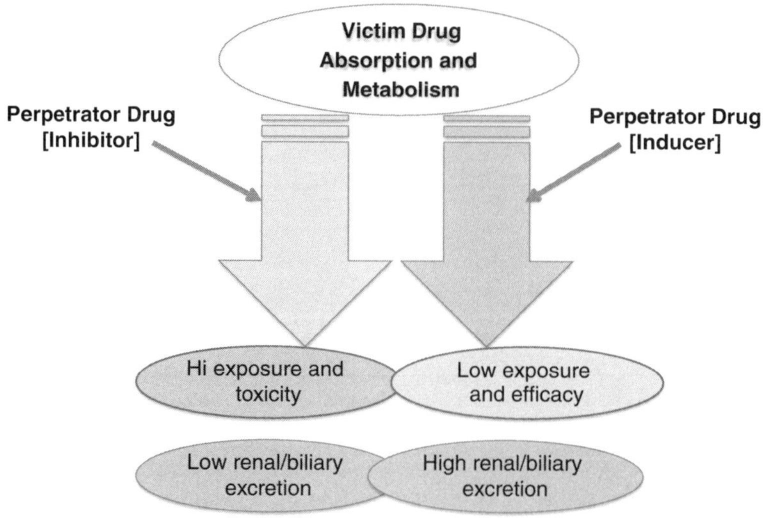

Figure 5.1 Effect of DDI on the disposition of victim target drug.

concentrations of victim/object drug and its perpetrator play a major role in causing effective interaction, and alteration of drug concentrations will lead to diminishing the DDI potential. On the other hand, DDI can result in altering the therapeutic effect, or sometimes altering the toxic effects of a medication by administration with another drug. As shown in Figure 5.1, inhibition or induction of the absorption, distribution, metabolism, or elimination of victim drug by a coadministered drug could result in altering blood/target organ levels and potential effects on efficacy and/or safety of the victim drug. Such DDIs are classified as pharmacokinetics (PK) interactions. There are other types of interactions, such as pharmacodynamics (PD) interactions that occur when one drug alters the pharmacologic effect (efficacy and/or safety) of another coadministered drug without affecting its PK.

Historically, the impact of DDI on drug disposition and safety was reported in several drug therapy programs during the last decade where unadequate DDI evaluation of drug candidates resulted in postmarketing withdrawal after drug approval [1]. For example, the calcium channel blocker, mibefradil, used in the medical management of hypertension and angina, produced dangerous and occasionally fatal interactions with sensitive substrates of P450 3A (CYP3A), such as the calcium channel blocker felodipine due to its absorption, distribution, metabolism, and excretion (ADME) profile as a strong mechanism-based inactivator of CYP3A and an inhibitor of the efflux transporter P-glycoprotein (P-gp). Mibefradil was voluntarily withdrawn in June 1998 within a year following approval, after United States Prescribing Information (USPI) recommended the need to administer the drug in concomitant use of 26 drugs. The overall benefit/risk ratio for mibefradil was unfavorable, as other therapeutic drugs with fewer safety issues were already on the market. This example illustrates the importance of early assessment of risk for drugs under investigation to produce DDIs with coadministered agents via effects on their PK (e.g., via metabolic inhibition/induction by other drugs).

Similarly, there are a few examples worth mentioning, such as the prokinetic agent cisapride, which was used for the management of gastroesophageal reflux disease, and the antihistamine drugs terfenadine and astemizole, which are inhibitors of the ion channel hERG and play a critical role in cardiac repolarization. These three drugs were cleared almost exclusively via metabolism by CYP3A. With coadministration of inhibitors of this enzyme, a clinical increase in their exposure occurs. Furthermore, there is an increased risk for the fatal cardiac arrhythmia *torsades de pointes* of these drugs clinically if they are coadministered with many common therapeutic agents, including antibiotics such as erythromycin and consumer products such as grapefruit juice.

This set of examples illustrates the importance of an adequate level of premarketing characterization of DDI risk, thus identifying the safety profile relative to the therapeutic index [2].

5.2 DDIS IMPLICATED WITH DRUG-METABOLIZING ENZYMES (DMES) AND DRUG METABOLISM

5.2.1 DDI Mediated by P450 Inhibition

As mentioned earlier, the PK DDI can occur when one drug alters the metabolism, by inhibition or induction, of a coadministered drug. The most significant PK DDI is emphasized by the metabolic routes of elimination, mostly of those occurring via the P450 enzymes, by inhibition with concomitant drug treatment leading to serious clinical DDI such as those cases briefly described above. Although P450 inhibitions are implicated in the majority of clinically relevant DDIs [3,4], there have been a few incidents of DDI with conjugated enzymes, as will be briefly mentioned below.

In a clinic setting, DDI mediated by P450 inhibition was observed by the increase of plasma concentrations of victim drugs when coadministrated with a potent inhibitor (perpetrator), as indicated in Figure 5.1. When ketoconazole, a potent CYP3A4 inhibitor, was administered with triazolam, CYP3A4 substrate, a 22-fold increase in triazolam exposure was observed [5]. As expected, these incidents of DDI resulted in dose adjustment, serious drug monitoring, or sometimes drug development termination of investigational drugs, especially when they involve a drug that has a narrow therapeutic range, such as warfarin, resulting in an increase in plasma concentration. Inadequate DDI investigations during late discovery or early drug development may result in overdrug exposure, and hence unwanted toxicity in some patients when the metabolism-mediated drug elimination is diminished by coadministrated inhibitor. A classic example of a drug interaction is with the antihistamine Seldane. (terfenadine) and the common antibiotic erythromycin [6]. When terfenadine was dosed along with erythromycin that inhibited CYP3A4 responsible for its metabolism, hence clearance, terfenadine was accumulated to extensive toxic blood levels and to a potentially fatal arrhythmia. The case resulted in a recall of Seldane..by the Food and Drug Administration (FDA) and recommendation that DDI investigations be assessed relatively early in drug development. Similarly, as mentioned earlier, mibefradil caused serious DDI with simvastatin and with β-blockers [7,8]. Subsequently, mibefradil was found to be a potent inhibitor of CYP3A4, CYP2D6, and P-gp and a potent time-dependent inhibitor (TDI) for CYP3A4/3A5 [9–11]. These serious adverse events resulted when mibefradil was coadministered with CYP3A4 substrates, likely because of time-dependent inhibition of CYP3A4, and mibefradil was withdrawn from the market a year after launch. Although CYP3A4-mediated DDI was responsible for several clinically relevant DDI and drug withdrawal, other P450 enzymes were also responsible for serious DDIs. One example is the recent withdrawal of rofecoxib, a cyclooxygenase-2-selective nonsteroidal anti-inflammatory drug (NSAID), in 2006. It caused moderate increased plasma concentrations of theophylline [12] and R-warfarin [13], the effect implicated in some cardiovascular events in treated patients. Similarly, rofecoxib increased the plasma concentration of tizanidine more than 10-fold due to the potent inhibition of CYP1A2-mediated metabolism and clearance by rofecoxib [14,15].

In drug development, the DDI mediated by CYP inhibition of a drug candidate can be assessed in two steps: (1) by using in vitro models and methodologies to estimate the potency of inhibition, and (2) by translating the in vitro information to clinical pharmacology investigation and determining the correlation and magnitude of interaction.

5.2.1.1 In Vitro P450 Inhibition Models and Methodologies The in vitro models, methodologies, strategies, and data interpretation to assess the potential inhibition of P450 activity by investigational drug candidate have been well established now to permit their routine integration into preclinical to clinical drug investigation programs [16–21]. The use of liver subcellular fraction microsomes was found to be the most simple and common approach to investigate the rate of disappearance or appearance of metabolites of the drug under investigation. Also, they are routinely used to determine not only the rate of overall P450 reaction and kinetics but also to specify these parameters of each isozyme. In addition, liver microsomes are used to determine the inhibition of P450 enzymes by coadministered drugs as well as the potential that a specific enzyme may be inhibited by the tested drug. Utilizing liver microsomes as an in vitro tool has been successfully applied to measure the extent of drug metabolism and its inhibition by other drugs within a large human population by using a pool of 10–50 liver microsome preparations from diversified normal human subjects. The activity of each P450 isoform is carried out in the presence of a prototype substrate of each specific enzyme at its K_m value and in the presence of various concentrations of tested drug to estimate IC_{50} values as measures of inhibitory potency. Clinically significant P450 inhibition of human drug metabolism and DDIs includes reported CYPs 1A2, 2B6, 2C8, 2C9, 2C19, 2D6, 2E1, and 3A4/5 (CYP3A) [16,17,20]. The selective substrates for these are universally recognized and are often phenacetin (1A2), diclofenac (2C9), S-mephenytoin (2C19), bufuralol (2D6), chlorzoxazone (2E1), and testosterone (3A4). These screens are now firmly established as selectivity screens.

TABLE 5.1 Relationship Between I/K_i for CYP Inhibition by an Object Drug and DDI Risk Based on Fold Increase in Systemic Exposure of an Orally Administered Object Drug Whose Clearance Depends on the Metabolism of Inhibited Enzyme.

$[I]/K_i$ Ratio	Estimated Risk
$[I]/K_i > 1$	Likely
$1 > [I]/K_i > 0.1$	Possibly
$0.1 > [I]/K_i$	Remotely

These studies are typically conducted at a substrate concentration of K_m of the index reaction such that the apparent K_i can be approximated as $0.5 \times IC_{50}$ under the conservative assumption of competitive inhibition.

When the IC_{50} estimation indicates potent inhibition of a particular P450 by a new drug, there is need for a further evaluation of the inhibitory mechanism and kinetics for definitive estimation of the apparent K_i. This kinetic mechanism of inhibition and values associated with the inhibition constant can definitively assess by extrapolation the magnitude of clinical interaction, as shown in Table 5.1 [21]. Such additional refinement may not be a critical requirement from a pragmatic point of view, though it may be useful when the underlying mechanism of inhibition is atypical or complex [22]. When possible, estimation of K_i from IC_{50} can be determined, as it is important to assess the risk of DDIs and to guide the development studies toward the optimal clinical DDI evaluation and its strategy [23,24].

As has been indicated in many translational preclinical and clinical investigations by several pharmaceutical researchers, it is important to minimize the extent of nonspecific microsomal binding in the design of in vitro CYP inhibition DDI studies, as it influences the determination of K_m and IC_{50} values, thus improving the accuracy of in vitro-in vivo scaling and prediction of drug clearance [25–28]. It has been shown that nonspecific binding of drug may lead to overestimation of its IC_{50}/K_i in vitro and the underestimation of its inhibitory potency to P450 enzymes [21,29–32]. Currently, human liver microsomal protein concentrations 0.1–0.2 mg/mL are used in DDI in vitro [20,23]; at this level, the extent of microsomal binding would likely be minimized for almost all drugs [e.g., microsomal unbound fraction ($f_{u,\,mic}$) ≥ 0.8]. The inhibitory potency measured by $K_{i_{unbound}}$ will be determined as follows:

$$K_{i_{unbound}} = f_{u,\,mic} \times \text{observed } K_i. \tag{5.1}$$

Microsomal binding in the calculation of K_i values for basic drugs is recommended in the European Medicines Agency (EMA) draft guideline [33] by measuring the in vitro unbound fraction of drug candidate at the microsomal concentration selected in the P450 inhibition studies.

Although the pooled human liver microsome is the most simple and commonly used system for determination of in vitro P450 inhibition, it should be noted that

P450 inhibition studies can also be conducted using human hepatocytes in a pooled cryopreserved suspension preparation. Such a test may be helpful specifically when extensive non-P450 or nonmicrosomal-mediated metabolism is expected or known. The use of hepatocytes for estimating inhibitory potency for in vitro P450 inhibition by tested drug may be considered as more representative of the clinical outcomes [34]. Some structural features are associated with highly potent CYP inhibitors, both reversible and to a lesser extent irreversible. The presence of unhindered nitrogen in a saturated ring system (pyridine, imidazole, triazole) may result in the lone pair of the nitrogen being able to form a ligand interaction with the heme of the CYP450. Many of the potent CYP450 inhibitors bind in this manner, and the interaction adds 6 kcal to the binding energy. This interaction is the basis for the action of azole antifungals and a number of aromatase inhibitors. As this interaction is commonplace and invariably leads to highly potent inhibitors, such functionality is best avoided from the outset. Imidazole ring systems are also prevalent in mechanism-based time-dependent CYP450 inhibitors, although the relationships are more complex and are not accompanied by a mechanistic understanding [35]. Sufficient information is available to allow in silico filtering of compound structures to determine possible avenues that may lead to this problem [36].

5.2.1.2 Translating In Vitro P450 Inhibition Data to Clinical DDI After the in vitro inhibitory potency (K_i or IC_{50}) is assessed, the following step is to translate the data to strategic clinical study design that takes into consideration safe drug exposure, adequate PK properties, and a suitable dose of drugs associated with the DDI. The selection of the clinical dose(s) to forecast the level of risk for DDI with substrates of the enzyme being inhibited is the most significant step to enable development of risk management plans in later phases of clinical drug development.

The in vitro–in vivo extrapolation (IVIVE) correlation, done prospectively to permit the prediction of exposure of victim drug, is the ideal strategy not only among the major patient populations but also within the specific population. To reach this high prediction confidence, retrospective studies during the last decade have made substantial progress in the IVIVE of P450-inhibitory DDI, with some recently published examples of successful predictions from large databases [20,22,37,38]. However, uncertainty still remains in key parameters that are critical to the prediction of DDI magnitude, such as enzyme-available inhibitor concentration. Consequently, P450 inhibition-mediated DDI remains an area under development, which can balance scientific precision and an adequate conservative prediction though the approaches currently recommended by the FDA draft DDI guidance [39]. The approaches recommended by the FDA involve a pragmatic ranking level of safety risk based on the $[I]/K_i$ ratio versus systemic maximum plasma concentration (C_{max}) of the objecting drug at the highest clinical dose/frequency (e.g., steady-state C_{max}). The $[I]/K_i$ versus fold increase in AUC of an orally administered substrate drug indicates that its clearance is 100% mediated by P450 metabolism catalyzed by enzyme inhibited due to drug treatment.

From Table 5.1, the magnitude of DDIs measured by the ratio $[I]/K_i$ cutoffs of < 0.1 (corresponding projected maximum increase in AUC of < 1.1-fold) and

$[I]/K_i > 1$ (corresponding projected maximum increase in AUC of greater than 2-fold) are to be used for qualitative risk level classification as "low" and "high," respectively, though these ratios/classifications are not intended to serve as quantitative predictions of DDI magnitude. The application of this arrangement is basically to abolish unnecessary follow-up clinical DDI studies for enzymes that are at low inhibition-mediated DDI risk and to sequentially prioritize the conduct of clinical DDI studies. The relationship between $[I]/K_i$ and fold increase in AUC of an orally administered substrate drug whose clearance is entirely mediated 100% via metabolism by the enzyme that is inhibited by the drug candidate was found to be hyperbolic, and estimated risk can be determined from the following equation:

$$\frac{AUC_{inhibited}}{AUC_{control}} = \frac{CL_{int,\ control}}{CL_{int,\ inhibited}} = 1 + \frac{[I]}{K_i}. \qquad (5.2)$$

For example, the anticancer drug Everolimus, which is used for the treatment of advanced renal cell carcinoma, inhibits CYP2D6 activity in vitro. In the clinic, Everolimus was found to have mean steady-state C_{max} at the recommended 10 mg daily dose ~12-fold below the CYP2D6 inhibitory K_i. Based on the DDI classified potential listed in Table 5.1, Everolimus USPI concludes that Everolimus effect on the metabolism of CYP2D6 substrates is unlikely [40]. This example illustrates how in vitro data, when clearly indicative of low DDI risk, can inform prescribing guidance without the need for unnecessary clinical DDI studies.

The approach to DDI risk assessment using the $[I]/K_i$ ratio alone (with [I] defined as the clinically observed systemic C_{max} of the new drug candidate), while it is pragmatic and straightforward, is not without limitations. That simple classification system has been addressed in respect to multiple considerations: metric of [I], route of administration of the victim drug, and potential for extrahepatic metabolism. If the goal is to make quantitative predictions of DDI magnitude that extend beyond risk, the exposure change due to inhibition of P450-mediated metabolism may be determined from the following equations.

For reversible competitive or noncompetitive inhibition:

$$\frac{CL_{in\text{-}vitro\ int}\ Control}{CL_{int}\ Inhibitor} = 1 + \frac{I}{K_i}. \qquad (5.3)$$

For mechanism-based inhibition:

$$\frac{CL_{int}\ Control}{CL_{int}\ Inhibitor} = 1 + \frac{K_{inact}}{K_I} * \frac{[I]}{K_{deg}}, \qquad (5.4)$$

where [I] is inhibitor concentration, K_{inact} is the rate constant of P450 inactivation, K_I is the half-maximal inactivation rate, and K_i is the dissociation constant of enzyme-inhibitor complex.

As mentioned before, DDIs should not be viewed as solely undesirable, as there have been several cases in which the PK of one drug is modulated by another via a well-planned design to improve the exposure, hence the efficacy of the affected drug [41]. Kaletra is a coformulation of lopinavir and ritonavir, whereby ritonavir-mediated CYP3A4 inhibition results in higher plasma levels of lopinavir and boosts its anti-HIV protease activity. Also, ketoconazole (KTZ), a potent CYP3A4 inhibitor, is commonly used in combination with cyclosporine A (CsA) to enhance the immunosuppressive properties of the latter by inhibiting its first pass metabolism mediated by CYP3A, which in turn results in increasing CsA bioavailability and an increase in CsA exposure and efficacy.

Several commonly used therapeutics have been found to inhibit in vitro UDP-glucuronosyl transferase (UGT) activity. For example, the drugs tacrolimus, cyclosporine, and diclofenac are among the most potent (K_i values range from 0.033 to 7.9 μM) with probenecid, troglitazone, and naproxen being less potent inhibitors (K_i values range from 20 to 172 μM) [42]. The compound 7,7,7,-triphenylheptyl-UDP has been reported to be a mechanism-based inhibitor of UGT [43,44]. DDIs involving glucuronidation seem to be less prevalent than those identified for CYP450s possibly for the following reasons described by Williams et al. (2004) [45]. UGTs typically have much higher substrate K_m values (300 μM and often much higher) than those of CYP450s (K_m typically around 3 μM) and are usually metabolized by multiple UGTs. Given that the in vivo concentrations of most drugs are usually below 10 μM, UGT-metabolized drugs rarely saturate their own metabolism. This along with the fact that K_i values for most UGT inhibitors are usually > 10 μM leads to the conclusion that, in general, the AUC_i/AUC ratio will be relatively low even in situations where the fraction of the drug metabolized by a single UGT is high. Consequently, as further pointed out by Williams et al. (2004) [45], DDIs involving UGT result in exposures that are 2-fold or less of substrates in the presence of coadministered UGT inhibitors. In turn, DDIs involving glucuronidation that result in toxicity are rare but have been observed. For example, lamotrigine coadministered with valproic acid increases the incidence of skin rash, which is a known side effect of lamotrigine [46]. A number of intrinsic and extrinsic factors are known to affect drug glucuronidation in humans, including age, cigarette smoking, diet, disease state, ethnicity, genetic factors, hormonal factors, and interaction with other drug therapies [47].

5.2.2 Mechanism-Based P450 Inactivation DDI

Mechanism of DDI-mediated P450 inhibition as discussed previously can be reversible or irreversible and always results in reduction of intrinsic clearance of the pathway that is inhibited [48]. Reversible inhibition, also known as direct inhibition, can be typically classified as competitive, uncompetitive, noncompetitive, or mixed, with competitive inhibition being the most commonly observed pathway of inhibition. Reversible inhibitors bind to enzymes through weak, noncovalent interactions such as hydrogen bonds, hydrophobic interactions, or ionic bonds. The sum of the multiple weak interactions between the inhibitor and the enzyme active site results in strong, specific but still reversible binding [49]. In contrast, irreversible or mechanism-based inhibitors (MBIs) cause enzyme inactivation through covalent or quasi-irreversible modification of the enzyme structure.

Many clinically significant PK-related DDIs result from impairment of metabolic clearance via MBIs of CYP enzymes. Catalytic bioactivation due to MBIs of a drug candidate by an enzyme is a type of inhibition that increases over time. When in vitro method was used to investigate this DDI mechanism, preincubation of enzyme and potential inhibitor (drug candidate) is conducted. MBIs of CYPs have been extensively investigated, and the presence of functional groups [50] such as aniline, nitrobenzene, hydrazine, benzyl/propargyl/cyclopropyl amine, hydantoin, thioureas, thiazole, furan, thiophene, epoxides, methylene dioxy, methyl indoles, alkyne, isothiocyanate, and terminal alkenes, on a new chemical entity (NCE) warrants immediate and early assessment of inactivation potential of the NCE to avoid severe DDI liability in late-stage development. When an NCE possesses a structural alert as those listed above, it is not implying that it will be a potent inhibitor. Distinguishing an MBI from a simple reversible inhibitor is critical in predicting a clinical DDI, since applying a reversible inhibition model to an MBI may result in significant underprediction of a DDI risk. This can be readily appreciated from an examination of the strong and moderate inhibitors of the major human DME CYP3A identified in the 2012 FDA draft guidance on drug interaction studies [39]. It has been found that 75% of identified clinically significant CYP3A inhibitors are either established or putative mechanism-based inactivators of the enzyme; they can be either food products such as grapefruit juice or be prescription drugs spanning several therapeutic classes, including antiretroviral agents (e.g., ritonavir, saquinavir), antibiotics widely used in general practice (e.g., clarithromycin, erythromycin), the calcium channel blockers diltiazem and verapamil, the antidepressant agent nefazodone, or anticancer agents, such as tamoxifen [50]. Shown in Figure 5.2 are example MBIs of drugs and chemicals by P450 enzymes such as CYP3A4, CYP2C, 2D6.

5.2.2.1 Translating the In Vitro Information to Clinical Pharmacology Investigation Several approaches were reported to accurately define the MBI and when it clinically becomes significant. Li et al. (2011) [51] modify the classic P450 IC_{50} shift assay for more accurately screening CYP3A TDIs. In contrast to the regular IC_{50} shift assay, in which only one pair of P450 inhibition curves is generated, the modified method generated two pairs of inhibition curves, one pair of curves created from human liver microsomal incubations with the test article in the presence or absence of NADPH (same as the traditional assay), and the other pair created from new microsomal incubations with extract (compound/metabolites) of previous incubations. To assess the true CYP3A time-dependent inhibition, the authors propose a new parameter, the vertical IC_{50} curve shift (VICS), represented by vertical shift difference between the two sets of curves divided by inhibitor concentration at which maximal vertical shift of curves \$ + / − \$NADPH is observed. As has been indicated, a shift in the curves \$ + / − \$NADPH could mean a time-dependent inhibition or formation of a more active inhibitory metabolite(s). This proposed approach promises a more reliable characterization of the shift as a result of a true TDI- or metabolite-mediated reversible inhibition. Nine known TDI drugs were evaluated using this refined shift assay. The authors showed that derived VICS values correlated well with the reported K_{inact}/K_I values derived via

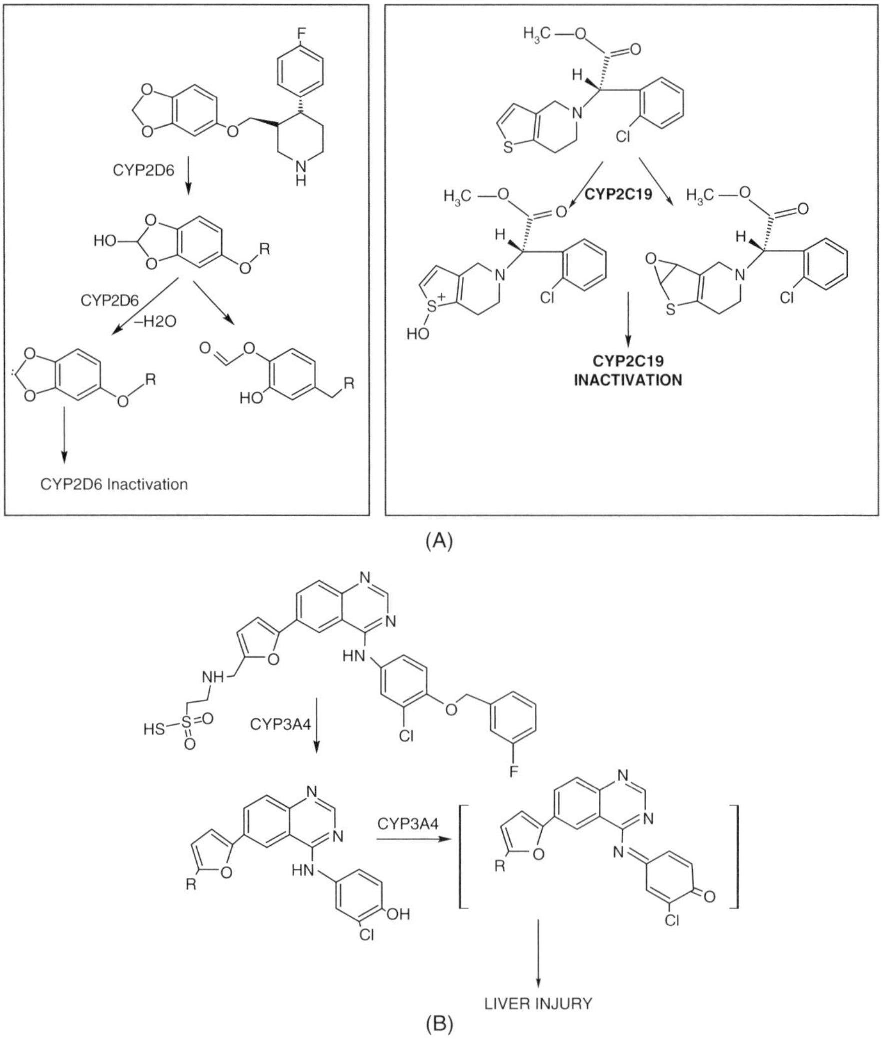

Figure 5.2 (A) Example of mechanism-based inhibition by CYP2D6 and CYP2C19. (B) Mechanism-based inhibition by CYP3A4 implicated in liver toxicity and injury.

the conventional dilution assay method [51]. Thus, the refined assay can be used to identify a true TDI and quantitatively assess the inactivation potential of TDIs in a high throughput fashion and can be invaluable to screen for true P450 TDIs in the early drug discovery. In a more recent review article by Orr et al. (2012) [50] that reviews MBIs of P450 enzymes, the authors discuss structure activity relationships (SARs) and discovery strategies to mitigate DDI risks of adverse, sometimes fatal, events in patients on multiple drug therapies that significantly are due to PK DDIs leading to elevated exposure to drugs with toxicity and eventually

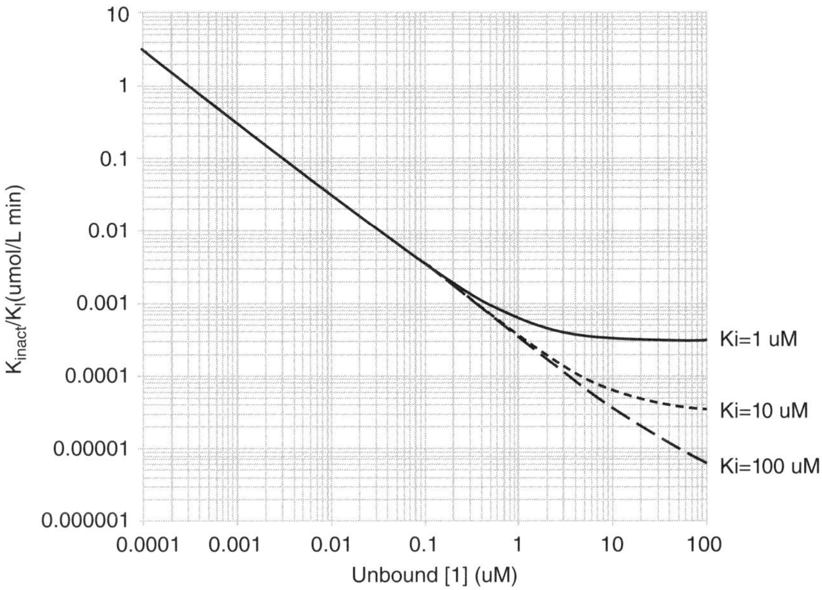

Figure 5.3 Relationship between K_{inact}/K_I and [I]. Adapted from Ref [50] with permission.

leading to withdrawal from the U.S. market (e.g., mibefradil in 1998). Irreversible inactivations generally involve metabolism of inactivators to reactive metabolites, which covalently modify the P450 enzymes and can result in loss of P450 activities. While the IC_{50} shift is able to provide early assessment in candidate selection stage for those that are potent inactivators, the precise prediction of in vivo DDI risks is not as simple of a task as had been described by others. A quantitative relationship between the magnitude of the IC_{50} shift and clinical DDI risk on human PKs was attempted by several pharmaceutical and academic investigators for tens of marketed drugs that identified as CYP3A4 inhibitors that were TDI positive based on IC_{50} shift cutoff 1.3-fold. Sekiguchi et al. (2009) [52] established a relationship between the observed IC_{50} shift, the ratio of unbound inactivator concentration at the steady state to competitive IC_{50}, and subsequent DDI risk. Also, for a relationship between the parameter K_{inact}/K_I, as shown in Figure 5.3, and estimated unbound plasma concentration of inactivator [I], the risk assessment can be made relative to an expected 2-fold interaction with lines at K_I value of 1, 10, and 100 μM. Compounds that fall below the line of 2-fold interaction would have low risk for DDI. Recently, good correlation has been reported for data of IC_{50} shift and K_{inact}/K_I assays and that obtained from K_{obs} for hundreds of CYP3A4 TDI drugs and compounds at single inactivator concentration of 10 μM [53]. The authors established that compounds that flagged as TDI positive are those that have K_{obs} with lower limit of 0.020 min^{-1}.

This approach, as indicated by Orr et al. (2012) [50], can be used in early discovery, once the risk is identified, to modify the structure of leads as an attempt to mitigate the risk prior to candidate selection.

For prediction of clinical DDI for MBI-TDI as identified in the late discovery stage, the strategy is to move forward with compounds that may demonstrate targeted efficacy at concentrations below the relevant DDI.

There have been numerous publications related to strategies for using in vitro CYP TDI data for prediction of clinical DDI, with approaches ranging from very simple (three input parameters) to more complex (four or more input parameters) [54–56]. In vitro–in vivo prediction approaches have in common three basic inputs: a measure of potency of inhibition (e.g., K_I), a measure of rate of inactivation (e.g., K_{inact}), and a measure of in vivo drug concentration (e.g., [I]). Approaches to predicting DDIs can be categorized into two general types, static and dynamic. Static approaches assume that the concentration of inhibitor does not change over time, while dynamic approaches incorporate changes in inhibitor concentration with time and may also incorporate other system dynamics.

For the static model, Mayhew et al. (2000) [54] adapted from Equation (5.4). The well-stirred model was adapted for extrapolation of in vitro intrinsic clearance (CL_{int}) to in vivo AUC, in addition to terms describing the effect of the CYP inactivator ($K_I$and K_{inact}) on the degradation rate of the enzyme. This relatively simple model requires one constant (K_{deg}, the degradation rate constant for the CYP isoform of interest) and three variable input parameters (in vitro K_I, K_{inact}, and [I]), as shown below, that may be used to determine the change in AUC of a probe substrate in the presence of TDI using the static model:

$$\frac{AUC'}{AUC} = \frac{CL_{int}}{CL'_{int}} = \frac{[E]_{ss}}{[E]'_{ss}} = \frac{1}{\dfrac{k_{deg}}{k_{deg} + \dfrac{[I] \times k_{inact}}{[I] + K_I}}}, \tag{5.5}$$

where AUC and AUC′ represent AUC of the object drug in absence and presence of inactivator, respectively; CL_{int} and CL'_{int} represent intrinsic clearance of object drug in absence and presence of inactivator, respectively; and [E]ss and [E]′ss represent concentration of clearing enzyme in absence and presence of inactivator, respectively. This model was validated by accurately predicting clinical DDI for three CYP3A4 inhibitors, such as clarithromycin, N-desmethyl diltiazem, and fluoxetine, after coadministration with CYP3A4 substrates. Using Equation (5.5), K_I of 0.04 min^{-1}, K_I of 2.4 μM, [I] of unbound plasma concentration of telithromycin at steady state of 0.17 μM and K_{deg} of 0.0005 min^{-1} predicted a 6.2-fold increase in AUC of midazolam in the presence of the tested inhibitor, which agreed well with the observed 6-fold increase of the substrate AUC in DDI study of midazolam and telithromycin in clinic. Some modification of the model described in Equation (5.5) was made to improve the prediction by incorporating the fraction of substrate metabolism by enzyme under inhibition by the tested inactivators and also by including the fraction of drug that escapes gut in the absence of inactivator, as reported in several recent publications [55,56]. It is worth noting that these static-based models [as shown in Equation (5.5)] address the extent of DDI under steady-state (equilibrium) conditions for average systemic blood/plasma concentration for specific human populations. However, when addressing variability of DDI among human populations, the dynamic model approach using physiologically based pharmacokinetic (PBPK) modeling and software can come into play, as discussed elsewhere in this book. Briefly, the dynamic approach allows one to link pharmacokinetics/pharmacodynamics (PK/PD) time course of predicted precipitant and liver and intestine exposure to the known P450 interaction mechanism. The PBPK software was developed to enable modeling and simulation of human PK, tissue distribution, and ADME variability in the presence of inactivator across human populations [57].

After reviewing these various approaches to determine MBI-TDI, the best approach to be selected in predicting the clinical DDI depends on the molecule under evaluation.

- For lead optimization, K_{inact}/K_I ranking may be appropriate.
- For candidate selection, quantitative estimates for DDI potential for positive MBI such as a static [Equation (5.5)] or PBPK dynamic model.

Finally, in consideration of using in vitro systems for MBI kinetic studies, in vitro MBI can be performed using recombinant expressed CYP enzyme isoforms, human liver microsomes (HLMs), or hepatocytes, though it should be noted that MBI kinetic parameters estimated using certain recombinantly expressed CYP preparations may not be reflective of those estimated using native HLMs [58]. Thus, caution should be exercised in their use for DDI risk assessment, as has been demonstrated for CYP3A inhibitory DDIs produced by macrolide antibiotics [59]. Better IVIVE based on in vitro inactivation kinetic parameters determined in primary human hepatocytes compared to HLM has been recently reported [60]. Nevertheless, when HLM is pooled to represent the "population average" distribution of enzymes, it is relatively convenient, and that introduced an opportunity for "standardization" of an experimental system as the most commonly utilized system for studies of MBI for DDI risk assessment and for estimation of kinetic parameters of inactivation, in the expert opinion of the Pharmaceutical Research and Manufacturers of America (PhRMA) [61].

However, if substantial non-CYP- and/or nonmicrosomal- mediated metabolism of the new molecular entity (NME) is expected, and if such metabolism modifies the inactivation effects of the NME, data from human hepatocytes rather than HLM is considered to be more translatable to the clinical setting, as has recently been demonstrated in studies for gemfibrozil, bupropion, and ezetimibe [62]. Clinically relevant inhibition of CYP2C8 by gemfibrozil through MBI of the enzyme by its acyl glucuronide was observed by using human hepatocytes. Also, clinically relevant metabolism-based inhibition of CYP2D6 by bupropion was produced of the enzyme by its metabolites erythro- and threo-hydrobupropion. As hepatocytes represent a more complete biotransformation system compared to HLM, potent inhibition of CYP2C8 and CYP2D6 by gemfibrozil and bupropion, respectively, is readily observed following preincubation of hepatocytes with these drugs but is not observed using HLM as the in vitro system, where direct evaluation of the metabolites is necessary to observe potent enzyme inhibition. In contrast, CYP3A4-mediated MBI of ezetimibe was determined in HLM, whereas when using hepatocytes as the in vitro system that was not detected, because ezetimibe represents an example of glucuronidation-dependent protection against metabolism-dependent inhibition of CYP3A4 [63]. Consequently, when drug candidates are extensively metabolized by non-CYP enzymes, it is recommended to use human hepatocytes in addition to HLM (or recombinant enzymes) to evaluate their abilities to inhibit CYP enzymes.

It is worth mentioning, as discussed previously, that considering the relatively large microsomal concentrations generally used in the preincubation step in in vitro kinetic studies of MBI, nonspecific microsomal binding can be significant. It may introduce bias in K_I estimates and jeopardize their applicability in subsequent IVIVE of DDI magnitude. Therefore, correction of apparent K_I estimates for microsomal binding can be important in getting unbiased estimates of inactivator.

5.2.3 DDI Mediated by P450 Induction

In addition to inhibition of DME-mediated DDI, induction of DMEs is also a key mechanism of clinically significant DDI. Induction that increases the content of enzyme responsible for the metabolism either by increasing its expression or protein stabilization can result in an increase in intrinsic clearance of metabolism by the induced enzyme. This process usually leads to decreased systemic exposure of one drug that is a substrate for the enzyme that is induced by another drug. Although this typically manifests itself as a reduction in therapeutic efficacy (due to decreased exposure) and increased dosage requirements, as in the case of increasing cyclosporine A dose in patients taking St. John's wort, which is an inducer of CYP3A4 responsible for the metabolism of cyclosporine A. In autoinduction, one drug induces its own metabolism, causing a drug-induced clinical event. In the setting of induction, increases in metabolism can also result in an alteration of the safety profile when dealing with drugs with active and/or toxic metabolites. Unlike reversible inhibition, induction DDIs are time dependent in their onset and offset, complicating their clinical management both after initiation of treatment with an inducer and in the deinduction period following cessation of treatment with the inducer.

5.2.3.1 In Vitro P450 Induction Models and Methodologies The most common model currently in use and recommended by the FDA guideline (2012) [39] to assess potential DDIs due to P450 induction is the in vitro P450 induction model. Induction of drug clearance can occur by increasing expression through increasing the transcription of DMEs and transport proteins (discussed in the next section of this chapter) by the inducer. Therefore, assessing the potential DDI mediated by induction should require the use of whole cell systems with intact transcriptional and translational machinery, in contrast to inhibition DDI studies that can be assessed by using subcellular fraction as HLMs. The major molecular mechanisms of induction DDI involve binding to and activation of receptors that are involved in the regulation of transcription of genes encoding DMEs and/or transporters [64,65]. Although many such receptors and transcriptional mechanisms have been identified as regulators of gene expression, the three receptors that mediate gene transcription that is induced in the majority of clinically relevant induction DDIs include (1) the aromatic hydrocarbon receptor (AhR), (2) the constitutive androstane receptor (CAR), and (3) the pregnane X receptor (PXR).

The most current and reliable in vitro system, which considered as the industrial standard and is recommended by the FDA, is the cultured human hepatocyte, either as primary cultures of fresh human hepatocytes or as attachable cryopreserved hepatocytes, as reflected in the recently published perspective of the PhRMA [66]. It is recommended for in vitro induction studies in human hepatocytes that the potential for an NME to produce induction of CYPs 1A2, 2B6, and 3A4 be evaluated, as the genes encoding these enzymes are considered to be representative sensitive targets that respond to induction via the AhR, CAR, and PXR, respectively. Significant overlap and cross talk between the CAR and PXR systems is well established, with coinduction of CYPs 2B6 and 3A4 by prototypic inducers such as rifampin.

Accordingly, it was originally considered to be sufficient to test NMEs for potential for induction of CYPs 1A2 and 3A4, to enable risk assessment for induction DDIs that may result from increased expression of AhR and CAR/PXR target genes, respectively. On the basis of recent findings of the molecular mechanisms of induction, it is now generally accepted that a comprehensive and definitive in vitro evaluation of risk for a new drug to produce DDIs via CYP/transporter induction should include analysis of the potential for induction of CYPs 1A2, 2B6, and 3A4 in human hepatocytes [66], or in an appropriately qualified cell line that maintains inducible regulation via the AhR, CAR, and PXR mechanisms as also noted in the 2010 EMA draft DDI guideline [33]. Although the primary human hepatocyte is recognized as the gold standard in vitro system for induction DDI risk assessment, limitations do exist in the availability of high-quality human hepatocytes as well as limitations in assessing interindividual variability in responsiveness to inducer as reproducibly performing biological reagents for in vitro induction assays. There are challenges related to supply limitations and the intermittent availability of fresh human hepatocytes that are substantially offset by the availability of cryopreserved human hepatocytes. Establishment of experimental approaches and biological reagents for in vitro induction studies with reproducible performance characteristics, for example, the use of human hepatocytes or hepatocyte-derived cell lines, has received considerable attention.

Successful use of the immortalized hepatocyte cell lines such as Fa2N-4 has been described in the in vitro assessment of enzyme induction for PXR target genes (e.g., CYP3A4, MDR1, CYP2C9) and AhR target genes (e.g., CYP1A2) [67], with promising in vitro–in vivo correlations to enable induction DDI predictions [68]. However, since this immortalized hepatocyte cell line does not express CAR, induction of CYP2B6 is not observed in response to CAR-selective inducers, and induction of CYP3A4 in Fa2N-4 cells following treatment with CAR-selective inducers is either blunted or absent when compared to cryopreserved human hepatocytes [69,70]. Therefore, Fa2N-4 cells may represent a good surrogate for primary human hepatocytes for evaluating the potential for an NME to produce AhR- or PXR-mediated induction, but not CAR-mediated induction, limiting their utility as a definitive in vitro model for CYP induction studies in a drug development setting [66]. Other cell lines, such as the differentiated hepatoma cell line HepaRG, appear to be a promising surrogate for human hepatocytes. Recent studies have demonstrated expression, activity, and inducibility of CYPs 1A2, 2B6, and 3A4 [71,72] consistent with expression of AhR, CAR, and PXR under optimally cultured conditions [73]. Unlike Fa2N-4 cells, where CYP2B6 induction is not observed following treatment with CAR-selective inducers, induction of CYP2B6 by CAR-selective inducers, such as phenytoin or phenobarbital, has been observed in HepaRG cells [72]. Additionally, an excellent correlation between parameters characterizing the maximum magnitude, potency, and efficiency of inducers (E_{max}, EC_{50}, and E_{max} : EC_{50} ratio, respectively) in primary human hepatocytes versus HepaRG cells has also been described [74].

In general, it is required to conduct extensive validation before conducting DDI-mediated induction for risk assessment either when using the primary hepatocytes or cell lines to qualify induction of CYPs 1A2, 2B6, and 3A4. The validation testing requires using appropriate positive controls that measure regulation of all three pathways (AhR-, CAR-, and PXR-mediated induction) and measure the inter- and intravariability among at least three lots (for human hepatocytes) or six preparations for cell lines.

Assays that measure binding to nuclear receptors can be useful in the drug discovery setting due to their relatively higher throughput compared to definitive hepatocyte-based induction assays. Furthermore, nuclear receptor assays can be useful in elucidating the molecular mechanism of the observed induction in hepatocytes and distinguishing the mechanism of induction, for example, PXR versus CAR activation. However, since the mechanisms of induction are complex and involve multiple nuclear receptors with cross talk between the mechanisms, these assays alone are not considered sufficient to support clinical DDI risk assessment.

In addition to the appropriate choice of model, an appreciation of induction methodology and data interpretation is required to determine an accurate risk assessment of potential DDIs in clinic. The general methodology with in vitro induction studies in human hepatocytes, or qualified cell lines with adequate applicability for drug investigation, involves treatment with the new drug candidate for 2–3 days, with daily replacement of the culture medium containing three concentrations selected within a range including the therapeutic systemic concentration (x), and two order of magnitude higher (10x, 100x, and 1000x—only if drug solubility is granted) to assess concentration response. As mentioned, the studies are conducted in hepatocytes from at least three different donor livers to determine the interindividual difference in response. At the end of the incubation period, the endpoints typically measured include (1) activity of CYPs 1A2, 2B6, and 3A4 using isoform-selecting reactions; and/or (2) mRNA expression of CYPs 1A2, 2B6, and 3A4 using techniques such as reverse transcription–polymerase chain reaction (RT–PCR) [66]. It has been reported that many CYP3A4 inducers are additionally TDIs of the enzyme's activity. It is important to measure mRNA in addition to enzyme activity assessments to aid appropriate mechanistic interpretation of the results and translate the potential DDI to the clinical setting, thus ensuring appropriate clinical DDI study design as indicated by the PhRMA perspective [66]. In addition, both the FDA 2012 guideline and EMA 2010 guideline mandate the inclusion of mRNA measurement in in vitro induction studies for the interpretation of study results [33]. A classical example of a drug producing complex simultaneous inhibition/induction effects is the HIV protease inhibitor ritonavir, with in vitro induction studies showing increases in mRNA expression of CYP3A4 due to PXR-mediated induction and decreased CYP3A4 activity due to time-dependent inactivation of the induced enzyme [75]. The clinical pharmacologic picture is consequently characterized by complex dose- and time-dependent interactions that are additionally dependent on the PK properties of the substrate drug. For example, in the case of the CYP3A substrate alprazolam, the net effect was a clinically significant level of inhibition of oral clearance (~ 2.5 – fold increase in AUC) following short-term low-dose treatment with four doses of 200 mg ritonavir administered BID, considered to be representative of a dosage schedule that may be used to initiate treatment with ritonavir [76]. In contrast, a mild and clinically insignificant level of induction of alprazolam oral clearance (12% decrease in AUC) was observed as the net effect following a 10-day treatment with the usual therapeutic dose of 500-mg BID of ritonavir as reflected in the ritonavir USPI [77]. Depending on the substrate drug's PK properties, the outcome of multiple-dose treatment with therapeutic doses of ritonavir can be variable, with substantial impairment of oral clearance (i.e., net inhibition rather than induction) observed even in the setting of a week-long treatment with high dose (500 mg BID) ritonavir for substrates such as sildenafil [78]. In these cases of DDI, it is not that simple to predict from the in vitro results alone, but it is important to measure both mRNA and activity in the in vitro induction studies, especially when there is a potential for concurrent inhibition. It is important to recognize that the use of CYP3A4 activity and/or mRNA measurement as an endpoint in in vitro induction studies is to serve as a sensitive marker of PXR-mediated induction; consequently if CYP3A4 induction is not observed in an in vitro induction study in human hepatocytes following treatment with an NME at concentrations up to ten times the mean systemic C_{max} at clinically relevant doses, one can not only conclude that the risk for the NME to produce DDIs via CYP3A induction is low but also that the risk for it to produce DDIs via induction of other coinduced PXR targets such as CYPs 2C8, 2C9, 2C19, and MDR1 P-gp is also low. With an NME that has properties such as ritonavir (i.e., PXR-mediated inducer and MBI of CYP3A), if activity data alone were measured and used in risk assessment, it could result in a false negative risk assessment for induction DDIs. The risk for induction DDI is low but may still be valid for potential interactions with CYP3A substrates, since the net effect of inhibition may likely predominate and the TDI studies would lead to a clinical DDI evaluation of the effect of the drug candidate on the PK of a sensitive CYP3A substrate to assess clinical relevance. However, a bigger impact is on risk assessment for induction of non-CYP3A targets of PXR, where a low risk for induction DDI with substrates of such enzymes or transporters (e.g., digoxin, warfarin) may be erroneously concluded in the absence of mRNA measurements that would be required for observing induction of gene expression, which would not be reflected by CYP3A4 activity measurements alone. Based on these clinical investigations, in order to minimize any potential DDI, it is critical that to measure mRNA as lack of CYP induction can be unbiased by factors such as concurrent inhibition or inactivation that could jeopardize the ability to pick up induction potential of drug candidates.

A recent analysis across CYP3A4 inducers showed that the maximum fold increase in CYP3A4 mRNA expression observed in cryopreserved human hepatocytes was generally greater than the corresponding maximum observed fold increase in CYP3A4 catalytic activity [79], similar to a previous study using fresh human hepatocytes [74], indicating that inhibition of CYP3A4 occurs by the inducer. It is important to point out that, when using mRNA expression of CYP3A4, one can evaluate the risk of induction by both PXR and/or CAR. This was revealed from the higher fold of induction on the mRNA of CYP3A4 by CAR-selective inducers such as phenytoin and phenobarbital observed than the fold induction in mRNA of the other PXR/CAR-regulated DMEs examined (CYPs 2B6, 2C9, 2C19, 3A5) [79].

5.2.3.2 Translating In Vitro P450 Induction Data to Clinical DDI The translation of the results of in vitro induction studies performed in human hepatocytes (or adequately qualified cell lines that are responsive to PXR-, CAR-, and AhR-mediated induction, as discussed earlier) to clinical PK DDI risk assessment is a growing area in the process of drug development and safety. As described in assessing the magnitude of risk assessment with inhibition-mediated DDI, empirical cutoffs for fold increase in mRNA or activity over vehicle control (usually dimethyl sulfoxide [DMSO]) and for the percentage of the observed induction by a strong inducer positive control (e.g., rifampin for CYP3A), tested alongside in the same experimental system with the tested drug, have been used for classifying in vitro positives versus negatives for induction. For instance, observation of > 40% of the rifampin as positive control for CYP3A means classifying the new drug candidate as an in vitro positive inducer for CYP3A4 and introduces the need for further clinical evaluation of DDI risk and vice versa [80], as shown in Equation (5.6):

$$\frac{(E_{drug} - E_{DMSO})}{(E_{positive} - E_{DMSO})} \times 100 = \%\ \text{induction of positive control.} \tag{5.6}$$

As mentioned earlier, the current FDA 2012 draft guidance recommends evaluating mRNA levels of CYP1A2, CYP3A4, and CYP2B6, followed by CYP2C9 in cases where CYP3A4 induction is observed [81]. However, caution should be used when the analysis depends on the mRNA (or activity) approach, as the data are not based on mechanistic study design and formal statistical optimization of the cut-off values used in data interpretation. However, when there is no change, or more than a 40% increase in enzyme activity and mRNA expression relative to positive control inducer, the results are easy to interpret.

Recently, 20 clinically significant CYP3A inducers and 15 noninducers in cryopreserved human hepatocytes were used to examine the in vitro–in vivo correlation based on only mRNA measurement [79]. This retrospective analysis concluded that a cutoff of 4-fold increase (not > 40%) in mRNA expression provided 98% sensitivity while maintaining a specificity of $\sim 70\%$. The cutoff of 40% of the positive control's effect was an unacceptably low sensitivity (i.e., high false negative rate) [79].

For determining the magnitude of induction prediction, the concentration-effect relationship for the observed in vitro increase of mRNA and/or increase in enzyme activity is typically accomplished using an E_{max} model from sigmoidal nonlinear regression fit of the log concentration of inducer against % maximal induction (e.g., IC_{50} plot):

$$E\,(\%) = \frac{(E_{max} \times C_{\gamma})}{(EC_{50\,\gamma} + C_{\gamma})}. \tag{5.7}$$

IVIVE is then performed assuming that the fold increase in enzyme expression or activity estimated from the in vitro concentration-effect relationship at the in vivo–relevant concentration of the inducer directly translates to fold increase in intrinsic clearance of metabolism via the induced enzyme in vivo, as reported [82].

For an orally administered drug, the following equation mechanistically allows the translation of the estimated in vitro concentration-effect relationship to a predicted interaction magnitude:

$$\frac{\text{AUC induced}}{\text{AUC Control}} = \frac{1}{\left(1 - f_{m\,(CYPi)} + \left(f_{m\,(CYPi)} \times \left(1 + \dfrac{E_{max}\,[IND]}{EC_{50} + [IND]}\right)\right)\right.}. \tag{5.8}$$

With enzyme activity as the endpoint in the in vitro induction assay in fresh human hepatocytes, use of unbound systemic steady-state maximum concentration of the inducer to estimate [IND], use of unbound EC_{50} of the inducer in vitro [calculated from the in vitro induction EC_{50}, as shown in Equation (5.7)], and the measured unbound fraction of the inducer in hepatocytes, Equation (5.8) has been successfully used to predict the magnitude of CYP3A induction DDIs [83], using enzyme activity and not mRNA expression as the in vitro endpoint. It was found with using the latter as endpoint, the performance of IVIVE was poorer, with a trend for overprediction of the magnitude of clinical DDIs, explained by a higher E_{max} for mRNA expression compared to enzyme activity. The higher fold of increase in mRNA due to induction (max induction) may not quantitatively equal the fold increase in enzyme content, as discussed earlier, since certain CYP3A inducers also produce some level of inhibition of the enzyme's activity either via reversible or TDI. Measuring mRNA endpoint is a true induction effect unbiased by effects of the inducer on enzyme activity, the IVIVE of which should ideally be performed using approaches for inhibition DDIs. When mRNA is used, fold increase in mRNA and fold increase in enzyme content/activity cannot be assumed. Thus, there is a need for introducing an empirical adjustment factor on E_{max} in Equation (5.8) that can be derived by calibrating approach using positive controls within the same experimental design [84].

In clinic, an induction-mediated DDI by a new drug can be significant even if the induced enzyme modestly contributes ($f_{m\,(CYPi)} < 25\%$) to the total clearance, which is in contrast to the inhibition-mediated DDI when an enzyme is a minor contributor (e.g., $f_{m\,(CYPi)} < 25\%$) to total clearance of the drug and the magnitude of an inhibitory

DDI will be relatively trivial or even complete inhibition. Consequently, the relative contribution of induced enzymes can be very small, for the risk for induction DDI is low (e.g., resulting in a < 25% decrease in exposure). It has been reported that even if the contribution of CYP3A to the overall clearance of an object drug is only 25%, the enzyme-inducing phenytoin and carbamazepine can be expected to produce ~ 40 – 60% decrease in exposures of the target drug. Also, the strong CYP3A inducer rifampin produces ~ 65% decrease in the target drug exposure [85].

5.3 INCIDENCE OF DDI DUE TO DRUG TRANSPORTERS

As discussed in previous chapters, the significant contributions of drug transporters is in the ADME processes and dispositions of new drugs [86–88]. The critical role of drug transporters in ADME and safety depends on their expressions and functions at major body organ transporters, such as small intestine, liver, and kidney [89–95], while transporters expressed in brain and placenta play a major role in drug distribution and protection from potentially toxic compounds in and off the brain and fetal tissues, respectively.

The expression, localizations, and functions of the various transporters are summarized in Figure 5.4 and Table 5.2. The functions, and thus their contributions to DDI of drug transporters, can vary depend on the expression patterns: tissues and localizations. For example, efflux transporters such as P-gp and the breast cancer resistance protein (BCRP), and the multidrug resistance protein 2 (MRP2) are localized

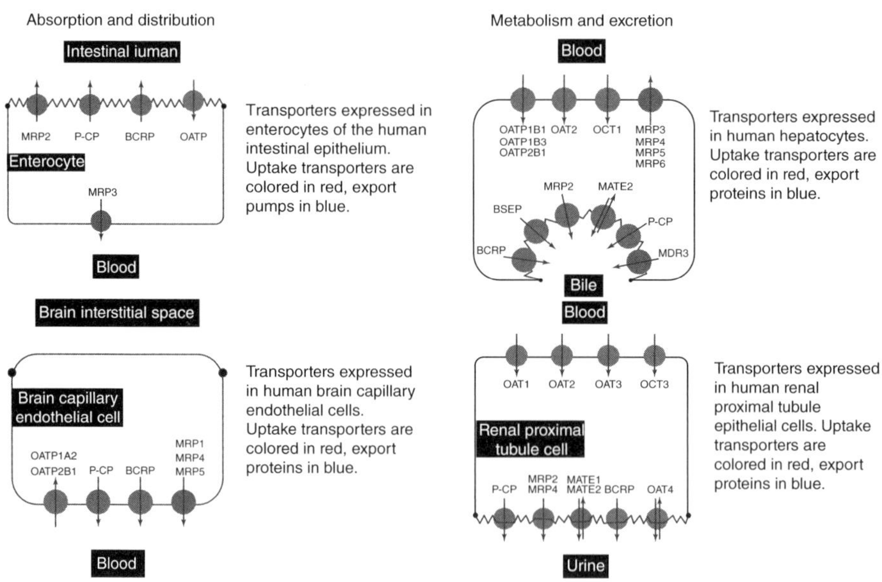

Figure 5.4 Drug transporters, localizations, and functions in major body organs influencing drug disposition. Adapted from Ref [90], [91] with permission.

TABLE 5.2 Adverse Drug Reactions of Selected Drugs in Association with Drug Transporter Function.

Drug	Adverse Drug Reaction	Transporter	Reference
Adefovir	Nephrotoxicity	OAT1 (uptake)	Ho et al. 2000
Bosentan	Cholestatic liver injury	BSEP (efflux)	Fattinger et al. 2001
Cidofovir	Nephrotoxicity	OAT1 (uptake)	Ho et al. 2000
Cisplatin	Nephrotoxicity	OCT2 (uptake)	Ciarimboli et al. 2010
Cyclosporine	Cholestatic liver injury	BSEP	Byrne et al. 2002
	Nephrotoxicity	P-gp	Zolk and Fromm 2011
Glibenclamide	Cholestatic liver injury	BSEP	Byrne et al. 2002
Irinotecan	Diarrhea	MRP2	de Jong et al. 2007
	Myelosuppression	BCRP	Cha et al. 2009
Loperamide	Respiratory depression	P-gp	Sadeque et al. 2000
Metformin	Hyperlactatacidemia, lactic acidosis	OCTs	Zolk 2012
Methotrexate	Nephrotoxicity	MRP2	Vlaming et al.,2009
Mycophenolate mofetil	Leucopenia, anemia, thrombocytopenia, diarrhea, nausea, vomiting, infection	OATP1B1	Michelon et al.,2010
Rifampin	Cholestatic liver injury	BSEP	Byrne et al. 2002
Simvastatin	Myopathy	OATP1B1 (uptake)	Brunham et al. 2012
Tacrolimus	Nephrotoxicity	P-gp	Zolk and Fromm 2011
Tenofovir	Nephrotoxicity	OAT1	Ray et al. 2006
Troglitazone	Cholestatic liver injury	BSEP	Funk et al. 2001

to the apical membrane of enterocytes (Figure 5.4), thereby limiting bioavailability of orally administered substrates, while their efflux functions at the apical canaliculi in liver or apical membrane in kidney can facilitate the elimination of drugs and metabolites to bile or urine, respectively. In the intestine, inhibition or induction of these efflux transporters by concomitantly administered drugs results in an increase or decrease, respectively, in the bioavailability of the victim drug, whereas in liver and kidney, inhibition or induction of these efflux transporters may reduce or enhance, respectively, the elimination of drug substrates, thus increasing the potential of drug-mediated toxicity or decreasing drug exposure, respectively (Figure 5.4).

5.3.1 DDI-Mediated Uptake Transporters

When discussing DDI in relation to drug transporters, the uptake transporters, organic anion transporting polypeptide (OATP, OATP1B1), that mediates statins drugs (see Table 5.2) into hepatocytes is one of the well-known clinically relevant adverse drug reactions.

Because OATP-mediated statin uptake in hepatocytes is required for their action as HMG-CoA reductase inhibitor, inhibition of this uptake may contribute to an increase in statin plasma concentrations with the risk of side effects such as myopathy and rhabdomyolysis. Multiple interaction studies [96,97] using statins as substrates for transporters have been conducted. A potent inhibitor of OATP-mediated statin transport is CsA. For OATP1B1-mediated pitavastatin the IC_{50} values of 0.24 μM [98] or rosuvastatin transport the IC_{50} values of 0.31 μM [98] have been calculated. Another study by Ho et al. (2006) [99] to determine the interaction between CsA and OATP indicated that CsA is a very potent inhibitor for the transport of rosuvastatin transport by OATP1B3 (IC_{50}value of 0.06 μM), and the effect of CsA in vitro can last hours even after eliminating CsA [100]. In a clinical setting, rosuvastatin and CsA interactions were investigated by Simonson et al. (2004) [101] in a study with patients who underwent heart transplantation and were taking 10 mg of rosuvastatin for 10 days. The AUC and the C_{max} increased 7.1- and 10.6-fold, respectively, over the control group that received rosuvastatin alone. By using in vitro models such as the oocytes expressed OATP1B1, the potency of CsA on inhibiting uptake of rosuvastatin was confirmed with an IC_{50} value of 2.2 μM [101]. The ability of CsA to inhibit the uptake of other statin drugs was revealed from the 4.5- and 6.6-fold increase of AUC and C_{max} by pitavastatin, respectively, in health volunteers after treatment with CsA [102]. Several in vitro studies using sandwich-cultured hepatocytes have indicated that rifampin is a potent inhibitor of OATP1B1 and OATP1B3 [103–105], confirmed in clinical study with healthy volunteers, where a single dose of rifampin (600 mg) raised the AUC of coadministered atorvastatin by more than 600% [106] by inhibiting its OATP-mediated hepatic uptake.

For bosentan, the endothelin receptor antagonist is metabolized by CYP2C9 and CYP3A4 and associated with OATP1B1 and OATP1B3 for its uptake [107]. Coadministration of CsA, ketoconazole, rifampin, or sildenafil increased the plasma concentrations of bosentan in clinical drug interaction studies [107,108]. Sildenafil inhibited OATP1B1- and OATP1B3-mediated bosentan uptake with IC_{50} values of 1.5 and 0.8 μM, respectively. Because sildenafil is not an inhibitor of CYP2C9 and CYP3A4, sildenafil inhibition of OATP1B1- and OATP1B3-mediated bosentan uptake may be the major determinant of this DDI. In general, for the interpretation of clinical DDI studies, it should be kept in mind that several inhibitors of OATPs (e.g., CsA) are also potent inhibitors not only of DMEs but also of other transport proteins (MRP2, P-gp).

In addition to OATP-mediated DDI, significant DDIs were observed in association with OAT inhibition in kidney [109], in particular OAT1 and OAT3, which facilitates the basolateral uptake of anionic drugs into the proximal renal tubular cells (Table 5.3 and Figure 5.4). Drugs associated with OAT-mediated DDI are those that identified as substrates for OAT, including drugs such as antibiotics (e.g., benzylpenicillin, K_m value of 54 μM [110,111]), antivirals (acyclovir and cidofovir, mainly by OAT1 as reported by Uwai et al. (2007) [112]), or H_2 -receptor antagonists [113]. A comprehensive list of OAT substrates and inhibitors was published in a recent review article [113]. For inhibition of OAT, both OAT1 and OAT3 are inhibited by probenecid, and because of its potency (K_i values of 4.3–12.1 for OAT1 and 1.3 – 9.0 μM for OAT3) it is rarely used in therapy but is used as a positive control for experimental

TABLE 5.3 Drug transporter, substrates, and in vitro/in vivo inhibitors with corresponding IC_{50}/K_i values.

Transporter	Selected Substrates	Selected Inhibitors	IC_{50} or K_i[a] μM
OATP1A2	Aliskiren	Naringin	75.5
	Fexofenadine	Apigenin	9.3
	Fexofenadine	Hesperidin	2.7
	Fexofenadine	Naringin	3.6–24.2
OATP2B1	Fexofenadine	Apigenin	13.9
	Rosuvastatin	Cyclosporine	0.07
	Rosuvastatin	Gemfibrozil	8
OATP1B1	Cerivastatin	Gemfibrozil	72
		Gemfibrozil-1-O-β-glucuronide	24
	Pitavastatin	Cyclosporine	0.24[a]
	Pitavastatin	Clarithromycin	8.26[a]
	Rosuvastatin	Gemfibrozil	25
	Rosuvastatin	Cyclosporine	0.31
OATP1B3	Pravastatin	Erythromycin	40% Inhibition by 10 μM
	Rosuvastatin	Cyclosporine	0.06
NTCP	Rosuvastatin	Cyclosporine	0.37
	Rosuvastatin	Gemfibrozil	23
OAT3	Methotrexate	Indomethacin	0.61–0.98
	Fexofenadine	Probenecid	1.3[a]
OCT1	Metformin	Cimetidine	104[a]
OCT2	Metformin	Cimetidine	124[a]
MATE1	Metformin	Cimetidine	3.8[a]
MATE2-K	Metformin	Cimetidine	6.9[a]
P-gp	Digoxin	Clarithromycin	4.1
	Digoxin	Cyclosporine	1.3
	Digoxin	Quinidine	2.2
	Digoxin	Telithromycin	1.8
	Digoxin	Verapamil	2.1

Note: [a]K_i value.

OAT-mediated DDI studies [114,115]. When the supplies of benzylpenicillin were low after World War II, probenecid was used to decrease its renal tubular secretion, which lead to an increase in serum concentrations [116]. Currently, probenecid has been used with other drugs to improve the PK of drugs that depend on OAT in their dispositions or to minimize OAT-mediated nephrotoxicity; for example, with methotrexate elimination [117], cidofovir [118], and furosemide [119].

In addition to probenecid, gemfibrozil and its metabolites were identified as potent inhibitors for OAT [120]. In vitro and in vivo studies showed that pravastatin uptake, which associates with OAT3, is inhibited by gemfibrozil. In healthy volunteers, gemfibrozil increased pravastatin plasma concentrations by inhibiting its hepatic uptake [121], and it is likely that gemfibrozil inhibition of OAT3 caused the decrease in renal pravastatin clearance by 43% [121,122].

Several in vitro and in vivo investigations reported DDI cases that associated with members of the OCT family and the multidrug and toxin extrusion protein (MATE) family in kidney and liver, as listed in Table 5.3. The most intensively investigated drug as substrate of OCTs and MATEs is the antidiabetic metformin. Because of its high pK_a (> 99.99% positively charged at pH 7.4) and its negative logP value, passive diffusion of metformin through cellular membranes is minimal, therefore its transport by drug transporters is pivotal for the permeation of metformin through cellular membranes [123,124], and selected inhibitors and DDIs are shown in Tables 5.2 and 5.3.

Most DDIs observed in humans that are attributed to inhibition of OCTs and/or MATEs are caused by cimetidine. In clinical study, Somogyi et al. (1987) [125] examined the interaction between cimetidine and metformin, which eliminated in humans mainly by renal excretion of the unchanged substance. Its renal clearance exceeds glomerular filtration rate by several-fold [126]. Coadministration of cimetidine (400 mg twice daily) in healthy subjects increased the metformin (250 mg once daily) C_{max} and $AUC_{0-24\,hours}$ by 81% and 50%, respectively, and decreased its renal clearance over 24 hours by 27% [125].

5.3.2 DDI-Mediated Efflux Transporters

Like DDI-mediated uptake transporters, numerous cases of adverse drug reactions were implicated with DDI-mediated efflux drug transporters (see Tables 5.2 and 5.3). The most recognized efflux drug transporter involved with DDI in humans is P-gp. The interactions were frequently seen with P-gp substrates, such as cardiac glycoside digoxin, human immunodeficiency virus protease inhibitors, immune suppressants, β blockers, and anticancer agents as reported by Ho and Kim (2005) and Shitara et al. (2005) [127,128]. The drugs transported by P-gp are usually hydrophobic molecules with cationic properties [86] and can either not be metabolized in humans (which are few, such as digoxin, dabigatran etexilate, fexofenadine, and talinolol [129,130]) or—the majority of drugs—are both P-gp substrates and can be metabolized by CYP3A4 and/or other DMEs. There were extensive in vitro and in vivo DDI investigations of efflux transporters in humans that evaluated the interactions between digoxin and inhibitors or inducers of P-gp. Fenner et al. (2009) [131] reported significant change in the digoxin PK (AUC) by P-gp inhibitors valspodar, followed by quinidine, cyclosporine, itraconazole, and clarithromycin. Coadministration of quinidine has been shown to increase digoxin bioavailability, which occurs either by (1) decreasing digoxin biliary elimination mediated by hepatobiliary P-gp efflux function, (2) reducing renal secretion mediated by P-gp efflux function of digoxin to urine [132], or (3) inhibition of P-gp pump that blocks the absorption of digoxin in the small intestine. In vitro studies using Caco-2 cells confirmed the latter [133].

In addition to the role of P-gp in DDI-mediated drug disposition and safety in liver, kidney, and intestine, P-gp plays a significant role in central nervous system (CNS) drug accumulation and life-threatening CNS toxicity. In humans, Sadeque et al. (2000) [134] showed that in healthy volunteers, inhibition of P-gp expressed in the blood-brain barrier by quinidine leads to respiratory depression caused by the opioid and P-gp substrate loperamide, whereas in cohort treated with loperamide alone didn't experienced this adverse event. Technologies such as quantitative whole-body autoradiography (QWBA) and positron emission tomography (PET) scanning have been used to assess the accumulation of P-gp substrates and interaction in the blood-brain barrier. By using this technology, Muzi et al. (2009) [135] showed that the inhibition of P-gp by cyclosporine increases CNS accumulation of the P-gp substrate verapamil.

Additionally, it is important to recognize that induction of transporters occurs in a pleiotropic manner via the nuclear receptors that regulate the expression of a battery of DME encoding genes. Accordingly, even though the relative contribution of CYP3A $\ll$ 25%, and if other PXR-inducible enzymes such as P-gp contribute additionally to the overall clearance of the drug, there is the possibility of clinically significant decreases in exposure following administration of strong PXR agonist.

Other efflux transporters are also involved in significant DDI in humans. Inhibition of bile salt export pump (BSEP) by certain drugs increases accumulation of hepatotoxic bile salts within the hepatocytes, leading to cholestatic liver injury. DDI cases with drugs such as bosentan, cyclosporine, and rifampin that competitively inhibit BSEP [136,137] can cause cholestatic liver injury.

5.4 CLINICAL DDI

In clinical DDI study design for a new drug to assess the potential and magnitude of inhibition or induction of P450 and/or drug transporters, important steps have to be taken. First is selection of a sensitive and selective probe substrate of the affected enzyme being investigated in vivo. It is important to keep in mind in selection of sensitive substrate that it is considered to be a sensitive substrate if its exposure has been shown to increase > 5-fold by a known inhibitor of that enzyme [82]. When evaluating clinical DDI associated with CYP3A, because of the expression of the enzyme both in the liver and in the small intestine, the sensitivity of a probe substrate depends on both the contribution of CYP3A to the overall hepatic metabolism ($f_{m\,(CYP3A)}$) and intestinal CYP3A contribution to the probe presystemic extraction following oral administration (F_G). In a study to compare the sensitivity of three substrates of CYP3A, midazolam, alprazolam, and buspirone indicated that the sensitivity of these three drugs to CYP3A-inhibitory DDIs is as follows: buspirone > midazolam > alprazolam against the effects of various CYP3A inhibitors such as ketoconazole, itraconazole, grapefruit juice, nefazodone, erythromycin, diltiazem, verapamil, and ritonavir. Those results were explained from the PK properties of these object drugs, and buspirone was explained to be more sensitive than midazolam due to the fact that buspirone has a smaller F_G than midazolam (0.21 versus 0.5), and midazolam more sensitive than alprazolam explained by differences in $f_{m\,(CYP3A)}$ (alprazolam: 0.74, midazolam: 0.93) as well as F_G (alprazolam: 0.99, midazolam: 0.5). Since alprazolam clearance in vivo highly depends on CYP3A4, then coadministration with strong inhibitors such as ketoconazole can result in clinically significant increases ~4-fold in alprazolam exposure [138], though alprazolam is not a sufficiently sensitive CYP3A substrate and is therefore not a suitable probe for evaluating the potential CYP3A inhibitory effect of new drug in a clinical DDI study. In contrast, both midazolam and buspirone represent sensitive probe substrates of CYP3A for use in clinical DDI studies with CYP3A inhibitors such as ketoconazole, ritonavir, or itraconazole and result in 5-fold higher magnitudes of DDI with these substrates.

When more than one enzyme contributes > 25% to the overall clearance of a new drug, DDI studies with strong inhibitors and inducers of each of these pathways are planned. A step-wise strategy can be implemented, with a DDI study to assess the effect of a strong inhibitor of the highest contributor enzyme, first followed by setting DDI studies with strong inhibitors for the lesser contributors. When a strong inhibitor produces a large and clinically significant increase in NME exposure for each contributing enzyme, further DDI studies with less potent inhibitors of that enzyme will be needed. If a contributing enzyme is polymorphic, with noncompetent enzyme activity poor metabolizer (PM), a PK study evaluating the effect of the polymorphism on the drug PK by comparing those parameters in extensive metabolizer (EM) versus PM can serve the purpose of a DDI study with a strong inhibitor, and no further DDI investigation with strong inhibitor will be recommended. An example for such an investigation is in the case of a drug that is cleared primarily by the metabolism with polymorphic enzyme CYP2D6 (80%) and the remainder by CYP1A2 (20%). It may appear that CYP2D6 EMs will not be at risk of DDIs via inhibition of CYP1A2, but clinically relevant increases in exposure of the drug can still be produced by fluvoxamine, a CYP1A2 inhibitor in the CYP2D6 PM. In general, the availability of clinical PK characteristics of any drug in EM and PM subpopulations (or at the preclinical stage) can be a powerful approach to DDI risk in such special populations and can support the design of DDI studies in such special populations or in populations that can have impaired or immature enzyme functions, such as pediatric populations.

5.4.1 DDI in Pediatric Patients

DDI is one cause of variability in drug response when one drug changes the effectiveness or toxicity of another drug in patients. As discussed in previous sections of this chapter, prediction of DDIs can be complex, as there are many factors that may contribute to the observed changes [139]. Also, as previously mentioned, most of the DDI studies in clinic of a specific drug are performed in healthy volunteers as part of the drug development process; however, any changes in biomedical and physiological conditions may alter the drug metabolism and thus the PK in patients, which may alter the magnitude of the potential DDI. For example, the adverse effects caused by DDI can be increased in patients with impaired renal [140] or hepatic functions [141], in poor metabolizers [142] for the noninhibited pathway, and in populations such as pediatric patients [143]. Data that describe the difference or extent of DDI in pediatric patients compared to adults is scarce, as carrying out DDI studies in younger children is challenging due to ethical consideration. One of the rare DDI studies in pediatrics was conducted to investigate the effect of felbamate therapy as adjunctive therapy for the treatment of epilepsy in 14 children (2–14 years) that already receiving valproic acid (VPA) [144]. Increase in VPA morning trough rose by 25% during the first week of felbamate therapy despite a 26% decrease in VPA dose on initiating the drug, which indicates that stronger interaction in pediatrics than adults and a more aggressive reduction in VPA dose in children will be needed. As it has been outlined previously, drug metabolism can be altered owing to the maturation of body functions, interaction with other medications, genetic variation, diet, and so forth. This can cause adverse drug reactions due to DDIs in pediatric populations that are not generally seen in the adult populations [145]. For instance,

- Altered metabolism of sodium valproate in children under 3 years caused higher incidence of hepatotoxicity.
- Impaired metabolism of chloramphenicol in neonates resulted in the gray baby syndrome that causes cyanosis and respiratory failure.
- Altered drug metabolism caused metabolic acidosis following the use of propofol in the critically ill children.
- Increased resistance to acetaminophen toxicity in children relative to adults was shown, apparently because of an increased capacity for sulfate conjugation early in life.

In a DDI study of nifedipine and cyclosporin (CsA) in pediatrics with smaller cohort age range from 3–18 years showed that the mean half-life of CsA was increased from 2.5 h to 4.1 h ($P < 0.04$) upon introducing nifedipine therapy, while in adults, no such effect was reported (literature reports). The contrast between DDI magnitude in children compared to adults revealed that nifedipine is capable of causing a serious DDI (nephrotoxicity) with CsA in children.

The difference in magnitude of interaction in a pediatric population versus adults might be influenced by ontogeny factors that affect the disposition of drugs, which lead to a higher or lower DDI potential compared to adults. Accordingly, the extrapolation of DDI in adults to pediatrics is not an adequate practice, and it is required [146] to conduct a DDI investigation in pediatrics to ensure a safe drug therapy.

In a recent report by Salem et al. (2013) [147], the authors summarized the literature data on pediatric DDIs and where possible compared the magnitude of reported DDIs in pediatrics with those in adult populations, with over a hundred reports of DDIs in the pediatric population age range from birth to 20 years. The magnitude of DDIs for 24 drug pairs from 31 different pediatric studies has been reviewed and used to compare with those DDIs in adults where corresponding data existed. Salem et al. reported the changes in exposure to the various substrates following drug interaction in pediatrics relative to the corresponding changes in exposure in adults (ratio of $AUC_{pediatric}/AUC_{adults}$). The results indicated that the extent of DDI, as revealed by the change in AUC and CL parameters in the presence and absence of inhibitor, were higher (> 1.25-fold), similar (0.8 – 1.25-fold), or lower (< 0.8-fold) than the corresponding ratio in adults in 10, 15, and 8 cases, respectively. The report concluded no age-related trend in the magnitude of DDIs could be established; however, the study highlighted the clear scarcity of the data in younger children < 2 years, and care should be exercised when applying the knowledge of DDIs from adults to those younger children.

5.4.2 Clinical DDI Study Designs

Clinical DDI study can be set up in several designs. The most common design is the randomized crossover, where subjects are randomized to one of two sequences. In one sequence a single test dose of the object drug for PK assessment in one period (1) is administered followed by administration of the object drug for PK during multiple dosing of the precipitant (at steady state) in a second period to assess the effect of the precipitant on the PK of the object drug. Dosing of the precipitant drug is continued through the period of PK assessment of the object drug. In the second sequence, the order of treatments is reversed such that the single-dose PK of the object drug is evaluated at the steady state of precipitant that was coadministered in period 1, followed by evaluation of the single-dose PK of the object drug in the absence of the precipitant in period 2. A long washout period separates periods 1 and 2, ensuring complete washout of both the object drug and the precipitant drug to reach baseline of metabolic activity at the start of the second period. The description of these DDI study designs is depicted in Figure 5.5(A) [148,149]. The period of washout varies depending on the DDI mechanism. For example, it is determined easily as five half-lives of the precipitant (and of any known circulating P450-oxidative metabolites) when the precipitant is a reversible CYP inhibitor. However, when the precipitant is a mechanism-based inhibitor or an inducer, the washout period can be longer, not only due to the half-life of the precipitant drug but also due to turnover half-life of the enzyme that is subject to inactivation or induction. Keep in mind that the washout period should not be longer than necessary but should be optimized using PBPK modeling and simulation of DDIs.

Other study designs for DDI investigation than the randomized crossover can be considered. For example, if one or both drugs possessed a very long half-life, then it would be impractical to incorporate a sufficiently long washout phase between the study periods. In these cases, a one-way crossover design (which is a fixed-sequence study design) is often used. In this approach, the PK of the object drug alone is determined before the PK assessment of the object drug administered with the precipitant [as shown in Figure 5.5(B)]. In this design, the washout period can be equivalent to the half-life of the object drug [147,148]. In reality, the fixed-sequence design was the most common (~70%) of all DDI studies evaluated [150]. However, if the DDI study includes measurement of PD endpoints (e.g., CNS depressant effects using a visual analog scale) in addition to PK endpoints, the randomized crossover design is recommended. It is also applicable in DDI studies that evaluate the effect of P-gp inhibition on the PK of new P-gp substrate-like drugs. In addition to increasing systemic drug exposure by an increase in the extent of oral absorption (via inhibition of intestinal efflux transport) and/or a decrease in renal and/or biliary clearance, it is possible that P-gp inhibition at the blood-brain barrier may additionally increase CNS drug availability beyond the extent that would be expected based on increases in systemic drug exposure alone. There is currently little clinical evidence to support the clinical significance of such DDIs that may increase human CNS drug distribution via inhibition of P-gp at the blood-brain barrier [151,134]. Because there is no

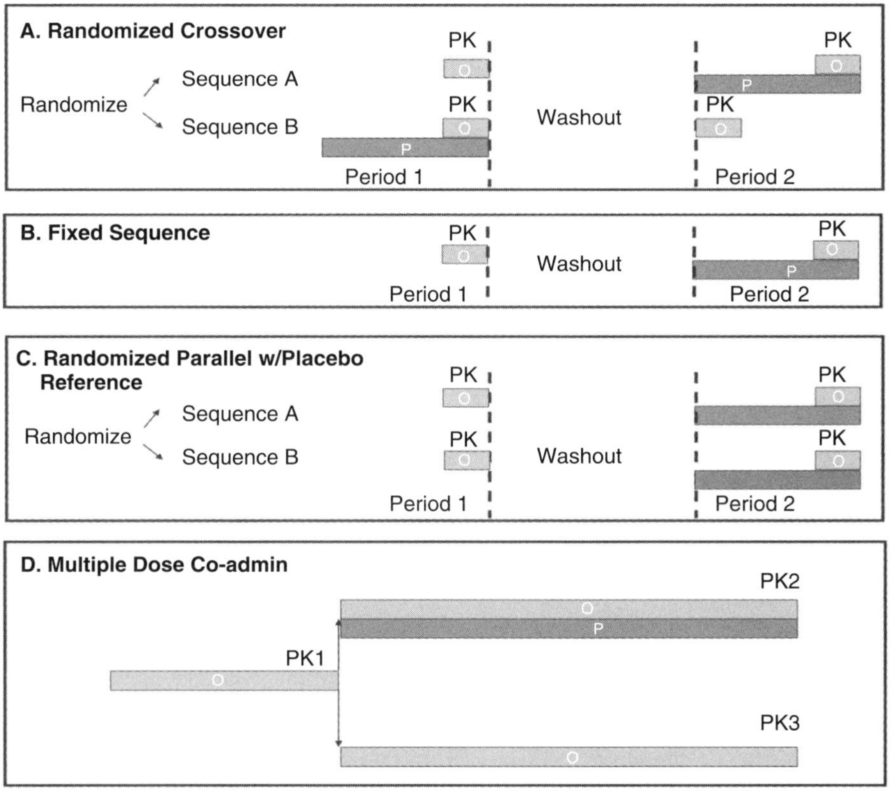

Figure 5.5 Possible clinical DDI study design objective drug (O) and precipitant drug (P) in relation to pharmacokinetic sampling (PK) and wash period (Washout). Adapted from Ref [148], [149] with permission.

direct assessment of any interaction for CNS drug distribution in DDI studies that evaluate the effects of a P-gp inhibitor, it may be useful to include CNS PD and/or safety evaluations in addition to systemic PK measurements in DDI studies of a P-gp inhibitor.

If neither the fixed-sequence design nor randomized crossover design are feasible, than an alternate design that may be used is the randomized parallel group design with a placebo reference [as depicted in Figure 5.5(C)] [148,149]. In this design, subjects are randomized into one of two groups. One group of subjects undergo PK characterization of the object drug alone in period 1 followed by PK characterization of the object drug in the setting of coadministration with the precipitant drug in period 2, and a second group of subjects undergo PK characterization of the object drug alone in period 1 and of the object drug following blinded administration of a placebo in period 2. Such a design provides the ability to separate out the effect of a potential DDI from any confounding period effects.

5.4.3 Statistical Approach in Clinical DDI Studies

As discussed in the next chapters, in any clinical study design including DDI study, the consideration of using an adequate statistical method with the appropriate sample size can enable an informative analysis of the resulting data to guide interpretation of the clinical significance of any observed interaction. In general, most DDI studies are designed with a sufficient number of subjects that can permit estimation of the magnitude of the interaction with a reasonable level of precision. That can be performed based on prior knowledge of the variability in PK of the object drug, from the variability of within-subject variance in the AUC and C_{max} of target drug for designs where each subject serves as his/her own control and undergoes PK characterization of the object drug alone as well as in the setting of coadministration of the precipitant drug.

PK data from a DDI study are typically analyzed using noncompartmental methods to calculate individual values of PK parameters AUC and C_{max} of the object drug when administered alone and when administered with the precipitant drug. Statistical analysis of PK parameters is performed using a mixed effects analysis of variance on log-transformed AUC and C_{max}, with estimation of the ratio of geometric mean AUC and C_{max} (object plus precipitant in reference to object alone) and associated 90% confidence intervals. This approach of estimation of interaction magnitude is preferred over hypothesis testing and use of p-values for the assessment of the statistical significance of DDI studies in new drug application (NDA) submissions to the FDA [150]. In statistical interpretation of clinical DDI, it is important to keep in mind that the p-values may not be the only parameter for the assessment of the statistical significance of DDI studies. For example, in a DDI study, estimated ratio of geometric mean AUCs is 89%, with a 90% confidence interval ranging from 85% to 95% and a p-value of < 0.05, although this study revealed a "statistically significant" interaction based on a p-value of < 0.05, which is a function of the low within-patient variability in the object drug's exposure, resulting in the ability of the study to demonstrate statistical significance in the minor decrease observed in object drug exposure (by 11% on average). However, the 90% confidence interval for the ratio of AUC is contained within the 80–125% range, and the observed mean decrease in exposure of 11% (101 – 90) would be of no clinical relevance in most cases. Therefore, the conclusion of statistical significance based on a p-value alone is not particularly meaningful or informative to guide the next steps in terms of whether the observed interaction requires risk management by increase in dose and/or exhibiting caution in drug labeling. Again, the p-value is of no value in reaching this conclusion; the ability to show statistical significance or the failure to demonstrate statistical significance should not drive interpretation of clinical DDI study results, and the use of a confidence interval approach (90% confidence interval about the geometric mean ratio of the AUC and C_{max} with and without the interacting drug) as opposed to testing for statistical significance is considered as appropriate in the analysis and interpretation of DDI study results. This is reflected in expert opinion articles [152] as well as the draft FDA and EMA DDI guidance documents [39,33].

5.5 CONCLUSION

DDI is a critical parameter that causes changes in the disposition of drug and can impact its PK and response (PD), including both the wanted therapeutic effects and the unplanned adverse effects. In this chapter, an overview was presented to describe the principles, translational approaches, and clinical strategies in humans that are recommended to successfully measure the PK DDI of new drugs during the preclinical and clinical stages of drug development. Furthermore, DDIs based on metabolic pathways by DMEs (mainly P450 enzymes) and nonmetabolic pathways such as transporter-mediated DDIs and DDIs observed in special populations were discussed. The quantitative translation of results from in vitro inhibition or induction studies to predict the clinical outcome in correspondence to the regulatory guidance (in the United States and the European Union) was described. Ultimately, the clinical relevance of PK changes resulting from DDIs will need to be assessed in the context of exposure-response understanding for the efficacy and safety characteristics of the object drug to contribute to product labeling. In this chapter, extensive review of the ongoing efforts to set up strategies for dealing with MBIs and P450 inactivation can be used to predict a clinically relevant DDI.

Last, despite the extensive translational efforts and progress that have been made within the pharmaceutical companies in collaboration with academic institutes, there is still uncertainty in the prediction of clinically relevant DDIs and identification of the therapeutic agent that will ensure more safe and effective treatment to all patient populations. However, with the advancement in using computational approaches and modeling, such as the use of PBPK and in silico prediction, a promised improvement in the prediction of preclinical and clinical DDIs is demonstrated by recent investigations such as those cited in chapter 8.

REFERENCES

1. Huang SM, Lesko LJ. Drug–drug, drug–dietary supplement, and drug–citrus fruit and other food interactions: what have we learned. J Clin Pharmacol 2004;44(6):559–569.
2. Huang SM, Hall SD, Watkins P, et al. Drug interactions with herbal products and grapefruit juice: a conference report. Clin Pharmacol Ther 2004;75(1):1–12.
3. Bachmann KA, Ring BJ, Wrighton SA. Drug–drug interactions and the cytochromes P450. In: Drug metabolizing enzymes: cytochrome P450 and other enzymes in drug discovery and development, ed. Lee J, Obach RS, Fisher MB. New York: Marcel Dekker, 2003, 311–336.
4. Thummel K, Kunze KL, Shen DD. Metabolically-based drug-–drug interactions: principles and mechanisms. Philadelphia: Lippincott Williams & Wilkins, 2000.
5. Varhe A, Olkkola KT, Neuvonen PJ. Oral triazolam is potentially hazardous to patients receiving systemic antimycotics ketoconazole or itraconazole. Clin Pharmacol Ther 1994;56:601–607.

6. Honig PK, Woosley RL, Zamani K, et al. Changes in the pharmacokinetics and electrocardiographic pharmacodynamics of terfenadine with concomitant administration of erythromycin. Clin Pharmacol Ther 1992;52:231–238.
7. Rogers RaP R. Profound symptomatic bradycardia associated with combined mibefradil and beta-blocker therapy. Med J Aust 1998;169:425–427.
8. Schmassmann-Suhijar D, Bullingham R, Gasser R, et al. Rhabdomyolysis due to interaction of simvastatin with mibefradil. Lancet 1998;351:1929–1930.
9. Ernst ME, Kelly MW. Mibefradil, a pharmacologically distinct calcium antagonist. Pharmacotherapy 1998;18:463–485.
10. Prueksaritanont T, Ma B, Tang C, et al. Metabolic interactions between mibefradil and HMG-CoA reductase inhibitors: an in vitro investigation with human liver preparations. Br J Clin Pharmacol 1999;47:291–298.
11. Wandel C, Kim RB, Guengerich FP, et al. Mibefradil is a P-glycoprotein substrate and a potent inhibitor of both P-glycoprotein and CYP3A in vitro. Drug Metab Dispos 2000; 28:895–898.
12. Bachmann K, White D, Jauregui L, et al. An evaluation of the dose-dependent inhibition of CYP1A2 by rofecoxib using theophylline as a CYP1A2 probe. J Clin Pharmacol 2003; 43:1082–1090.
13. Schwartz JI, Bugianesi KJ, Ebel DL, et al. The effect of rofecoxib on the pharmacodynamics and pharmacokinetics of warfarin. Clin Pharmacol Ther 2000;68:626–636.
14. Bachmann KA. Inhibition constants, inhibitor concentrations and the prediction of inhibitory drug drug interactions: pitfalls, progress and promise. Curr Drug Metab 2006; 7:1–14.
15. Karjalainen MJ, Neuvonen PJ, Backman JT. Rofecoxib is a potent, metabolism-dependent inhibitor of CYP1A2: implications for in vitro prediction of drug interactions. Drug Metab Dispos 2006;34:2091–2096.
16. Bjornsson TD, Callaghan JT, Einolf HJ, et al. The conduct of in vitro and in vivo drug–drug interaction studies: a pharmaceutical research and manufacturers of America (PhRMA) perspective. Drug Metab Dispos 2003;31(7):815–832.
17. Bourrie M, Meunier V, Berger Y, et al. Cytochrome P450 isoform inhibitors as a tool for the investigation of metabolic reactions catalyzed by human liver microsomes. J Pharmacol Exp Ther 1996;277(1):321–332.
18. Newton DJ, Wang RW, Lu AY. Cytochrome P450 inhibitors. Evaluation of specificities in the in vitrometabolism of therapeutic agents by human liver microsomes. Drug Metab Dispos 1995;23(1):154–158.
19. Tucker GT, Houston JB, Huang SM. Optimizing drug development: strategies to assess drug metabolism/transporter interaction potential—towards a consensus. Br J Clin Pharmacol 2001;52(1):107–117.
20. Venkatakrishnan K, von Moltke LL, Greenblatt DJ, et al. Drug metabolism and drug interactions: application and clinical value of in vitro models. Curr Drug Metab 2003; 4(5):423–459.
21. Walsky RL, Obach RS. Validated assays for human cytochrome P450 activities. Drug Metab Dispos 2004;32(6):647–660.
22. Brown HS, Galetin A, Halifax D, et al. Prediction of in vivo drug–drug interactions from in vitro data : factors affecting prototypic drug–drug interactions involving CYP2C9, CYP2D6 and CYP3A4. Clin Pharmacokinet 2006;45(10):1035–1050.

23. Galetin A, Ito K, Halifax D, et al. CYP3A4 substrate selection and substitution in the prediction of potential drug–drug interactions. J Pharmacol Exp Ther 2005;314(1):180–190.
24. Obach RS, Walsky RL, Venkatakrishnan K, et al. The utility of in vitro cytochrome P450 inhibition data in the prediction of drug–drug interactions. J Pharmacol Exp Ther 2006;316(1):336–348.
25. Obach RS, Walsky RL, Venkatakrishnan K, et al. In vitro cytochrome P450 inhibition data and the prediction of drug–drug interactions: qualitative relationships, quantitative predictions, and the rank-order approach. Clin Pharmacol Ther 2005;78(6): 582–592.
26. Obach RS. The importance of nonspecific binding in in vitro matrices, its impact on enzyme kinetic studies of drug metabolism reactions, and implications for in vitro–in vivo correlations. Drug Metab Dispos 1996;24(10):1047–1049.
27. Obach RS. Nonspecific binding to microsomes: impact on scale-up of in vitro intrinsic clearance to hepatic clearance as assessed through examination of warfarin, imipramine, and propranolol. Drug Metab Dispos 1997;25(12):1359–1369.
28. Obach RS. Prediction of human clearance of twenty-nine drugs from hepatic microsomal intrinsic clearance data: an examination of in vitro half-life approach and nonspecific binding to microsomes. Drug Metab Dispos 1999;27(11):1350–1359.
29. Venkatakrishnan K, von Moltke LL, Obach RS, et al. Microsomal binding of amitriptyline: effect on estimation of enzyme kinetic parameters in vitro. J Pharmacol Exp Ther 2000; 293(2):343–350.
30. Margolis JM, Obach RS. Impact of nonspecific binding to microsomes and phospholipid on the inhibition of cytochrome P4502D6: implications for relating in vitro inhibition data to in vivo drug interactions. Drug Metab Dispos 2003;31(5):606–611.
31. Tran TH, von Moltke LL, Venkatakrishnan K, et al. Microsomal protein concentration modifies the apparent inhibitory potency of CYP3A inhibitors. Drug Metab Dispos 2002; 30(12):1441–1445.
32. Walsky RL, Obach RS, Gaman EA, et al. Selective inhibition of human cytochrome P4502C8 by montelukast. Drug Metab Dispos 2005;33(3):413–418.
33. EMA. Draft Guideline on the Investigation of Drug Interactions, 2010. [Online]. Available: http://www.ema.europa.eu/docs/en_GB/document_library/Scientific_guideline/2010/05/WC500090112.pdf.
34. Oleson FB, Berman CL, Li AP. An evaluation of the P450 inhibition and induction potential of daptomycin inprimary human hepatocytes. Chem Biol Interact 2004;150:137–147.
35. Smith DA, Jones BC, Walker DK. Design of drugs involving the concepts and theories of drug metabolism and pharmacokinetics. Med Res Rev 1996;16(3):243–266.
36. Trunzer M, Faller B, Zimmerlin WA. Metabolic soft spot identification and compound optimization in early discovery phases using MetaSite and LC-MS/MS validation. J Med Chem 2009;52(2):329–335.
37. Einolf HJ. Comparison of different approaches to predict metabolic drug–drug interactions. Xenobiotica 2007;37(10–11):1257–1294.
38. Fahmi OA, Hurst S, Plowchalk D, et al. Comparison of different algorithms for predicting clinical drug–drug interactions, based on the use of CYP3A4 in vitro data: predictions of compounds as precipitants of interaction. Drug Metab Dispos 2009;37(8):1658–1666.
39. FDA. Drug Interaction Studies—Study Design, Data Analysis and Implications for Dosing and Labeling. Draft guidance, 2006. [Online]. Available: http://www.fda.gov/downloads/Drugs/GuidanceComplianceRegulatoryInformation/Guidances/ucm072101.pdf.

40. USPI. Afinitor, 2010. [Online]. Available: http://www. pharma.us.novartis.com/product/pi/pdf/afinitor.pdf.
41. Bohnert T, Gan L-S. The role of drug metabolism in drug discovery. In: Enzyme inhibition in drug discovery and development: the good and the bad, ed. Lu C, Li AP. Hoboken, NJ: Wiley, 2010, 91–176.
42. Kiang TKL, Ensom MHH, Chang TKH. UDP-glucuronosyltransferases and clinical drug–drug interactions. Pharmacol Ther 2005;106:97–132.
43. Said M, Noort D, Magdalou J, et al. Selective and potent inhibition of different hepatic UDP-glucuronosyltransferase activities by omega, omega, omega-triphenylalcohols and UDP derivatives. Biochem Biophys Res Commun 1992;187:140–145.
44. Zamek-Gliszczynski MJ, Hoffmaster KA, Nezasa K-I, et al. Integration of hepatic drug transporters and phase II metabolizing enzymes: mechanisms of hepatic excretion of sulfate, glucuronide, and glutathione metabolites. Eur J Pharm Sci 2006;27:447–486.
45. Williams JA, Hyland R, Jones BC, et al. Drug–drug interactions for UDP-glucurmonosyltransferase substrates: a pharmacokinetic explanation for typically observed low exposure (AUC_I/AUC) ratios. Drug Metab Dispos 2004;32(11): 1201–1208.
46. Garnett WR. Lamotrigine-interaction with other drugs. In: Anti-epileptic drugs, 5th ed., ed. Levy RH, Mattson RH, Meldrum BS. Philadelphia: Lippincott Williams & Wilkins, 2002, 300–331.
47. Miners JO, Mackenzie PI. Drug glucuronidation in humans. Pharmacol Ther 1991; 51:347–369.
48. Zhang ZY, Wong YN. Enzyme kinetics for clinically relevant CYP inhibition. Curr Drug Metab 2005;6:241–257.
49. Takakusa H, Wahlin MD, Zhao C, Hanson KL, New LS, Chan EC, Nelson SD. Metabolic intermediate complex formation of human cytochrome P4503A4 by lapatinib. Drug Metab. Dispos 2011;39:1022–1030.
50. Orr ST, Ripp SL, Ballard E, Henderson JL, Scott DO, Obach RS, Sun H, Kalgutkar AS. Mechanism-based inactivation (MBI) of cytochrome P450 enzymes: structure-activity relationships and discovery strategies to mitigate drug–drug interaction risks. J Med Chem 2012;55(11):4896–4933.
51. Li P, Lu C, Balani SK, Gan LS. A refined cytochrome P450 IC_{50} shift assay for reliably identifying CYP3A time-dependent inhibitors. Drug Metab Dispos 2011 39:1054–1057.
52. Sekiguchi N, Higashida A, Kato M, Nabuchi Y, Mitsui T, Takanashi K, Aso Y, Ishigai M. Prediction of drug–drug interactions based on time-dependent inhibition from high throughput screening of cytochrome P450 3A4 inhibition. Drug Metab Pharmacokinet 2009;24:500–510.
53. Zimmerlin A, Trunzer M, Faller B. CYP3A time-dependent inhibition risk assessment validated with 400 reference drugs. Drug Metab Dispos 2011;39:1039–1046
54. Mayhew BS, Jones DR, Hall SD. An in vitro model for predicting in vivo inhibition of cytochrome P450 3A4 by metabolic intermediate complex formation. Drug Metab Dispos 2000;28:1031–1037.
55. Wang YH, Jones DR, Hall SD. Prediction of cytochrome P4503A inhibition by verapamil enantiomers and their metabolites. Drug Metab Dispos 2004;32:259–266.
56. Brown HS, Ito K, Galetin A, Houston JB. Prediction of in vivo drug–drug interactions from in vitro data: impact of incorporating parallel pathways of drug elimination and inhibitor absorption rate constant. British J Clin Pharmacol 2005;60:508–518.

57. Chen Q, Ngui J S, Doss GA, Wang RW, Cai X, DiNinno FP, Blizzard TA, Hammond ML, Stearns RA, Evans DC, Baillie TA. Cytochrome P450 3A4-mediated bioactivation of raloxifene: irreversible enzyme inhibition and thiol adduct formation. Chem Res Toxicol 2002;15:907–914.
58. Polasek TM, Miners JO. In vitro approaches to investigate mechanism-based inactivation of CYP enzymes. Expert Opin Drug Metabol Toxicol 2007;3(3):321–329.
59. Polasek TM, Miners JO. Quantitative prediction of macrolide drug–drug interaction potential from in vitro studies using testosterone as the human cytochrome P4503A substrate. Eur J Clin Pharmacol 2006;62(3):203–208.
60. Xu L, Chen Y, Pan Y, et al. Prediction of human drug–drug interactions from time-dependent inactivation of CYP3A4 in primary hepatocytes using a population-based simulator. Drug Metab Dispos 2009;37(12):2330–2339.
61. Yang J, Jamei M, Rowland-Yeo K, et al. Kinetic values for mechanism-based enzyme inhibition: assessing the bias introduced by the conventional experimental protocol. Eur J Pharm Sci 2005;26(3–4):334–340.
62. Parkinson A, Kazmi F, Buckley DB, et al. System-dependent outcomes during the evaluation of drug candidates as inhibitors of cytochrome P450 (CYP) and uridine diphosphate glucuronosyltransferase (UGT) enzymes: human hepatocytes versus liver microsomes versus recombinant enzymes. Drug Metab Pharmacokinet 2010;25(1):16–27.
63. Hewitt NJ, Lecluyse EL, Ferguson SS. Induction of hepatic cytochrome P450 enzymes: methods, mechanisms, recommendations, and in vitro–in vivo correlations. Xenobiotica 2007;37(10–11):1196–1224.
64. Pascussi JM, Gerbal-Chaloin S, Duret C, et al. The tangle of nuclear receptors that controls xenobiotic metabolism and transport: crosstalk and consequences. Annu Rev Pharmacol Toxicol 2008;48:1–32.
65. Urquhart BL, Tirona RG, Kim RB. Nuclear receptors and the regulation of drug-metabolizing enzymes and drug transporters: implications for interindividual variability in response to drugs. J Clin Pharmacol 2007;47(5):566–578.
66. Chu V, Einolf HJ, Evers R, et al. In vitro and in vivo induction of cytochrome P450: a survey of the current practices and recommendations: a pharmaceutical research and manufacturers of America perspective. Drug Metab Dispos 2009;37(7):1339–1354.
67. Mills JB, Rose KA, Sadagopan N, et al. Induction of drug metabolism enzymes and MDR1 using a novel human hepatocyte cell line. J Pharmacol Exp Ther 2004;309(1): 303–309.
68. Ripp SL, Mills JB, Fahmi OA, et al. Use of immortalized human hepatocytes to predict the magnitude of clinical drug–drug interactions caused by CYP3A4 induction. Drug Metab Dispos 2006;34(10):1742–1748.
69. Hariparsad N, Carr BA, Evers R, et al. Comparison of immortalized Fa2N-4 cells and human hepatocytes as in vitro models for cytochrome P450 induction. Drug Metab Dispos 2008;36(6):1046–1055.
70. Kenny JR, Chen L, McGinnity DF, et al. Efficient assessment of the utility of immortalized Fa2N-4 cells for cytochrome P450 (CYP) induction studies using multiplex quantitative reverse transcriptase-polymerase chain reaction (qRT-PCR) and substrate cassette methodologies. Xenobiotica 2008;38(12):1500–1517.
71. Antherieu S, Chesne C, Li R, et al. Stable expression, activity, and inducibility of cytochromes P450 in differentiated HepaRG cells. Drug Metab Dispos 2010;38(3): 516–525.

72. Kanebratt KP, Andersson TB. HepaRG cells as an in vitro model for evaluation of cytochrome P450 induction in humans. Drug Metab Dispos 2008;36(1):137–145.
73. Aninat C, Piton A, Glaise D, et al. Expression of cytochromes P450, conjugating enzymes and nuclear receptors in human hepatoma HepaRG cells. Drug Metab Dispos 2006;34(1):75–83.
74. McGinnity DF, Zhang G, Kenny JR, et al. Evaluation of multiple in vitro systems for assessment of CYP3A4 induction in drug discovery: human hepatocytes, pregnane X receptor reporter gene, and Fa2N-4 and HepaRG cells. Drug Metab Dispos 2009;37(6):1259–1268.
75. Luo G, Cunningham M, Kim S, et al. CYP3A4 induction by drugs: correlation between a pregnane X receptor reporter gene assay and CYP3A4 expression in human hepatocytes. Drug Metab Dispos 2002;30(7):795–804.
76. Greenblatt DJ, von Moltke LL, Harmatz JS, et al. Alprazolam-ritonavir interaction: implications for product labeling. Clin Pharmacol Ther 2000;67(4):335–341.
77. USPI. Norvir, 2010. [Online]. Available: http://www.accessdata.fda.gov/drugsatfda_docs/label/2010/020945s028lbl.pdf.
78. Muirhead GJ, Wulff MB, Fielding A, et al. Pharmacokinetic interactions between sildenafil and saquinavir/ritonavir. Br J Clin Pharmacol 2000;50(2):99–107.
79. Fahmi OA, Kish M, Boldt S, et al. Cytochrome P450 3A4 mRNA is a more reliable marker than CYP3A4 activity for detecting pregnane X receptor-activated induction of drug-metabolizing enzymes. Drug Metab Dispos 2010;38(9):1605–1611.
80. LeCluyse EL. Human hepatocyte culture systems for the in vitro evaluation of cytochrome P450 expression and regulation. Eur J Pharm Sci 2001;13(4):343–368.
81. U.S. Department of Health and Human Services. Food and Drug Administration draft guidance for industry. Drug interaction studies—study design, data analysis, and implications for dosing and labeling recommendations. FDA, 2012.
82. Kato M, Chiba K, Horikawa M, et al. The quantitative prediction of in vivo enzyme-induction caused by drug exposure from in vitro information on human hepatocytes. Drug Metab Pharmacokinet 2005;20(4):236–243.
83. Shou M, Hayashi M, Pan Y, et al. Modeling, prediction, and in vitro in vivo correlation of CYP3A4 induction. Drug Metab Dispos 2008;36(11):2355–2370.
84. Fahmi OA, Maurer TS, Kish M, et al. A combined model for predicting CYP3A4 clinical net drug–drug interaction based on CYP3A4 inhibition, inactivation, and induction determined in vitro. Drug Metab Dispos 2008;36(8):1698–1708.
85. Ohno Y, Hisaka A, Ueno M, et al. General framework for the prediction of oral drug interactions caused by CYP3A4 induction from in vivo information. Clin Pharmacokinet 2008;47(10):669–680.
86. Giacomini KM, Huang SM, Tweedie DJ, Benet LZ, Brouwer KL, Chu X, Dahlin A, Evers R, Fischer V, Hillgren KM, et al. International Transporter Consortium Membrane transporters in drug development. Nat Rev Drug Discov 2010;9:215–236
87. Klaassen CD, Aleksunes LM. Xenobiotic, bile acid, and cholesterol transporters: function and regulation. Pharmacol Rev 2010;62:1–96.
88. Zolk O, Fromm MF. Transporter-mediated drug uptake and efflux: important determinants of adverse drug reactions. Clin Pharmacol Ther 2011;89:798–805.
89. Shitara Y, Sato H, Sugiyama Y. Evaluation of drug–drug interaction in the hepatobiliary and renal transport of drugs. Annu Rev Pharmacol Toxicol 2005;45:689–723.

90. Müller F, Fromm MF. Transporter-mediated drug–drug interactions. Pharmacogenomics 2011;12:1017–103.

91. König J, Müller F, Fromm MF. Transporters and Drug–drug Interactions: Important Determinants of Drug Disposition and Effects Pharmacol Rev 2013;65:944–966

92. Zhang L, Huang SM, Lesko LJ. Transporter-mediated drug–drug interactions. Clin Pharmacol Ther 2011;89:481–484.

93. Lepist EI, Ray AS. Renal drug–drug interactions: what we have learned and where we are going. Expert Opin Drug Metab Toxicol 2012;8:433–448

94. Morrissey KM, Stocker SL, Wittwer MB, Xu L, Giacomini KM. Renal transporters in drug development. Annu Rev Pharmacol Toxicol 2013;53:503–529.

95. Yoshida K, Maeda K, Sugiyama Y. Hepatic and intestinal drug transporters: prediction of pharmacokinetic effects caused by drug–drug interactions and genetic polymorphisms. Annu Rev Pharmacol Toxicol 2013;53:581–612.

96. Nies AT, Koepsell H, Damme K, Schwab M. Organic cation transporters (OCTs, MATEs), in vitro and in vivo evidence for the importance in drug therapy. Handb Exp Pharmacol 2011;201:105–167.

97. Neuvonen PJ. Drug interactions with HMG-CoA reductase inhibitors (statins): the importance of CYP enzymes, transporters and pharmacogenetics. Curr Opin Investig Drugs 2010;11:323–332.

98. Hirano M, Maeda K, Shitara Y, Sugiyama Y. Contribution of OATP2 (OATP1B1) and OATP8 (OATP1B3) to the hepatic uptake of pitavastatin in humans. J Pharmacol Exp Ther 2004;311:139–146

99. Ho RH, Tirona RG, Leake BF, Glaeser H, Lee W, Lemke CJ, Wang Y, Kim RB. Drug and bile acid transporters in rosuvastatin hepatic uptake: function, expression, and pharmacogenetics. Gastroenterology 2006;130:1793–1806.

100. Shitara Y, Takeuchi K, Nagamatsu Y, Wada S, Sugiyama Y, Horie T. Long-lasting inhibitory effects of cyclosporin A, but not tacrolimus, on OATP1B1- and OATP1B3-mediated uptake. Drug Metab Pharmacokinet 2012;27:368–378.

101. Simonson SG, Raza A, Martin PD, Mitchell PD, Jarcho JA, Brown CD, Windass AS, Schneck DW. Rosuvastatin pharmacokinetics in heart transplant recipients administered an antirejection regimen including cyclosporine. Clin Pharmacol Ther 2004;76:167–177.

102. Hasunuma T, Nakamura M, Yachi T, Arisawa N, Fukushima K, Ijima H, Saito Y. The drug–drug interactions of pitavastatin (NK-104), a novel HMG-CoA reductase inhibitor and cyclosporine. J Clin Ther Med 2003;19:381–389.

103. Yanni SB, Augustijns P, Benjamin DK Jr, Brouwer KLR, Thakker DR, Annaert PP. In vitro investigation of the hepatobiliary disposition mechanisms of the antifungal agent micafungin in humans and rats. Drug Metab Dispos 2010;38:1848–1856,

104. Vavricka SR, Van Montfoort J, Ha HR, Meier PJ, Fattinger K. Interactions of rifamycin SV and rifampicin with organic anion uptake systems of human liver. Hepatology 2002;36:164–172.

105. Tirona RG. Molecular mechanisms of drug transporter regulation. Handb Exp Pharmacol 2011;201:373–402.

106. Lau YY, Huang Y, Frassetto L, Benet LZ. Effect of OATP1B transporter inhibition on the pharmacokinetics of atorvastatin in healthy volunteers. Clin Pharmacol Ther 2007;81:194–204.

107. Treiber A, Schneiter R, Häusler S, Stieger B. Bosentan is a substrate of human OATP1B1 and OATP1B3: inhibition of hepatic uptake as the common mechanism of its interactions with cyclosporin A, rifampicin, and sildenafil. Drug Metab Dispos 2007;35:1400–1407.
108. van Giersbergen PL, Halabi A, Dingemanse J. Single- and multiple-dose pharmacokinetics of bosentan and its interaction with ketoconazole. Br J Clin Pharmacol 2002;53:589–595.
109. Lepist EI, Ray AS. Renal drug–drug interactions: what we have learned and where we are going. Expert Opin Drug Metab Toxicol 2012;8:433–448
110. Deguchi T, Kusuhara H, Takadate A, Endou H, Otagiri M, Sugiyama Y. Characterization of uremic toxin transport by organic anion transporters in the kidney. Kidney Int 2004;65:162–174.
111. Vanwert AL, Bailey RM, Sweet DH. Organic anion transporter 3 (Oat3/Slc22a8) knockout mice exhibit altered clearance and distribution of penicillin G. Am J Physiol Renal Physiol 2007;293:F1332–F1341.
112. Uwai Y, Ida H, Tsuji Y, Katsura T, Inui K. Renal transport of adefovir, cidofovir, and tenofovir by SLC22A family members (hOAT1, hOAT3, and hOCT2). Pharm Res 2007;24:811–815.
113. Burckhardt G, Burckhardt BC. In vitro and in vivo evidence of the importance of organic anion transporters (OATs) in drug therapy. Handb Exp Pharmacol 2011;201:29–104.
114. Hashimoto T, Narikawa S, Huang XL, Minematsu T, Usui T, Kamimura H, Endou H. Characterization of the renal tubular transport of zonampanel, a novel alpha-amino-3-hydroxy-5-methylisoxazole-4-propionic acid receptor antagonist, by human organic anion transporters. Drug Metab Dispos 2004;32:1096–1102.
115. Tahara H, Kusuhara H, Endou H, Koepsell H, Imaoka T, Fuse E, Sugiyama Y. A species difference in the transport activities of H2 receptor antagonists by rat and human renal organic anion and cation transporters. J Pharmacol Exp Ther 2005;315:337–345.
116. Robbins N, Koch SE, Tranter M, Rubinstein J. The history and future of probenecid. Cardiovasc Toxicol 2012;12:1–9.
117. Lilly MB, Omura GA. Clinical pharmacology of oral intermediate-dose methotrexate with or without probenecid. Cancer Chemother Pharmacol 1985;15:220–222.
118. Izzedine H, Launay-Vacher V, Deray G. Antiviral drug-induced nephrotoxicity. Am J Kidney Dis 2005;45:804–817.
119. Vree TB, Van Den Biggelaar-Martea M, Verwey-Van Wissen CP. Frusemide and its acyl glucuronide show a short and long phase in elimination kinetics and pharmacodynamic effect in man. J Pharm Pharmacol 1995;47:964–969.
120. Nakagomi-Hagihara R, Nakai D, Tokui T. Inhibition of human organic anion transporter 3 mediated pravastatin transport by gemfibrozil and the metabolites in humans. Xenobiotica 2007;37:416–426.
121. Kyrklund C, Backman JT, Neuvonen M, Neuvonen PJ. Gemfibrozil increases plasma pravastatin concentrations and reduces pravastatin renal clearance. Clin Pharmacol Ther 2003;73:538–544
122. Watanabe Y, Nakajima M, Ohashi N, Kume T, Yokoi T. Glucuronidation of etoposide in human liver microsomes is specifically catalyzed by UDP-glucuronosyltransferase 1A1. Drug Metab Dispos 2003;31:589–595.
123. Damme K, Nies AT, Schaeffeler E, Schwab M. Mammalian MATE (SLC47A) transport proteins: impact on efflux of endogenous substrates and xenobiotics. Drug Metab Rev 2011;43:499–523.

124. Nies AT, Herrmann E, Brom M, Keppler D. Vectorial transport of the plant alkaloid berberine by double-transfected cells expressing the human organic cation transporter 1 (OCT1, SLC22A1) and the efflux pump MDR1 P-glycoprotein (ABCB1). Naunyn Schmiedebergs Arch Pharmacol 2008;376:449–461.

125. Somogyi A, Stockley C, Keal J, Rolan P, Bochner F. Reduction of metformin renal tubular secretion by cimetidine in man. Br J Clin Pharmacol 1987;23:545–551.

126. Graham GG, Punt J, Arora M, Day RO, Doogue MP, Duong JK, Furlong TJ, Greenfield JR, Greenup LC, Kirkpatrick CM, et al. Clinical pharmacokinetics of metformin. Clin Pharmacokinet 2011;50:81–98.

127. Ho RH, Kim RB. Transporters and drug therapy: implications for drug disposition and disease. Clin Pharmacol Ther 2005;78:260–277.

128. Shitara Y, Sato H, Sugiyama Y. Evaluation of drug–drug interaction in the hepatobiliary and renal transport of drugs. Annu Rev Pharmacol Toxicol 2005;45:689–723.

129. European Medicines Agency. Guideline on the investigation of drug interactions. London: EMA, 2012.

130. Oswald S, Terhaag B, Siegmund W. In vivo probes of drug transport: commonly used probe drugs to assess function of intestinal P-glycoprotein (ABCB1) in humans. Handb Exp Pharmacol 2011;201:403–447.

131. Fenner KS, Troutman MD, Kempshall S, Cook JA, Ware JA, Smith DA, Lee CA. Drug–drug interactions mediated through P-glycoprotein: clinical relevance and in vitro–in vivo correlation using digoxin as a probe drug. Clin Pharmacol Ther 2009; 85:173–181.

132. Angelin B, Arvidsson A, Dahlqvist R, Hedman A, Schenck-Gustafsson K. Quinidine reduces biliary clearance of digoxin in man. Eur J Clin Invest 1987;17:262–265.

133. Wandel C, Kim RB, Kajiji S, Guengerich P, Wilkinson GR, Wood AJ. P-glycoprotein and cytochrome P-450 3A inhibition: dissociation of inhibitory potencies. Cancer Res 1999;59:3944–3948.

134. Sadeque AJ, Wandel C, He H, Shah S, Wood AJ. Increased drug delivery to the brain by P-glycoprotein inhibition. Clin Pharmacol Ther 2000;68:231–237.

135. Muzi M, Mankoff DA, Link JM, Shoner S, Collier AC, Sasongko L, Unadkat JD. Imaging of cyclosporine inhibition of P-glycoprotein activity using ^{11}C-verapamil in the brain: studies of healthy humans. J Nucl Med 2009;50:1267–1275.

136. Pauli-Magnus C, Meier PJ, Stieger B. Genetic determinants of drug-induced cholestasis and intrahepatic cholestasis of pregnancy. Semin Liver Dis 2010;30:147–159.

137. Russmann S, Jetter A, Kullak-Ublick GA. Pharmacogenetics of drug-induced liver injury. Hepatology 2010;52:748–761.

138. Greenblatt DJ, Wright CE, von Moltke LL, et al. Ketoconazole inhibition of triazolam and alprazolam clearance: differential kinetic and dynamic consequences. Clin Pharmacol Ther 1998;64(3):237–247.

139. Rostami-Hodjegan A, Tucker GT. In silico simulations to assess the in vivo consequences of in vitro metabolic drug–drug interactions. Drug Discov Today Tech. 2004; 1(4):441–448.

140. Zhao P, Vieira Mde L, Grillo JA, et al. Evaluation of exposure change of nonrenally eliminated drugs in patients with chronic kidney disease using physiologically based pharmacokinetic modeling and simulation. J Clin Pharmacol. 2012;52(1 Suppl): 91S–108S.

141. Orlando R, De Martin S, Pegoraro P, Quintieri L, Palatini P. Irreversible CYP3A inhibition accompanied by plasma protein- binding displacement: a comparative analysis in subjects with normal and impaired liver function. Clin Pharmacol Ther 2009;85(3): 319–326.

142. Collins C, Levy R, Ragueneau-Majlessi I, Hachad H. Prediction of maximum exposure in poor metabolizers following inhibition of nonpolymorphic pathways. Curr Drug Metab. 2006;7(3):295–299.

143. Johnson TN, Rostami-Hodjegan A. Resurgence in the use of physiologically based pharmacokinetic models in pediatric clinical pharmacology: parallel shift in incorporating the knowledge of biological elements and increased applicability to drug development and clinical practice. Paediatr Anaesth. 2010;21(3):291–301.

144. Delgado MR. Changes in valproic acid concentrations and dose/level ratios by felbamate coadministration in children. Ann Neurol. 1994;36(3):538–539.

145. Yanni SB, Allegaert K. Drug metabolism and interaction in pediatric patients. Encyclopedia of Drug Metabolism, 2011, V6 Wiley & Sons: New York.

146. Committee for Human Medicinal Products. Guideline on the investigation of drug interactions. European Medicines Agency, 2010.

147. Salem F, Rostami-Hodjegan A, PharmD, Johnson TN. Do children have the same vulnerability to metabolic drug–drug interactions as adults? A critical analysis of the literature. J Clin Pharmacol 2013;53(5):559–566.

148. Venkatakrishnan K. DDI risk assessment and evaluation in pharmaceutical development: interfacing drug metabolism and clinical pharmacology. In: Encyclopedia of drug metabolism and interactions, ed. Lyubimov AV. Hoboken, NJ: Wiley, 2012, 10:1–62.

149. Venkatakrishnan K, Pickard MD, von Moltke LL. A quantitative framework and strategies for management and evaluation of metabolic drug–drug interactions in oncology drug development: new molecular entities as object drugs. Clin Pharmacokinet 2010; 49(11):703–727.

150. Huang SM, Lesko LJ, Williams RL. Assessment of the quality and quantity of drug–drug interaction studies in recent NDA submissions: study design and data analysis issues. J Clin Pharmacol 1999;39(10):1006–1014.

151. Khaliq Y, Gallicano K, Venance S, et al. Effect of ketoconazole on ritonavir and saquinavir concentrations in plasma and cerebrospinal fluid from patients infected with human immunodeficiency virus. Clin Pharmacol Ther 2000;68(6):637–646.

152. Huang SM, Temple R, Throckmorton DC, et al. Drug interaction studies: study design, data analysis, and implications for dosing and labeling. Clin Pharmacol Ther 2007;81(2): 298–304.

6

GENERAL TOXICOLOGY: PRINCIPLE, METHODS, AND APPLICATIONS[1]

6.1 INTRODUCTION: THE HISTORY OF TOXICOLOGY

Toxicology, from the Greek "toxikos" meaning poisonous, is the study of the adverse effects of xenobiotics [1]. Due to the nature and evolution of *the* field, toxicology overlaps with a wide range of disciplines such as biology (including micro, molecular, and biochemical), chemistry, medicine, environmental science, epidemiology, pharmacology, and public health, to name a few.

The history of toxicology dates back to the earliest humans; in fact, the Ebers papyrus, an ancient Egyptian medical papyrus dating back to 1550 BC, listed a number of poisons including hemlock and aconitum, used in Chinese culture as an arrow poison, as well as a number of metals including lead, copper, and antimony. Knowledge and use of poisons have had numerous roles throughout various aspects of human history and helped shape much of our current understanding of therapeutics and experimental medicine. Toxicants have played a part in many human historic events, from the use of hemlock in the execution of Socrates and the toxic concoction in Cleopatra's suicide, to the wide use of arsenicals in the Middle Ages used to dethrone kings and monarchs [1], to their use in biological warfare and agriculture in modern times.

Much of what we consider to be fundamentals of toxicology came into practice due to Philippus Aureolus Theophrastus Bombastus von Hohenheim, who later became

[1] Also contributed by Ingrid L. Druwe PhD and Samuel C. Suarez PhD

Translational ADMET for Drug Therapy: Principles, Methods, and Pharmaceutical Applications, First Edition. Souzan B. Yanni.

known simply as Paracelsus. Paracelsus (1493–1541) was a physician and alchemist in the late Middle Ages and is known as "the Father of Toxicology" [2]. Paracelsus promoted many ideas that were controversial and contradicted the medical and scientific consensus at that time, but today his ideas are integral to the study and practice of the field of toxicology [3]. Some of these ideas include (1) experimentation is essential in the examination of response to chemicals; (2) one should make a distinction between the therapeutic and toxic properties of chemicals (therapeutic vs. toxic dose—Paracelsus believed that the properties between the therapeutic and toxic were sometimes but not always indistinguishable except by dose); and (3) one can establish a degree of specificity of chemicals and their therapeutic or toxic effects [4].

One of the most famous concepts credited to Paracelsus is that "all substances are poisons; there is none which is not a poison. The right dose differentiates a poison from a remedy" [1,2,4]. Today, this quote by Paracelsus is considered to be a cornerstone of toxicology.

Another pioneer in the field of toxicology was Bernardino Ramazzini, an Italian physician. Ramazzini contributed to the field of toxicology with his book on occupational diseases, *De Morbis Artificum Diatriba* (Diseases of Workers) [1,5]. In his book Ramazzini outlined the health hazards of chemical, dust, metals, and other disease causative agents encountered by workers in numerous occupations at that time [1,5]. Following Ramazzini's contributions, Sir Percivall Pott was the first scientist to demonstrate an association between chimney soot and a high incidence of scrotal cancer in chimney sweep workers [5]. This was the first occupational link to cancer [5]. These observations were later the foundation for animal studies by Kennaway, who in 1933 identified benzo[a]pyrene as a chemical carcinogen by painting coal tar onto the skin of mice [6]. Benzo[a]pyrene occurs in high concentrations in smoke and chimney soot [6].

The early twentieth century saw a movement to regulate various toxicants for the sake of consumer safety in a number of industries. After Upton Sinclair published the novel *The Jungle* describing numerous health violations in the meatpacking industry, the United States passed the 1906 Pure Food and Drug Act. This law was one of the first of its kind, and it regulated not only the meat industry but also the marketing of drugs. It required accurate labeling of dosage and contents. Prior to this, many drug makers advertised unsubstantiated benefits from "secret" ingredients, and this new law served as the driving force for the creation of the U.S. Food and Drug Administration Federal Drug Agency (FDA) [7]. The law stayed in place until 1938, when the deaths of 100 people exposed to diethylene glycol in sulfanilamide medicine prompted the replacement of the Pure Food and Drug Act with the Federal Food, Drug, and Cosmetic (FD&C) Act [1]. The FD&C Act required companies to perform safety testing and obtain FDA approval prior to marketing new drugs [7].

In 1910, the first pesticide control law was enacted in the United States [8]. The purpose of the law was to provide consumer protection from ineffective pesticide products and labeling. In 1947 the U.S. Congress passed the Federal Insecticide, Fungicide and Rodenticide Act (FIFRA) [8]. FIFRA established registration procedures for pesticides with the U.S. Department of Agriculture and established labeling provisions for pesticides [8]. The focus of the law was still on consumer protection against nonefficacious pesticides rather than safety and regulation of pesticide use. As pesticide use in agriculture became more popular, concerns about its use and effect on human health and the environment began to surface among farmers and environmentalists [9,10]. However, it was the work of Rachel Carson, an American marine biologist and conservationist, that brought the toxicological issues of pesticide use and its effects on human health and the environment to the American people [11,12]. Much of Carson's research on the environmental effects of dichlorodiphenyltrichloroethane (DDT) was published in 1962 in the book *Silent Spring* [11–13]. The book was met with fierce criticism and opposition from politicians and chemical companies, yet it resulted in a reversal of national pesticide policies and lead to a nationwide ban on DDT and the creation of several agencies including the U.S. Environmental Protection Agency (EPA) and the National Institute for Occupational Safety and Health (NIOSH) [11–13].

In 1972, 10 years after *Silent Spring* was published, FIFRA was fundamentally rewritten, amended by the Federal Environmental Pesticide Control Act (FEPCA) [13]. Since then, the law has evolved through numerous amendments from a law ensuring pesticide efficacy to a law that allows the U.S. EPA to regulate the use and sale of pesticides to protect human health and preserve the environment [13].

In addition to Carson's contributions, the 1960s issued in many advances in the field of toxicology, starting with what became known as the "thalidomide tragedy." Thalidomide was a sedative and antiemetic commonly prescribed to elderly patients. By 1960 the drug was widely popular in Europe and had already been approved for use in Canada and a number of African countries. Although thalidomide was originally intended for use in elderly patients, due to its effectiveness in relieving nausea and absence of significant side effects in the elderly, a number of physicians began prescribing thalidomide to pregnant women to ease nausea and vomiting associated with pregnancy. In 1960, the drug company Richardson-Merrell submitted a new drug application to the FDA for thalidomide under the trade name Kevadon [14,15]. Richardson-Merrell was confident that their application would be approved by the U.S. FDA and even lobbied for a quick approval of the drug. Additionally, Richardson-Merrell dispensed millions of thalidomide pills to over a thousand doctors across the United States, as was common permitted practice at the time for drugs pending approval [14,15]. To the surprise and dismay of the drug company, the reviewing FDA pharmacologist, Frances Kelsey, denied the thalidomide application numerous times, citing lack of substantial testing, and required further testing be performed before approval for the drug would be granted [14–16]. The drug company applied much pressure to the FDA and insisted that the thalidomide application be approved, but Kelsey stood firm against the approval of the drug [14–16]. In 1962, it was discovered that thalidomide had the ability to cross the placental barrier and inhibited long limbs from forming, resulting in numerous children being born with phocomelia (malformation of the limbs). In addition, many of these children had malformations of the eyes that resulted in blindness, as well as malformations of the heart and alimentary and urinary tracts [15]. The exact mechanism of action behind thalidomide's teratogenic toxicity is theorized to be due to thalidomide's ability to inhibit angiogenesis, in particular inhibition of bFGF or VEGF [17]. The "thalidomide tragedy," as it has come to be known, affected approximately 10,000 children worldwide, and only about 50% of these children survived [15]. Due to this tragedy, the U.S. Congress passed the 1962 Kefauver Harris Amendment in order to strengthen drug regulation. The new drug testing reforms required stricter limits of the testing of new drugs in order to protect the safety of the public. For example, the Kefauver Harris Amendment required that drug companies be able to demonstrate the efficacy of new drugs prior to marketing them, report any observed adverse effects to the FDA, and request consents from patients undergoing clinical trials [18].

Progress in the area of occupational health safety was also made in the 1960s due to the work of Irving Selikoff and Christopher Wagner. Both of these physicians independently identified an association between occupational asbestos exposure and the development of various lung diseases, including mesothelioma. Their combined work and research were instrumental in pressuring the Occupational Safety and Health Administration (OSHA) to make changes and limit occupational asbestos exposure [19].

Much progress in the field of toxicology was also due to the advances made in technology, which allowed for more thorough testing of chemicals for toxicity [1,20]. Much of the groundwork that made these advances possible was due to the work of Elizabeth and James Miller. Together they impacted the field of toxicology when they discovered that synthetic and naturally occurring chemicals can be metabolized by an organism into carcinogenic metabolites. They showed that these active metabolites had the ability to interact with macromolecules such as DNA and produce DNA adducts [20]. Based on this work, Bruce Ames developed a simple inexpensive assay to test the carcinogenesis potential of a chemical. The Ames assay, as it came to be called, employed the use of a histidine mutant strain of Salmonella. The bacteria is grown on histidine-containing agar plates. The chemical of interest is mixed with rat liver homogenates containing metabolizing enzymes and then added to the plate containing the mutant Salmonella. The plates are coated overnight and the bacteria growth is measured. Chemicals that are able to induce reverse mutations at the histidine loci cause bacterial growth and therefore are potential carcinogens. The Ames assay led to a better understanding of the underlying mechanisms of chemical-induced carcinogenicity [1]. Together with the Millers, Bruce Ames helped to propel the field of toxicology forward, as their work added to the understanding of molecular and genetic mechanisms in carcinogenesis, as well as enzymatic metabolism of xenobiotics [1].

The knowledge and new awareness that many human diseases were directly or indirectly related to chemical exposures caused much concern about the health effects of chemical agents in the environment [19]. In response to these concerns, Congress formed the National Toxicology Program (NTP) in 1978. Today this interagency program is run by the U.S. Department of Health and Human Services and is headquartered at the National Institute of Environmental Health Sciences (NIEHS) in Research Triangle Park, NC. The goal of the NTP is to coordinate federal toxicology testing among different governmental agencies, strengthen the scientific base in toxicology, develop and validate improved testing methods, and inform various stakeholders such as scientific and medical communities as well as the public about potentially toxic chemicals [21].

6.2 THE MULTIFACETED FIELD OF TOXICOLOGY

Toxicology is a broad and multifaceted field. Toxicity studies of chemicals are performed not only in the medical field in order to identify new molecules that can be used for diagnostic, preventative, and therapeutic purposes but also in agriculture and in the chemical industry, where toxicants may be used in the production of food, solvents, components, and additives that have become part of our modern lives. Toxicology is also concerned with the health effects of metals, such as those produced in mines and smelters and petroleum products [22–26]. A large function of the field of toxicology is to provide information for risk assessment and policy-making decisions. This function involves a careful balance between assessing the toxicity of a substance against the potential benefits that it may provide. Each of these toxicity studies seeks to accomplish a different goal, and therefore toxicology has a number of different subdisciplines [22].

6.2.1 Various Disciplines in Toxicology

- *Analytical toxicology:* To identify and measure toxic agents in biological and environmental samples, analytical toxicology is conducted to determine the extent of exposure [22].
- *Clinical toxicology:* The study of the harmful effects of chemicals, drugs, pesticides, and other substances on humans through the clinical diagnosis of symptoms and biological poisoning; may be involved in the treatment and management of intoxicated patients [22].
- *Environmental toxicology:* The study of the harmful effects of environmental exposure to chemical, physical, and biological agents on living organisms as well as the effects of these on ecosystems [22].
- *Forensic toxicology:* Specializes in the study of alcohol, legal and illicit drugs, and poisons. Forensic toxicologists study chemical composition, preparation, and identification. They also study the absorption, distribution, and elimination characteristics of chemicals and substances in the body, as well as the way in which the body responds to them and the factors that determine drug safety and effectiveness [22].
- *Occupational toxicology:* Involves the study of the harmful effects of substances used in the workplace to determine a safe or acceptable level of exposure to workers, as well as appropriate control measures to limit worker exposure [27].
- *Regulatory toxicology:* Has the primary role of ensuring public health and safety from the use of chemicals, drugs, and pesticides by identifying potential health risks posed by exposure to such substances. Regulatory toxicologists provide advice to governments, health professionals, politicians, and the public

on potential risks associated with chemical exposure so that appropriate risk management strategies are put in place, such as setting laws, regulations, and standards to limit or suspend the use of extremely toxic chemicals and or define conditions for the use of others in order to protect the health of workers and the public. In order to be able to set significant regulations and safety standards toxicologists must have extensive toxicity data profiles for the chemical(s) in question. Such data profiles can only be established through a wide variety of relevant toxicological studies that involve the use of conventional toxicology as well as mechanistic toxicology, which provides information on the mechanisms of action and enhances the toxicological evaluation [22–26].

6.2.2 Principles of Toxicology

In order to understand what deleterious effects may arise from any exposure to a toxicant, we must comprehend a variety of parameters about that exposure. These parameters include which class the toxicant falls into; duration, formulation, and route of exposure; venue of exposure; available dose the organism is exposed to; and finally the context of previous exposures that may be related. Each of these parameters informs what can be determined as to the potential harm of exposure.

6.3 CHARACTERISTICS OF TOXICANTS, TOXINS, AND EXPOSURES

Potentially harmful substances are characterized in a multitude of different ways, each depending on some aspect of the exposure, the substances that do harm (i.e., toxicants), or the extent of harm (adverse effect). Toxic substances can be classified according to their physical states, stability or reactivity, chemical makeup and structure, or potential poisonous effect. They can also be examined by target organ system, venue of use, source, and adverse outcome. Overarching classification by biochemical mode of action can also serve to encompass more than one of these individual classification modes. Classification also depends on the interests of the individual describing the toxic agent. No single classification system is sufficient to cover all needs in describing potential toxicity of substances, so classifiers often choose to combine these systems in order to maximize utility for their desired purpose. Classification systems that take into account both biological and chemical properties as well as exposure characteristics of the agent are most likely to be useful for general toxicology, as well as regulatory, control, and prevention purposes [23].

The simplest form of classification when discussing toxicology is the distinction between the terms "toxicant" and "toxin." Toxicant is the broadest descriptor of a potentially harmful substance, defined as "any chemical, of natural or synthetic origin, capable of causing a deleterious effect on a living organism.' Toxins are a subset of these chemicals, defined as "a toxicant produced metabolically by a living organism," and thus the terms are not interchangeable. Confusion often occurs when discussing toxic agents introduction by both geologic and biologic activity. Polycyclic aromatic hydrocarbons can result from combustion as a result of activities both natural (e.g., forest fires) and anthropogenic (e.g., fossil fuel combustion). Metal pollution can exist as a natural contaminant of groundwater [23] or as the result of mining activity. Even though biological activity can be the proximate cause of the release of these substances, they lack the metabolism component necessary for classification as toxins, and thus are characterized as toxicants [23].

6.3.1 Use Classes

Organisms are not exposed to one chemical at a time but are rather exposed to a matrix of chemicals that exist to form their environment, each of which can come from multiple use classes. Use class description may not be sufficient for mechanistic consideration because of this fact. It is essential, however, for understanding multiple branches of applied toxicology as well as the scope of general toxicology. It also is indispensable for description of the likely characteristics of any mixture of toxicants present in a particular exposure scenario.

Metals comprise over 75% of the periodic table of the elements, and a further 8 elements are considered metalloids [24]. Most metals occur in the environment in ores, rocks, and soil. Most are widely dispersed, or those deposits that are concentrated are usually buried. Anthropogenic activities can increase these concentrations near inhabitants of these areas, resulting in potential doses high enough to produce toxic effects [26]. Some metals perform vital functions in living systems, and certain dietary levels are required to maintain homeostasis. Thus, toxic effects can be observed from deficiency as well as excess. The chemical properties of some nonessential toxic metals can mimic those of essential metals and can cause toxicity when misincorporated into biological molecules, disrupting their function [24]. Nonessential toxic metals can interact with DNA, disrupt organelles of individual cells, disrupt the nervous system, and damage multiple individual organs and organ systems. Some of these metals show multiple of these effects. Each depends on the identity and chemistry of the individual elements [26]. Another group of metals to consider are those that are nonessential, but are used in medicinal therapies [24].

Naturally occurring toxins produced by organisms have mostly evolved as defense mechanisms in order to deter or kill predators or pathogens, although a smaller subset of animals have developed venoms to assist in subduing or killing their prey. Organisms producing toxins range from as small as bacteria, algae, and fungi to as large as plant and animal species. The actions of biologically produced toxins have been recognized throughout human history and have been put to use for both therapeutic and nefarious purposes. We continue to discover and understand the potential of natural toxins as investigations into their possible use progress [26].

Polycyclic aromatic hydrocarbons are often released due to the incomplete combustion of organic materials and are found in the smoke arising from many sources, as well as some overcooked foods. They are also present in coal and crude oil. Some are well described as carcinogens due to metabolite interaction with DNA. This metabolic activation has been studied extensively, and this model underpins our understanding of chemical carcinogenesis [26].

Agricultural chemicals are deployed to control pests that damage crops. Deliberate use makes them unusual among environmental toxicants, as their purpose is the hope of killing some form of life. In an ideal situation, the pesticidal properties of any compound would be highly selective in controlling the desired pest. However, there is rarely a pesticide that does not display toxicity to some undesired organism. This requires that their use must be weighed against the potential for degradation of environmental quality as well as potential adverse effects on human health. Environmental contamination may lead to the disruption of food webs as well as aquatic adulteration. Pesticides may also persist in the environment and bioaccumulate, leading to effects long after the initial application. Advances in the understanding of how adverse effects resulting from pesticide application have fundamentally changed the products used and have led to the production of safer, more effective chemistries [26].

Therapeutic drugs and drugs of abuse comprise a large number of chemicals with the potential for toxic action. Pharmacology is the study of the therapeutic properties of chemicals, and toxicology overlaps with pharmacology due to the fact that all therapeutic drugs can be toxic at some dose. The risks of drugs must be weighed against potential benefits, taking into account the gravity of the condition under treatment. Other drugs are classified not as therapeutics but as drugs of abuse due to potential misuse at higher doses than necessary for therapy. Drugs of abuse also have higher potential for harmful physical effects, physical dependence, and potentially fatal overdoses. Both classifications of drugs present an important category of potential toxicants [26].

6.3.2 Characteristics of Exposure

Toxic effects of chemicals are not observed unless the organism in question is exposed to the toxicant and the toxin is present in its active form, either in its native form or through metabolic activation. The active form of the toxicant must be present at the site of biological relevance for a sufficient duration and necessary concentration to produce toxic effect. Thus, the toxic effects of any exposure are determined by the properties of the chemical itself, the exposure, metabolic activation by the organism, the dose that arrives active at the site of biological relevance, and the overall susceptibility of the organism. To fully understand the potential for harm of any substance, we must know what properties the chemical possesses, the nature of the exposure, and the distribution of the exposure in an organism. The duration and frequency of the exposure are important to examine, as well as the route of exposure [23].

6.3.3 Length of Exposure

One of the descriptors to define exposure by is length. In laboratory toxicity tests, acute exposure generally refers to a single exposure of less than 24 h. Subacute exposures consist of repeated administration over 1 month or less. Subchronic exposures are characterized by duration of 1–3 months, with chronic exposures lasting 3 months to multiple years. Human or environmental exposures are not often as controlled as laboratory studies but still use similar terminology. Acute exposures are single events, subchronic events occurring multiple times over weeks or months, while chronic exposures occur on time horizons of months to years [23].

Acute toxicity usually manifests immediately following acute exposure, while chronic toxicity is evidenced over longer periods. It is important to distinguish between acute versus chronic exposure in contrast to acute versus chronic effect. Acute exposures can have both acute and chronic effects. Chronic exposures may show some acute effects in addition to longer-term chronic toxicity. It is also noteworthy that all chemicals can have acutely toxic effects if administered in sufficiently high dose. Exposure to high dosages may mask the potential for chronic toxicity. Differing experimental designs must be used when examining acute and chronic toxicity in order to determine the potential hazards posed by each type of exposure [28].

6.3.4 Routes of Exposure

In each of the components of the planet is some opportunity to be exposed to toxicants. The air contains compounds that can be harmful if inhaled. Soil contamination is also a concern for not only the possibility of contamination of food supplies but for potential movement by rainfall transferring toxicants from the soil into water supplies. Water sources can be contaminated from nonpoint source (soil runoff) and point source (industrial release) pollution, as well as erosion of mineral deposits in the ground. Ingestion of contaminated water is a serious hazard to human health, as water is essential to life. Inhalation, ingestion, and dermal exposure account for exposure possibilities from these sources [27]. When considering exposure from all toxicants, including pharmaceuticals and other drugs, the greatest and most rapid effects are achieved from exposures by the following routes in approximate decreasing order of effectiveness: intravenous, inhalation, intraperitoneal, subcutaneous, intramuscular, oral, and dermal [22].

In addition to the components of environmental habitats mentioned above, two other venues must be discussed as possible settings of exposure. The domestic environment provides daily living space, and modern advances have increased health and longevity. These advances also form opportunities for exposure to substances that may have toxic effects [27]. The occupational environment forms the second locality associated with toxic exposures. Industrial activity involves many potential toxicants, and care must be taken by workers to avoid harmful doses. This also includes taking preventative measures to limit exposures to toxicants of unknown risk and implementing surveillance programs to detect and limit unexpected effects in susceptible individuals [25].

6.3.5 Dose Response

Potential toxicity between individuals is modulated by a large number of factors, including age, diet, interactions with other chemicals, and individual phenotypic and genotypic variation [26]. In order to measure the potential acute hazard of a chemical, it is useful to examine the range of potential responses of a population of exposed individuals. By plotting the response of a population of organisms to differing doses, information is determined about the entire population. If the curve has sufficient dose-response information, we extrapolate useful values, such as threshold dose and LD_{50}. Threshold dose determines the point where effect is first observed and implies that doses below this threshold will show no effect. This no observed effect level (NOEL) is an important concept, as it implies there is some level of intake below which no effects will be observed and can be considered safe. However, some chemicals that have carcinogenic or other chronic effects may not show acute toxicity, and thus for regulatory purposes must be treated as a no-threshold phenomenon [29]. LD_{50} is another measure often used to give information about a population and represents a dose lethal to 50% of the population. If some other effect is measured, the effective dose for 50% of the population is referred to as the ED_{50}. These dose-response testing results can be used in risk assessment and identification of doses to be used in other sublethal, chronic toxicity tests, as well as providing insight into the mode of action of acute toxicity [28].

6.3.6 Tolerance

Organisms adapt to their environment through a variety of mechanisms. They are also able to adapt to doses of toxicants, whether or not the doses produce deleterious effects. These prior doses can provide a stimulus for some biochemical change to ameliorate or exacerbate the effects of subsequent doses of the same or closely related chemicals [22]. These adaptations fall into two categories: alteration of disposition of a toxicant to the site of biological effect (pharmacokinetics) and adaptive changes at the site of biological effect (pharmacodynamics) [29]. Each plays a role in affecting the toxic action on organisms, and previous exposure should be taken into account where possible to understand differences in dose response among individuals to a toxicant.

6.4 ADVERSE DRUG REACTIONS: IDIOSYNCRATIC AND DRUG-INDUCED LIVER INJURY (DILI)

6.4.1 Idiosyncratic Drug Reactions (IDRs)

IDRs are drug-induced toxicity and are adverse reactions (ARs) that are not related to known pharmacological properties of the drug and occur in only a small percentage of the population. Based on the total AR, IDRs account for $< 10\%$ of all ARs. Most of IDRs show no dose-response relationship, and they are unpredictable, often causing serious and life-threatening ARs. Obviously, the IDRs pose a major health concern in clinical drug development and after drug approval for public use. Several drugs that have been on the market for clinical usages are recognized to cause IDRs. In contrast to drugs that cause toxicity due to direct exposure, drugs that induce IDRs typically have a delayed onset of reaction following their administration and may involve various organs, including liver, skin, bone marrow, or perhaps a generalized systemic response.

One of the most challenging aspects facing pharmaceutical companies that develop drugs is whether the new candidate is causing IDRs, as the mechanism is not clearly understood. Several theories have been proposed, and IDRs can be generally categorized mechanistically as either (1) immune-mediated, or (2) nonimmune-mediated. There are some hypotheses for the mechanism behind IDRs, including:

- Hapten hypothesis: generates foreign protein from native by covalent binding [30].
- Danger hypothesis: proposed that not all covalent binding leads to immune response and second signal may be required, for example, cell death [31].
- Nonmetabolism-dependent immune response: mediate T cell stimulation with sulfamethoxazole [32]

In addition, there are hypotheses such as:

- Metabolic idiosyncrasy hypothesis
- Pharmacological interaction concept
- Mitochondrial injury hypothesis
- Inflammagen hypothesis

For both immune-mediated and nonimmune-mediated mechanisms, there is circumstantial evidence that reactive metabolites play a role in IDRs. In general, most drugs that cause idiosyncratic reactions are not sufficiently reactive in themselves, but most are metabolized to chemically reactive metabolites that may covalently bind to cellular proteins. Recent investigations continue to support the association of reactive metabolites and immune-mediated responses with IDRs [33–35]. As expected, IDRs can lead to drug failure due to the difficulty in predicting the clinical risk of the events and due to the ineffectiveness in identifying drugs that may potentially cause an IDR in preclinical toxicity studies in animals. Similar to the situation in humans, unpredictability of IDRs in animals limits the success of drug studies in in vivo preclinical toxicity studies. However, IDRs in animals do share some characteristics and mechanisms observed in humans, supporting the need for further study of animal models. Examples of animal models that most closely reflect human IDRs include sulfonamide hypersensitivity in dogs, nevirapine rash in rats, propylthiouracil (PTU)-induced autoimmunity in cats, and d-penicillamine autoimmunity in brown Norway (BN) rats. Although attempts with other animal models, such as procainamide-induced lupus and felbamate-induced liver toxicity and aplastic anemia, have been unsuccessful to date, they are examples of compounds where reactive metabolites are thought to play a role in ADRs and may provide some insight into possible mechanisms that may be relevant to IDRs. As evidenced from the many hypotheses that have been proposed, the mechanisms involved in the occurrence of IDRs are complex, and the development of appropriate animal models will be key to gaining a better understanding of the basic underlying mechanisms involved in IDRs. The development of chimeric mice that carry human immune system might be one model under investigation to potentially identify IDRs before human use of new drugs.

6.4.2 Drug-Induced Liver Injury

As mentioned, adverse drug reactions (ADRs) continue to be the major challenge for health providers and the most concerning issue related to drug therapy, as the safety of drug therapy sometimes can't be fully assessed. As a result, serious ADRs often emerge only after approval of a drug by the FDA [36]. Because of adverse reactions, 10% of approved drugs are either withdrawn or include a black box warning, as seen in Table 6.1 [37]. The liver, which plays an important role in the metabolism of drugs, is a frequent target of ADRs. As discussed in several chapters of this book, hepatotoxic drug reactions or metabolism-induced drug toxicity are often categorized as one of two types:

- *Type A,* which covers 75% of the observed drug-related toxicities. It is characterized by being:
  - Predictable from the known pharmacology or toxicology
  - Dose dependent
  - Drug metabolism may be involved
- *Type B,* hypersensitive reaction and covers 25%. It is characterized by the following:
  - Not predictable from the known pharmacology of the drug
  - Not simple dose dependency
  - Can be severe (Stevens Johnson Syndrome/TEN)
  - Drug metabolism is presumed to be involved

TABLE 6.1 Example of Drugs Associated with DILI and the Degulatory Actions, Withdrawals, Warnings, or Second on line to Withdrawal or Warnings.

Actions		
Withdrawals	Second Tiers	Warnings
Bromfenac	Tolcapone	Acetaminophen
Pemoline	Trovafloxacine	Nefazodone
Troglitazone	Felbamate	Rifampin
		Valproic Acid
		Interfron 1b-1b and 1a
		Bosentan
		Saquinavir
		Zifirlukast
		Telithromycin

As mentioned in previous sections, preclinical safety evaluation is first conducted by studies in a rodent and nonrodent species prior of testing new molecular entities in humans. The appropriate translation of preclinical safety risks to humans requires not only an understanding of the preclinical safety findings but also an appreciation of the sensitivity of the preclinical species to the identified toxicity risk in relation to man. Results from these preclinical data often showed poor correlation to human toxicities; the poorest correlation between human and animal toxicities have been found in the areas of hypersensitivity/dermal and hepatotoxicity. These types of toxicities resulted in a high termination rate of drug development programs.

By reviewing the clinical incidents of type A—dose-related reactions and type B—idiosyncratic reactions, type A reactions are dose dependent and most likely occur in overdosed individuals. A typical example is an AR due to acetaminophen overdose [38,39], causing hepatotoxicity. Hepatotoxicity or DILI is the most common reason for safety-related market withdrawal or regulatory actions. As mentioned in the previous section, idiosyncratic ADRs differ from type A reactions in that they are unpredictable and occur in a minority (often a quite small percentage) of patients at therapeutic doses, while DILI is dose related and many drugs have been associated with DILI reactions that occur in multiple forms:

- Acute hepatocellular necrosis
- Acute hepatitis
- Steatosis
- Cholestasis
- Chronic active hepatitis
- Fibrosis
- Chronic cholestasis
- Granulamatous hepatitis
- Vascular disorders

Drug metabolism is implicated in many but not all forms. As shown in Figure 6.1, when reactive metabolite occurs, it causes, for example, necrosis, and that triggers the elevation of liver enzymes such as elevations of serum aminotransferase (ALT) levels, ALT $> 3 -$ fold the upper limit of normal ($3 \times$ ULN). Several investigators have studied serum ALT and other biomarkers for DILI following the treatment of drugs that are known to cause liver injury [40–43]. The FDA has established certain guidelines and recommendations to avoid incidents of hepatotoxicity in drug development. The recommendations are

1. Avoid certain molecular structures that may form reactive metabolite.
2. Target daily dose to $< 10\ \text{mg/day}$.
3. In vitro testing indicates low covalent binding in liver microsomes.
4. The phase II metabolism indicates low production of glutathione conjugates.
5. Low incidence ($< 5\%$) of ALT $> 3 \times$ ULN in clinical trials.

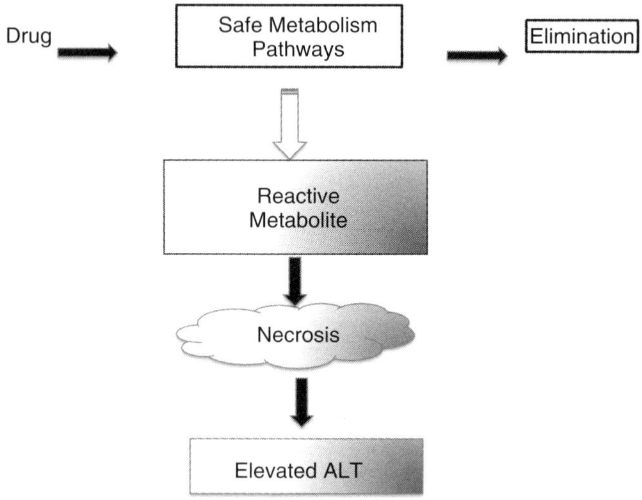

Figure 6.1 Pathway of drugs that undergo safe disposition mediated by metabolism vs. pathway of drug associated with DILI.

DILI can result in permanent disability or death and lead to withdrawals of many drugs or obligated dose restrictions in the use of efficacious drugs. An example of a drug that caused idiosyncratic hepatotoxicity is troglitazone (TGZ), which was withdrawn from the market by the FDA in 2000 [44]. TGZ is a peroxisome proliferator-activated receptor (PPAR)-γ agonist that was used to treat type 2 diabetes. The main reason for its withdrawal was that TGZ was associated with the development of acute liver failure [45], and it accounted for 10% of all IDRs between 1998 and 2001 [44]. Although TGZ was beneficial in reducing insulin resistance, 94/1.92 million patients reported liver failure [46] and the majority of these patients died, whereas others required liver transplants.

Similar to IDRs, DILI reactions are difficult to predict, and the mechanism remains absent and is generally not reproducible in traditional animal models used in preclinical safety evaluation. For example, oral administration of TGZ at large doses (60- to 120-fold larger than the therapeutic dose) to monkeys for 52 weeks did not increase levels of serum liver enzymes and had little gastrointestinal, hematologic, or hepatic effects [47]. The lack of preclinical animal models causes mechanisms of DILI to remain poorly understood, so that appropriate action cannot be taken to prevent or treat these reactions.

As mentioned above, elevation of liver enzymes such as ALT has been viewed as a biomarker for DILI; however, novel biomarkers such as sorbitol dehydrogenase (SDH), glutamate dehydrogenase (GLDH), and miRNA (miR-122) have been reviewed as novel biomarkers to detect DILI. SDH is localized in cytoplasm and mitochondria of liver and kidney. Its changes are parallel to ALT with drugs that are known to cause DILI such as enoxaparin, adomiparin, and dalteparin. For GLDH, it is

TABLE 6.2 Fold Changes in Biomarker Data Shown as the Fold Difference for the Average Biomarker Values of Two Dose Groups and Time Point When Compared to Control (vesicle) Data at Day 29.

Days	ALT	GLDH	SDH	miR-122
	Fold Diff.	Fold Diff.	Fold Diff.	Fold Diff.
		15 mg/kg		
Day 7	10.8	10.4	4.9	78.3
Day 14	7.6	3.0	2.1	54.1
Day 29	11.8	3.6	3.6	6.1
		100 mg/kg		
Day 7	29.6	14.6	9.6	105.7
Day 14	29.1	10.8	5.0	43.57
Day 29	26.2	9.3	4.0	18.0

characterized as high sensitivity and tissue specificity. The change is dose dependent but prolonged persistence compared to ALT. It localizes in mitochondria matrix of liver more specific than in kidney. miR-122 is expressed specifically in liver and measured in serum. Both dose and exposure cause changes in expression and earlier than any detected changes in ALT. Table 6.2 shows the correlation between ALT and GLDH, SDH, and miR-122 [48].

6.5 IN VITRO DETERMINATION OF REACTIVE METABOLITE FORMATION, OXIDATIVE STRESS, MITOCHONDRIAL DAMAGE, AND NEPHROTOXICITY

The incorporation of in vitro methods for the detection of *reactive metabolites*, oxidative stress, and mitochondrial permeability/damage early in the lead optimization phase of drug development provides significant data and the decision of candidate selection with the highest probability of success in the preclinical safety studies. As discussed in other chapters, compounds undergoing metabolic activation may form soft or hard electrophiles. Soft electrophiles such as quinones, epoxides, quinone imines, and quinone methides can be trapped by glutathione (GSH) [49]. Hard electrophiles such as iminium species and aldehydes do not bind well to GSH but can be trapped using potassium cyanide (KCN) and the products detected by mass spectrometry or LC/MS/MS [50]. Monitoring GSH depletion, instead of GSH-adduct, offers an indirect method for determining bioactivation of new drug candidates, by conducting a 30- to 60-min incubation reaction with a standard reaction mixture containing potassium phosphate buffer (100 mM; pH 7.4), hepatic microsomes (0.5 – 1 mg/mL), NADPH (1 mM), test drug (10 μM), and GSH (1 mM), and initiation with NADPH. The reaction is stopped with either an equal volume of ice-cold acetonitrile or 10% trichloroacetic acid added in a volume that is 15% of the total reaction volume. After centrifugation the supernatant is processed, and the GSH depletion can be monitored [51]. Azathioprine is a purine synthesis inhibitor that results in the loss of adenosine triphosphate (ATP). It is also known to disrupt mitochondrial function, leading to oxidative stress and depletion of GSH. Leflunomide treatment caused a significant loss in cellular GSH levels at low exposure concentrations. It is important to note that in each of these examples, cell viability remained high, as determined by membrane integrity (MemTox). Endpoints that indicate mitochondrial effects (3-(4,5-Dimethylthiazol-2-yl)-2,5 diphenyltetrazolium bromide (MTT) or ATP) and cell proliferation can be affected by exposure concentrations and times that do not cause cell death. MTT can be reduced intracellularly by reduction reactions in mitochondria, cytosol, endoplasmic reticulum, and lysosomes. Therefore, endpoints such as GSH depletion and depletion of markers for energy production as indicators of mitochondria function such as ATP or mitochondrial membrane potential are commonly used as markers for assessing cytotoxicity for in vitro toxicity assessment, though the interpretation of effect is greatly influenced when more than one endpoint is used. Thus, it is important to use multiple endpoints in the in vitro model for correct interpretation of the data obtained [52].

Phenothiazines, such as thioridazine and chlorpromazine, are often administered together with antidepressants to treat many disorders including psychotic depression and depression associated with schizophrenia. Despite the clear benefits of these drugs, there have been reported cases of rare, but severe, hepatic injury [53]. The hepatic injury associated with these drugs was labeled as idiosyncratic in nature. Recent studies have demonstrated that many of the idiosyncratic drug toxicities in liver can be traced to the formation of reactive metabolites [54]. The latter drugs were found to undergo extensive "first-pass" metabolism. The primary metabolites generated from phase I P450 enzymes are the products of hydroxylation, N-dealkylation, N-oxidation, and S-oxidation. The 7-hydroxy form of both the drugs can undergo additional oxidation to form an electrophilic quinone imine. The enzymes involved in the formation of the reactive intermediates are CYP2D6 and CYP1A2, with the former being more active than the latter. One possible explanation for the idiosyncratic nature of the liver toxicity associated with these drugs could be that CYP2D6 represents a small percentage of the total hepatic P450 enzymes under normal conditions. Therefore, in order for toxicity to occur, a larger amount of drug would need to enter the liver following absorption from the gastrointestinal (GI) tract because CYP2D6 is in very low levels in the intestine. Metabolism in normal individuals by CYP3A4 in the intestine to nonreactive metabolites could greatly reduce the amount of parent drug that the liver would receive and hence reduce the formation of the reactive intermediate mediated by CYP2D6. In case of coadministered drug that inhibits CYP3A4 metabolism in both the intestine and the liver, the amount of drug reaching the liver would be greater. Accumulated drug is shunted to secondary pathways that produce toxic metabolites and a higher tissue concentration of the toxic intermediates, which results in liver injury. GSH binding can serve as a surrogate for binding to more complex proteins.

In addition, some metabolites lead to the generation of *oxidative stress* via an increase in reactive oxygen species (ROS) or nitrogen species, which would also not bind directly to GSH. Therefore, in addition to determining bioactivation potential, it is important to monitor the formation of ROS (superoxide anion and hydrogen peroxide), lipid peroxidation, and redox enzymes (GSH peroxidase, GSH reductase, catalase, and superoxide dismutase). Drug-induced changes in mitochondrial membrane potential can cause a release of superoxide anion and hydrogen peroxide into the cytosol. As a result, it is also important to monitor compound effects on mitochondrial membrane potential. The accumulation of ROS can be measured using the fluorescent probe 5-(and-6)- carboxy-2′,7′-dichlorofluorescein diacetate (CM-DCFDA) [55]. In this assay, nonfluorescent DCFDA readily crosses the cell membrane. Once in the cell, the acetate moiety is cleaved by endogenous esterases, which prevents leakage out of the cells and activates dichlorofluorescin (DCF) for binding to ROS (e.g., hydroperoxides and lipid peroxides that produce a fluorescence signal) [36]. The cleaved molecule cannot cross cell membranes and is therefore trapped inside the cell, thus rat hepatoma (H4IIE) cells can be used as a model here. The formation of ROS occurs rapidly, so it is important to conduct this experiment over shorter exposure periods (e.g., 1–4 h). The loading concentration of the DCFDA is typically $2-10$ μM for 60 min. An increase in fluorescence due to ROS binding is measured at an excitation wavelength of 503 nm and an emission wavelength of 529 nm. Two hydroperoxides, t-butyl hydroperoxide and cumene hydroperoxide, produced rapid increases in the DCF signal. Rotenone inhibits mitochondrial respiration, which leads to a release of superoxide anion. This process requires more time, which is detected by the assay as a delay in the increase in fluorescence. In comparison, a chemical like menadione, which requires metabolism via CYP1A1/2 to the reactive quinone, requires more time to increase the oxidative stress signal.

Mitochondrial damage can also cause an increase in ROS. This can occur when the membrane potential is disrupted, which allows superoxide anion to leak into cell cytosol. Therefore, it is also important to evaluate the effects of new drug candidates on mitochondrial membranes. A commonly used assay is based on the uptake of the fluorescent dye 5,5′, 6,6′-tetrachloro-1,1′, 3,3′-tetraethylbenzimidazolylcarbocyanine iodide (JC-1). JC-1 is taken up by healthy mitochondria, where it forms aggregates as the inner membrane becomes polarized. The ratio of the monomer to aggregate JC-1 is measured and used to determine the drug-induced changes in mitochondrial membrane potential. Results are collected by measurements at an excitation wavelength of 490 nm, then quantifying the monomer at 535 nm, and the aggregate at 595 nm. The ratio of aggregate to monomer fluorescence determines the effect on mitochondrial membrane potential [55,56].

For *nephrotoxicity*, a recent developing model by Brown and collaborators (2008) [57] has demonstrated the use of proximal tubular cells, isolated from human and rat kidney, as in vitro cell models to assess the potential nephrotoxicity early before screening drugs in preclinical animal models or before drugs reach the clinical trials stage. Because recent reports suggest that around 50% of drugs that successfully pass animal safety tests fail in subsequent human trials, this can become a huge financial burden to the pharmaceutical industry. Furthermore, 20% of drug attrition during development is attributable to drug-induced nephrotoxicity—the toxic effects of new chemical entity (NCE) to kidney structure and functions. Brown and his colleagues

(2015) [60] proposed that identifying the NCEs that have toxicity potential, or those that missed in recognized toxicity potential in animal models before human use, can offer several benefits, including saving human lives and a company over billion of dollars in development costs.

The kidney, in particular the first part of the nephron known as the proximal tubule, is very important in removing drug molecules from the body. The proximal tubule achieves this through a series of specialized proteins that transport drug molecules from the blood into the urine. Brown and collaborators developed novel in vitro cell culture models of the kidney [see Figure 6.2(A) for the preparation scheme] as tools to understand this process and identify why some drug molecules or combinations of these molecules result in kidney toxicity. Developing highly differentiated cell culture models for the human and rat proximal tubule (Proximate cells), which mimic the physiology of an in vivo proximal tubule, promised a successful screen for nephrotoxicity [57–60]. These models will provide substantial new mechanistic and species-specific data about the nephrotoxic potential of new drug molecules. In essence, the strategy would replace ineffective preclinical screening of compounds in animals with effective screening in advanced cultured cell models. This would reduce the number of animals entering mandatory preclinical drug safety testing by

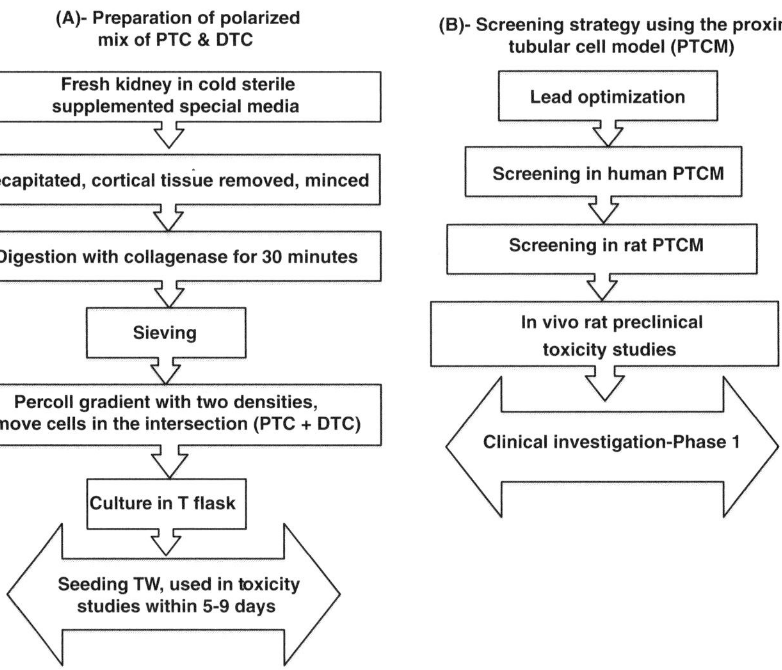

Figure 6.2 Preparation of in vitro model to study and predict the renal drug disposition and toxicity (A) and screening strategy (B) for nephrotoxicity using human PTCM and rat PTCM prior to in vivo animal screening and human clinical trials.

identifying the compounds with the potential to generate human toxicity at the in vitro screening stage. A highly predictive in vitro human and rat model of the primary proximal tubule cell (HPTCM and RPTCM) is important for identifying the candidates that can be evaluated in preclinical animal models and subsequently in humans. The screening strategy is depicted in Figure 6.2(B).

6.6 PRESENT AND FUTURE FOR ASSESSING TOXICITY IN DRUG DISCOVERY AND DEVELOPMENT

An essential component of any drug discovery program is to understand the potential adverse effects associated with new drug candidates. It is clear that extensive presystemic metabolism can reduce bioavailability, and production of reactive metabolites can increase the risk of toxicity. Many new compounds can be enzymatically converted to reactive intermediates capable of binding to cellular proteins or DNA. In vitro methods for characterizing metabolism, bioactivation, and cytotoxicity are now included in the screening paradigm. Emphasis has been placed on the relationship between the in vitro concentration that causes adverse effects and the expected in vivo maximum therapeutic plasma concentration. Using in vitro models helps to identify potential toxicity and relate it to the observed effects to the in vivo parameter.

Primary hepatocytes in sandwich culture is considered one of the critical models to determine drug metabolism and toxicity in vitro. Hepatocytes cultured between two layers of collagen or Matrigel provide sustained metabolic capacity of phase I and phase II metabolism and hepatobiliary transporters. This model provides a means of measuring metabolic stability, reactive metabolite formation by measuring GSH conjugates and depletion of GSH, and oxidative stress using DCFDA.

In general, cell culture models should be used with multiple endpoints that measure cell health and a broad range of exposure concentrations. These include ATP levels, cell proliferation, reduction of MTT, and membrane integrity [52]. These parameters combined provide a potential adverse event by new drug candidates in vivo. Hepatocytes from rat, dog, and human can provide insight into potential species differences and may guide the selection of the nonrodent species required in preclinical drug safety studies. In a study by Harrill et al. (2014) [48] with NP260, a drug candidate that was designed as a first-in-class selective antagonist of α4-subtype $GABA_A$ receptors that had promising efficacy in animal models of pain, epilepsy, psychosis, and anxiety was complicated following a 28-day safety study in dogs in which there were pronounced elevations of serum ALT levels. There was no histopathological indication of hepatocellular injury. Analysis of serum samples from the 28-day dog study for liver specific biomarkers miR-122 as well as enzymatic biomarkers GLDH and SDH indicated liver necrosis. Cytotoxicity assessments were conducted in hepatocytes derived from dog, rat, and human liver samples to address the species specificity of the liver response to NP260. All biomarkers, except ALT, returned toward baseline by day 29 despite continued drug treatment, suggesting adaptation to the initial injury. In vitro analysis of the toxicity potential of NP260 to hepatocytes indicated a relative sensitivity of dog > human > rat. These results may explain, in part, the liver effects that were not evident in the rodent safety studies and a diagnostic biomarker approach, coupled with sensitive in vitro screening strategies such as measuring the changes in ATP and GSH, may facilitate interpretation of toxicity potential when an adaptive event masks the underlying toxicity.

In addition to using hepatocytes, PTCM from human and rat kidney can also be used to screen for nephrotoxicity, as 20% of drug-induced toxicity is attributable to nephrotoxicity before in vivo investigation in animal models and before drug candidates administered in humans within clinical studies, as indicated in Figure 6.2 [57–60].

Furthermore, in 2007 the National Academy of Sciences (NAS) published a report noting that the toxicology testing methods being used at that time were descriptive in nature. Methods included using a large number of laboratory animals and making a number of assumptions and extrapolations. As an example, the report noted that many of the tests in use exposed laboratory animals to higher doses of a chemical than were physiologically relevant for human exposure, and many of these tests did not take into account species differences [61]. In addition, with the urgency to have complete toxicity profiles for these compounds, traditional approaches were not an option. The NAS report recommended (1) the use of high throughput techniques to study biologic activity of thousands of chemicals, (2) the use of a systems biology approach to understand how changes in a biological pathway result in a disease state, and (3) the use of cells and tissues of human origin [37]. In order to address the concerns identified by the 2007 NAS report and implement the recommendations, a collaboration was formed between the NIEHS, the NTP, the EPA, the FDA, and the National Human Genome Research Institute. The goal of this collaboration is to take the recommendations of the NAS and advance the state of toxicity testing using high throughput screening methods enabled by advances in computing, robotics, and cell-based analytical approaches [62]. Additionally, this collaboration seeks to identify new mechanisms of chemical activity in cells to better predict human response to toxic substances. The strategy being used by the Tox21 collaborators uses a tiered approach with an integrated testing strategy to chemical toxicity testing. In the first tier, chemicals of interest are tested at multiple concentrations using high throughput screening assays, which employ a multiwell system using in vitro cell models, ideally of human origin [61]. The effects of these chemicals on cell viability and ability to induce apoptosis, as well as activation and or inhibition of key signaling pathways associated with disease, are tested. Chemicals with "positive hits" are then further tested in the second tier using a medium throughput screening system with alternative model systems such as zebrafish in multiwell platforms. Chemicals that have positive hits would then be tested in laboratory animals (see Figure 6.3).

In order to further address the recommendations of the NAS report the Adverse Outcome Pathways (AOP) tool was created. The idea behind the tool is to provide a simplified mechanistic representation of the existing knowledge linking chemical interactions with biological processes at varying levels of complexity and an adverse human or environmental health risk [63–65]. The overall goal of an AOP is to provide a framework that connects the events of the chemical interaction that starts a molecular initiating event (MIE) and through a series of key events leads to an adverse

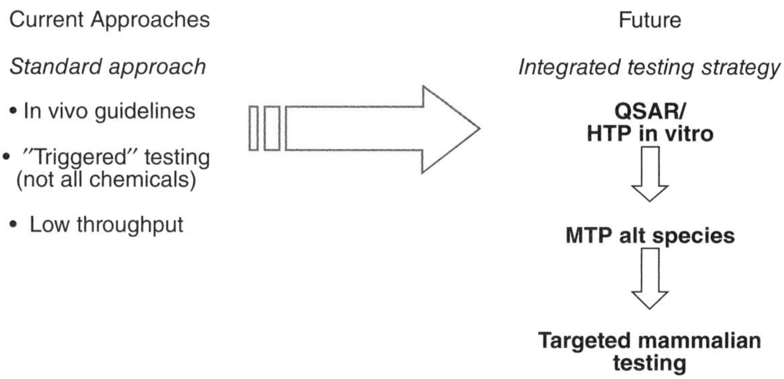

Figure 6.3 Integrated testing strategy as recommended by National Research Council using tiered testing approach employing QSAR and high throughput in vitro testing, medium throughput testing (MTP) using alternative species, and targeted mammalian testing.

outcome (AO) relevant to risk assessment. The information used to feed each of the building blocks of an AOP may be the result of an in-depth survey of the relevant scientific literature and/or results from experimental studies. However, other relevant information such as structural data, "omics" data, or in vivo and in vitro data may also be used to build or strengthen an AOP. An AOP is composed of a single MIE and results in a single AO but may involve an infinite number of intermediary steps. Each of the intermediary steps in an AOP should be scientifically sound and encompass knowledge of the normal physiological processes that are perturbed and cause the AOP (Figure 6.4). Within the intermediary steps, key events (KE) may be identified.

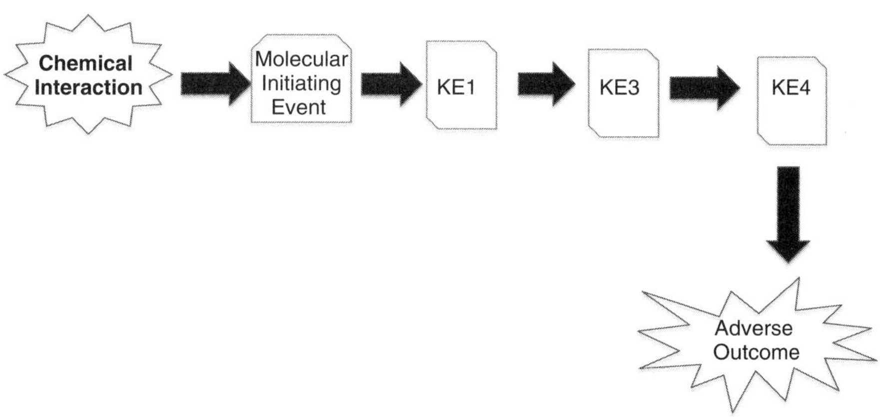

Figure 6.4 Adverse outcome pathway (AOP) structure. An AOP begins with a chemical interaction leading to a molecular initiating event, then through a series of key events (KE) leading to an adverse outcome (AO).

REFERENCES

1. Gallo M. History and scope of toxicology. In: Casarett and Doull's toxicology: the basic science of poisons, ed. Casarett LJ, Doull J, Klaassen CD. New York: McGraw-Hill, 2001, chap. 1.
2. Debus AG. A philosophical path for paracelsian medicine: the ideas, intellectual context, and influence of Petrus Severinus (1540–1602) (review). Bull Hist Med 2005;79:806–808.
3. Pagel W. Paracelsus: an introduction to philosophical medicine in the era of the renaissance. Basel: Karger, 1982.
4. Debus AG. Paracelsus, five hundred years: three American exhibits. Bethesda, MD: U.S. National Library of Medicine, 1998, 4.
5. Franco G. Ramazzini and workers' health. Lanct 1999;354:858–861.
6. Waller RE. 60 years of chemical carcinogens: Sir Ernest Kennaway in retirement. JRSM 1994;87:96–97.
7. Swann JP. FDA history, part 1. FDA, 2009.
8. Environmental Protection Agency. Federal Insecticide, Fungicide, and Rodenticide Act (FIFRA), 2012. [Online]. Available: http://www.epa.gov/agriculture/lfra.html.
9. Bicknell F. Chemicals in food and in farm produce: their harmful effects. London: Faber and Faber, 1960.
10. Murray B. Our synthetic environment. New York: Knopf, 1962.
11. Paull J. The Rachel Carson letters and the making of Silent Spring. SAGE Open 2013;July–Sept:1–12.
12. Linda JL. Rachel Carson: witness for nature. New York: Holt, 1997.
13. Environmental Protection Agency. The Guardian: Origins of the EPA, 2014. [Online]. Available: http://www2.epa.gov/aboutepa/guardian-origins-epa.
14. Fintel B, Samaras AT, Carias E. The Thalidomide Tragedy: Lessons for Drug Safety and Regulation, 2009. Available: http://helix.northwestern.edu/article/thalidomide-tragedy-lessons-drug-safety-and-regulation.
15. Teo SK, Colburn WA, Tracewell WG, Kook KA, Stirling DI, Jaworsky MS, Scheffler MA, Thomas SD, Laskin OL. Clinical pharmacokinetics of thalidomide. Clin Pharmacokinet 2004;43(5):311–327.
16. Frances Oldham Kelsey Chemical Heritage Foundation, 2014. Available: http://www.chemheritage.org/discover/online-resources/chemistry-in-history/themes/public-and-environmental-health/food-and-drug-safety/kelsey.aspx.
17. D'Amato RJ, Loughnan MS, Flynn E, Folkman J. Thalidomide is an inhibitor of angiogenesis. Retina 1996;16:268.
18. Swann JP. FDA history, part 3. FDA, 2009.
19. National Institute of Environmental Health Sciences. Benchmarks in toxicology. 2014.
20. Kadlubar FF. In memoriam: James A. Miller (1915–2000). Chem Res Toxicol 2001;14(4):335–337.
21. National Toxicology Program. About NTP, 2014. [Online]. Available: https://ntp.niehs.nih.gov/about/index.html.
22. Kacew S, Lee B-M. Lu's basic toxicology: fundamentals, target organs, and risk assessment. Boca Raton, FL: CRC Press, 2012.

23. Eaton DL, Steven GG. Principles of toxicology. In: Casarett and Doull's toxicology: the basic science of poisons, ed. Casarett LJ, Doull J, Klaassen CD. New York: McGraw-Hill, 2001, chap. 2.

24. Liu J, Goyer RA, Waalkes MP. Toxic effects of metals. In: Casarett and Doull's toxicology: the basic science of poisons, ed. Casarett LJ, Doull J, Klaassen CD. New York: McGraw-Hill, 2001, chap. 23.

25. Thorne PS. Occupational toxicology. In: Casarett and Doull's toxicology: the basic science of poisons, ed. Casarett LJ, Doull J, Klaassen CD. New York: McGraw-Hill, 2001, chap. 33.

26. Cope WG, Hodgson E. Classes of toxicants: use classes. In: A textbook of modern toxicology. Hoboken, NJ: Wiley, 2010, chap. 4.

27. Cope WG. Exposure classes, toxicants in air, water, soil, domestic, and occupational settings.' In: A textbook of modern toxicology. Hoboken, NJ: Wiley, 2010, chap. 3.

28. LeBlanc GA. Acute toxicity. In: A textbook of modern toxicology. Hoboken, NJ: Wiley, 2010, chap. 10.

29. Hodgson E, Mailman RB, Chambers JE. Dictionary of toxicology. New York: Van Nostrand Reinhold, 1988.

30. Riley RJ, Leeder JS. In vitro analysis of metabolic predisposition to drug hypersensitivity reactions. Clin Exp Immunol 1995;99(1):1–6.

31. Pirmohamed M, Naisbitt DJ, Gordon F, Park BK. The danger hypothesis—potential role in idiosyncratic drug reactions. Toxicology 2002;181–182:55–63.

32. Engler OB, Strasser I, Naisbitt DJ, Cerny A, Pichler WJ. A chemically inert drug can stimulate T cells in vitro by their T cell receptor in non-sensitized individuals. Toxicology 2004;197(1):47–56.

33. Harrill AH, Roach J, Fier I, Eaddy JS, Kurtz CL, Antoine DJ, Spencer DM, Kishimoto TK, Pisetsky DS, Park BK, Watkins PB. The effects of heparins on the liver: application of mechanistic serum biomarkers in a randomized study in healthy volunteers. Clin Pharmacol Ther 2012;92(2):214–220.

34. Kroh EM, Parkin RK, Mitchell PS, Tewari M. Analysis of circulating microRNA biomarkers in plasma and serum using quantitative reverse transcription-PCR (qRT- PCR). Methods 2010;50(4):298–301.

35. Lagos-Quintana M, Rauhut R, Yalcin A, Meyer J, Lendeckel W, Tuschl T. Identification of tissue-specific microRNAs from mouse. Curr Biol 2002;12(9):735–739.

36. O'Brien PJ, Slaughter MR, Polley SR, Kramer K. Advantages of glutamatedehydrogenase as a blood biomarker of acute hepatic injury in rats. Lab Anim 2002;36(3):313–321.

37. Wojcinski ZW, Uetrecht JP. Idiosyncratic drug reactions. In: Encyclopedia of drug metabolism and interactions, ed. Lyubimov AV. Hoboken, NJ: Wiley, 2012, 8:1–29.

38. Dieckhaus CM, Thompson CD, Roller SG, et al. Mechanisms of idiosyncratic drug reactions: the case of felbamate. Chem-Biol Interact 2002;142:99–117.

39. Watkins PB. Idiosyncratic liver injury: challenges and approaches. Toxicol Pathol 2005;33(1):1–5.

40. Zou W, Roth RA, and Ganey PE. Animal models of idiosyncratic, drug-induced liver injury: emphasis on the inflammatory stress hypothesis. Idiosyncratic drug reactions. In: Encyclopedia of drug metabolism and interactions, ed. Lyubimov AV. Hoboken, NJ: Wiley, 2012, 8:1–25.

41. Lasser KE, Allen PD, Woolhandler SJ, et al. Timing of new black box warnings and withdrawals for prescription medications. JAMA 2002;287:2215–2220.

42. Amar PJ, Schiff ER. Acetaminophen safety and hepatotoxicity—where do we go from here? Expert Opin Drug Saf 2007;6(4):341–355.

43. Larson AM, Polson J, Fontana RJ, et al. Acetaminophen-induced acute liver failure: results of a United States multicenter, prospective study. Hepatology 2005;42(6):1364–1372.

44. Ostapowicz G, Fontana RJ, Schiodt FV, et al. Results of a prospective study of acute liver failure at 17 tertiary care centers in the United States. Ann Intern Med 2002;137:947–954.

45. Chojkier M. Troglitazone and liver injury: in search of answers. Hepatology 2005;41:237–246.

46. Graham DJ, Green L, Senior JR, et al. Troglitazone-induced liver failure: a case study. Am J Med 2003;114:299–306.

47. Rothwell C, McGuire EJ, Altrogge DM, et al. Chronic toxicity in monkeys with the thiazolidinedione antidiabetic agent troglitazone. J Toxicol Sci 2002;27:35–47.

48. Harrill AH, Eaddy JS, Rose K, Cullen JM, Ramanathan L, Wanaski S, Collins S, Ho Y, Watkins PB, Lecluyse EL. Liver biomarker and in vitro assessment confirm the hepatic origin of aminotransferase elevations lacking histopathological correlate in beagle dogs treated with GABAA receptor antagonist NP260. Toxicol Appl Pharmacol 2014;277(2):131–137.

49. Prakash C, Sharma R, Gleave M, et al. In vitro screening techniques for reactive metabolites for minimizing bioactivation potential in drug discovery. Curr Drug Metab 2008;9(9):952–964.

50. Kalgutkar AS, Vaz AD, Lame ME, et al. Bioactivation of the nontricyclic antidepressant nefazodone to a reactive quinone-imine species in human liver microsomes and recombinant cytochrome P450 3A4. Drug Metab Dispos 2005;33(2):243–253.

51. Griffith OW. Determination of glutathione and glutathione disulfide using glutathione reductase and 2-vinylpyridine. Anal Biochem 1980;106(1):207–212.

52. McKim JM Jr. Building a tiered approach to in vitro predictive toxicity screening: a focus on assays with in vivo relevance. Comb Chem High Throughput Screen 2010;13(2):188–206.

53. Guo JJ, Wigle PR, Lammers K, et al. Comparison of potentially hepatotoxic drugs among major US drug compendia. Res Soc Adm Pharm 2005;1(3):460–479.

54. Williams DP, Park BK. Idiosyncratic toxicity: the role of toxicophores and bioactivation. Drug Discov Today 2003;8(22):1044–1050.

55. Qu B, Li QT, Wong KP, et al. Mechanism of clofibrate hepatotoxicity: mitochondrial damage and oxidative stress in hepatocytes. Free Radic Biol Med 2001;31(5):659–669.

56. MKim JM Jr. In vitro toxicity screening in early drug discovery: importance of metabolism and reactive metabolites. In: Encyclopedia of drug metabolism and interactions, ed. Lyubimov AV. Hoboken, NJ: Wiley, 2012, 7:1–28.

57. Brown CD, Sayer R, Windass AS, Haslam IS, De Broe ME, D'Haese PC, Verhulst A. Characterisation of human tubular cell monolayers as a model of proximal tubular xenobiotic handling. Toxicol Appl Pharmacol 2008;233(3):428–438.

58. Verhulst A, Sayer R, De Broe ME, D'Haese PC, Brown CD. Human proximal tubular epithelium actively secretes but does not retain rosuvastatin. Mol Pharmacol 2008;74(4):1084–1091.

59. Verhulst A, D'Haese PC, De Broe ME. Inhibitors of HMG-CoA reductase reduce receptor-mediated endocytosis in human kidney proximal tubular cells. J Am Soc Nephrol 2004;15(9):2249–2257.

60. Brown C. Testing for toxicity. International Innovation 2015; issue 167:98–99.

61. Krewski D, Acosta D Jr, Andersen M, Anderson H, Bailar JC 3rd, Boekelheide K, Brent R, Charnley G, Cheung VG, et al. Toxicity testing in the 21st century: a vision and a strategy. J Toxicol Environ Health B Crit Rev 2010;13(2–4):51–138.

62. National Institutes of Health, National Center for Advancing Translational Sciences. Toxicology in the 21st Century, 2014. Available: http://www.ncats.nih.gov/research/reengineering/tox21/tox21.html.

63. Ankley GT, Bennett RS, Erickson RJ, Hoff DJ, Hornung MW, Johnson RD, Mount DR, Nichols JW, Russom CL, Schmieder PK, Serrrano JA, Tietge JE, Villeneuve DL. Adverse outcome pathways: a conceptual framework to support ecotoxicology research and risk assessment. Environ Toxicol Chem 2010;29(3):730–741.

64. Vinken M. The adverse outcome pathway concept: a pragmatic tool in toxicology. Toxicol 2013;312:158–165.

65. Organisation for Economic Co-operation and Development. Proposal for a template and guidance on developing and assessing the completeness of adverse outcome pathways. OECD, 2012.

7

TOXICOKINETICS AND TOXICITY TESTING IN DRUG DEVELOPMENT[1]

7.1 INTRODUCTION: TOXICOKINETICS AND ITS RELATIONSHIP WITH PHARMACOKINETICS AND ADME IN PRECLINICAL DEVELOPMENT

As illustrated in previous chapters of this book, the relationship between drug dose, exposure, plasma, and body tissue concentration with time occurs as a result of the absorption, distribution, metabolism, and excretion (ADME) process and pharmacokinetics (PK) of the drug. The response of drug treatment in relation to time and the drug concentration in the body is described by the pharmacodynamics (PD) [1].

Toxicokinetics (TK) defines the toxicity effects as a function of drug exposure. Hence, both PK and TK can be viewed, according to Varma et al. (2012) [2], as biomarkers of drug exposure as well as markers of efficacy and safety. As mentioned, the determinants of the PK/TK of a drug are the ADME parameters, and the modern practice of late drug discovery and early development requires that both PK and TK analyses be determined as accurately as possible [3]. This assessment is helpful in (1) selection of dose for preclinical toxicity testing and dosage route, (2) understanding drug exposure, and (3) identifying dose for first in human (FIH) studies and the follow-up clinical trials.

The objective of this chapter is to provide a basic review of drug toxicity measurements in the in vivo preclinical setting and introduce fundamentals of PK and TK

[1] Also contributed by Gabriel A. Knudsen, PhD

Translational ADMET for Drug Therapy: Principles, Methods, and Pharmaceutical Applications, First Edition. Souzan B. Yanni.

and their applications in drug discovery and development. The genotoxicity, oncogenicity, reproductive toxicity versus toxicogenomics, and biomarkers in preclinical species are discussed.

First, as an introduction, toxicity testing is the way to measure the toxic effects that may be produced after exposure to chemicals or xenobiotics. Several toxicity assessment studies (e.g., acute, single, repeat-dose studies, GLP) in both rodent and nonrodent species are used to better refine safety margins and PK/PD modeling and set appropriate dosages before starting FIH clinical trials. Critical parameters that define the PK/TK profile are area under the concentration curve (AUC), maximal plasma concentration (C_{max}), time of maximum concentration (T_{max}), clearance (CL), apparent volume of distribution (V_d) apparent volume of distribution at steady state (V_{dss}), half-life ($t_{1/2}$), and bioavailability (F%). Toxicity tests may not assess the safety of xenobiotics. A tiered approach is typically taken that may incorporate tests from some or all of the types of studies described in subsequent sections, depending on the guidelines described by the regulatory body charged with jurisdiction (e.g., Food and Drug Administration, Environmental Protection Agency, etc.). The proliferation of new "omics" technologies has yielded a call for the use of more and more in vitro tests to take advantage of pathway-based approaches and move away from apical endpoints (target organ damage, mutagenesis, teratogenesis, etc). Now there is an emerging and increased utility of toxicogenomics to better define drug toxicity/exposure to support compound selection and safety risk assessment. This new path was outlined in 2007 by a joint effort of the National Academy of Sciences and the National Research Council [4], but validation of the methods still remains a significant hurdle to implementation of high-throughput toxicity screening. Major advances in this area have been made by the U.S. Tox21 interagency collaboration in developing and evaluating cellular models for the evaluation of environmental chemicals using high-throughput methods [5]. An excellent series of protocols with descriptions is available from the International Conference on Harmonization of Technical Requirements for Registration of Pharmaceuticals for Human Use (ICH), along with information on the recent changes to harmonize toxicity studies in the United States, Europe, and Japan (ICH guideline M3) [6, 7].

7.2 TYPES OF PRECLINICAL DOSING THAT SUPPORT TOXICOKINETICS

Several toxicology studies are conducted during drug discovery and all the way to the late stages of drug development before a new drug application (NDA) filing is made. Despite comprehensive toxicity assessment in early and late-stage discovery, attrition of drug candidates in clinical studies due to adverse drug reaction or general drug-related toxicity is common, owing to disconnect in predictions of risk in humans on the basis of preclinical data obtained from cell culture and animal models. However, extensive preclinical assessment and appropriate scaling and modeling tools have showed an improvement in the predictions [8]. In general, the correlation between human and animal toxicities is good for conditions such as cardiovascular, hematological, and gastrointestinal diseases, with the poorest correlation for adverse drug reactions such as idiosyncratic reactions, skin rash, hypersensitivity, and hepatotoxicity.

As mentioned, toxicology testing in drug discovery is initiated by the high-throughput screening, which is followed up by definitive tests. Screening refers to the methods that yield rapid and comprehensive data, which often use in vitro tools. The origin of any toxicological or safety outcome is multifactorial and complex and thus demands use of sophisticated systems for definitive assessment. Thus, in vivo animal-based toxicology studies have been conducted earlier, quite often in the lead optimization stage. Extensive and appropriate toxicology studies of varying duration ranging from acute, single dose to chronic, repeat dose in rodent and nonrodent species are requested to establish safe human clinical trials. Acute drug plasma concentration toxicity (single-dose ranging) studies in preclinical species are performed to support selection of a drug candidate for potential advancement to repeat-dose toxicology studies and ultimately to enable initial FIH clinical trials.

7.2.1 Single-Dose Toxicity Studies

The objective of such studies, especially acute toxicity studies, is conducted to determine the lethal dose (e.g., LD50) and to identify a dose at which the major adverse effects are observed. These studies are usually carried out in rodents, following a single dose up to a limit of 2000 mg/kg. The information obtained may be translated to select the dose levels for the FIH studies and also to give an indication of potential effects of acute overdose in humans. These studies are likely to be carried out before bioanalytical methods appropriate for toxicokinetic modeling have been developed. In such studies, it is advisable that biological samples and excreta be collected and stored for subsequent analyses. Analyte stability in the various matrices is required in this scenario. For disposition studies, evaluation of the effects of a range of doses is advisable. These doses may be bounded at the upper level by the maximum tolerated dose/minimum toxic dose and at the lower end by a no observed effect level (NOEL) or other practicable limit [9]. Single-dose toxicity studies underpin the basis of most, if not all, subsequent toxicity and PK studies.

7.2.2 Repeated-Dose Toxicity Studies

Early drug development starts with candidate compound selection. Subacute dosing regimens are used for range finding typically using two to four doses for durations of 14 or 28 days, after which clinical chemistries and histopathology are performed. Subchronic studies largely aim to establish the no observed adverse effect level (NOAEL) and identify or characterize organ-specific toxicity. A useful lowest observed adverse effect level (LOAEL) may also be described in such a study, with implications for reference dose and benchmark dose establishment for regulatory agencies.

Subchronic studies are typically conducted using at least two species, often rats and mice of both sexes, with administration by the expected route, most commonly oral. Subchronic studies are typically 90 days in duration and will employ a minimum of three doses, one of which is expected to elicit a toxic response that produces fatalities in less than 10% of subjects, a dose that evokes no apparent toxic response, and a midrange dose. Statistical power calculations may be used to establish a sufficient number of subjects in each dose group, typically 10–20 for rodent species.

Chronic exposure studies follow the same general format as subchronic studies but may be conducted for 6 months and up to 2 years, depending on the nature of the chemical and expected human exposures. International Conference on Harmonization of Technical Requirements for Registration of Pharmaceuticals for Human Use (ICH) S4 guidance for pharmaceuticals recommends studies of 6 months in rodent species but 9 months in nonrodent species. Environmental contaminants or food additives, which result in lifelong exposure to humans, are often tested using the 2 years guideline.

Dose selection must ensure that significant numbers of subjects do not succumb to premature mortality due to chronic toxicities. The highest dose in a chronic study will likely be the maximum tolerated dose, as required by most regulatory agencies. The various definitions for the maximum tolerated dose, as described by Haseman (1985) [10], may be a function of weight loss induced in the subchronic study, PK, metabolites in urine, or other indicator [11]. The carcinogenic potential of a chemical may also be considered in a chronic toxicity test. These studies of the genotoxic or oncogenic potential require special sets of parameters. Repeat-dose toxicity studies (7–14 days in duration) in both rodent and nonrodent species are used to better refine safety margins and PK/PD modeling, as well as set appropriate dosages for the subsequent good laboratory practice (GLP) 1-month general toxicology and safety pharmacology (i.e., cardiovascular testing in a nonrodent; central nervous system [CNS] and respiratory function tests in a rodent) studies that proceed the investigational new drug (IND) application before starting FIH clinical trials. TK assessment is based on the multiple samples obtained throughout the duration of the study along with the PK data. Such data are critical in defining NOAEL and the projected plasma concentrations achieved in humans. It is generally considered that a 100-fold safety factor (rodent-to-human exposure ratio) from the most sensitive species NOAEL provides good safety margin in clinical studies. However, the enhanced capability of understanding interspecies sensitivity and detecting more and more subtle effects may warrant a more flexible approach. The toxicology assessment profile can include the maximum tolerated dose, safety margins and therapeutic index, target organ toxicities, most sensitive preclinical species, and reversibility of an effect/toxicity. Biomarkers characterization and preclinical to clinical translation can also be investigated in these GLP toxicology studies.

Known PD and PK parameters should be used to determine subsequent treatment regimens (doses, formulation, route[s] of administration, dosing frequency) and species for repeat-dose studies. These parameters may consist of exposure profile monitoring at the beginning and end of the initial repeat-dose study. These data will aid in tailoring appropriate measures to be taken in subsequent studies. Alterations may include extension, reduction, or modification of the monitoring protocol for an individual chemical where problems or uncertainties were noted in prior studies.

7.3 PHARMACOKINETIC PARAMETERS IN SUPPORT OF TOXICOKINETIC ASSESSMENTS

It has been shown that poor PK is one of the leading causes for failure of compounds in preclinical and clinical drug development [12]. Poor PK profile is usually for compounds that have low oral systemic plasma exposure and high interindividual variability, which limits their therapeutic utility [13]. A better understanding of the PK profile early can improve its safety and efficacy. The PK/TK assessments of drug candidates in late discovery and early development may have several positive outcomes, such as

1. Support for the selection of compounds with the maximum potential of reaching the target and determination of the appropriate route of administration/delivery (typically oral)
2. Development of the relationship between level of drug in the systemic circulation (exposure) and its efficacy or toxicity in order to choose efficacious and safe doses
3. Enabling appropriate dose selection for rodent and/or nonrodent species in toxicology testing and drug safety evaluation
4. Making the optimum decision on the frequency and duration of dosing in order to maintain adequate drug concentration at target for disease modification and accurately predict the PK in human profile before clinical studies

A PK/TK study involves dosing animals or humans with new drug candidates and collecting blood samples at predefined time points. After sample preparation and quantification, a concentration–time profile is generated. In drug discovery, preliminary PK studies are usually conducted in rodents to evaluate the extent of drug exposure in vivo. This is commonly followed by studies in nonrodents, such as dogs or monkeys, to better characterize the PK profile of the compound and to support safety risk assessment studies. PK *allometry scaling* is used to predict human PK profile using preclinical data. Accurate prediction of human PK profile is imperative to minimize drug failure in development because of the PK liability. Briefly, the PK parameters that support the TK are described in the following sections.

7.3.1 Area Under the Curve (AUC)

AUC is a primary measure of the extent of drug availability to the systemic circulation; that is, it reflects the total amount of unchanged drug that reaches the systemic circulation following intravenous or extravascular administration. Mathematically, area under the plasma (or blood) concentration–time curve can be calculated from the obtained concentration–time profile. The unit for AUC is concentration per unit time (e.g., ng h/mL). AUC is determined using a simple integration method or often by a linear trapezoidal method. The area of each trapezoid is calculated using the following equation:

$$\mathrm{AUC}\ t_1 \rightarrow t_2 = \frac{(C_2 + C_1)}{2} \times (t_2 - t_1).$$

The extrapolated area from t_{last} to ∞ is estimated as

$$\mathrm{AUC}_{t_{last} \rightarrow \infty} = \frac{C_{last}}{K_e},$$

where C_{last} is the last observed concentration at t_{last} and K_e is the slope obtained from the terminal portion of the curve, representing the terminal elimination rate constant. The total AUC ($\mathrm{AUC}_{0 \rightarrow \infty}$) is determined as

$$\mathrm{AUC}_{0 \rightarrow \infty} = \mathrm{AUC}_{0 \rightarrow t_{last}} + \mathrm{AUC}_{t_{last} \rightarrow \infty}.$$

AUC is used in the calculation of clearance (CL), apparent volume of distribution, and bioavailability (see following sections) and reflects the general extent of exposure over time.

7.3.2 Maximum Plasma Concentration (C_{max}) and Time of Maximum Concentration (T_{max})

C_{max} is defined as the maximum observed drug concentration in the plasma concentration–time profile following intravenous or oral dosing. Most commonly, C_{max} is obtained by direct observation of the plasma concentration–time profile as shown for the AUC. For some drugs, the pharmacological effect is dependent on the C_{max}; for example, aminoglycosides (gentamicin) need to achieve a C_{max} to obtain a clinical response > 90%. The unit of C_{max} is concentration unit (e.g., ng/mL). On the other hand, T_{max} is the time required to reach C_{max}. As with C_{max}, T_{max} is usually determined from direct observation of the plasma concentration–time profile. The unit of T_{max} is time (e.g., h).

7.3.3 Clearance

CL is defined as the irreversible elimination of a drug from the systemic circulation. It is equivalent to the volume of blood or plasma that is totally cleared of its content of drug per unit time. Accordingly, it measures the rate and not the extent of removal of drug from blood or plasma. Thus the units are given as mL/min or mL/min/kg (normalized to body weight).

The most widely used approach to evaluate plasma (total) CL involves intravenous administration of a single dose and measuring its plasma concentration at different time points:

$$CL_{tot} = \frac{Dose_{iv}}{AUC_{iv}}.$$

In general, a drug is either eliminated unchanged through excretion in the urine and/or bile or by metabolic conversion into more polar metabolite(s) that can be readily excreted in urine and/or bile. Since all these CL mechanisms operate independently, total body CL is a sum of all CLs by individual mechanisms and can be expressed as

$$CL_{tot} = CL_{hep} + CL_{ren} + CL_{bil},$$

where CL_{tot} is the total body CL from all different organs and mechanisms, CL_{hep} is the hepatic CL, CL_{ren} is the renal CL, and CL_{bil} is the biliary CL.

About 75% of the top 200 prescribed drugs in the United States are primarily cleared by hepatic metabolism [14].

7.3.4 Apparent Volume of Distribution (V_d)

Volume of distribution (V_d) is a proportionality factor that relates the amount of a drug in the body to its blood or plasma concentrations:

Amount of drug in the body at time t = $V_d \times Ct$.

Similar to CL, V_d is a primary PK parameter and is represented as volume by unit body weight (e.g., L/kg). V_d is used to assess the extent of drug distribution in the body. This is usually achieved by comparing the drug V_d to the total body water. If the drug has a V_d smaller than the total body water (human total body water = 42 L per 70 kg human body weight), the drug has limited tissue distribution (e.g., naproxen has a V_d = 11 L per 70 kg human body weight).

Being proportionality constant, V_d ranges from 3 to more than 40,000 L per 70 kg human body weight and therefore is usually referred to as apparent volume of distribution.

7.3.5 Apparent Volume of Distribution at Steady State (V_{dss})

V_{dss} is the volume of distribution that is determined when plasma concentrations are measured at steady state and in equilibrium with the drug concentration in the tissue compartment:

$$V_{dss} = \frac{\text{Amount of drug in the body at equilbrium condition}}{\text{Steady} - \text{state plasma concentration (Css)}}.$$

Although V_{dss} is a steady-state parameter, it can be calculated using non-steady-state data as:

$$V_{dss} = CL \times MRT,$$

where MRT is mean residence time, which is the average time for all drug molecules to exist.

7.3.6 Half-Life ($t_{1/2}$)

Half-life, usually expressed in hours, is the time that is required for the amount (or concentration) of a drug in the plasma to decrease by one-half:

$$t_{1/2} = \frac{0.693 \times V_d}{CL}$$

Half-life is a dependent PK parameter that is determined by the independent PK parameters CL and V_d. It is the most widely reported PK parameter since it may constitute a major determinant of the duration of action after single and multiple dosing. In addition, half-life plays a key role in determining the time that is required to reach steady state following multiple dosing and the frequency with which doses can be given.

7.3.7 Bioavailability (F%)

As indicated in Figure 7.1, bioavailability (F%) is the rate and extent to which an active moiety is absorbed from a pharmaceutical form and becomes available in the systemic circulation [2]. It is usually determined by calculating the respective AUC after oral and intravenous administrations as

$$\text{Absolute bioavailability} = \frac{AUC_{po}}{AUC_{iv}} \times \frac{Dose_{iv}}{Dose_{po}}.$$

Oral bioavailability is determined by the fraction of dose absorbed (F_a) in the gastrointestinal tract and fraction of dose that does not undergo metabolism in the intestinal tract (F_g) and liver (F_h), as indicated in Figure 1.1. It is expressed as

$$F = F_a \cdot F_g \cdot F_h.$$

F_h is calculated using the following equation:

$$F_h = 1 - E_h = 1 - \frac{CL_h}{Q},$$

where Q is hepatic flow and Eh is the hepatic extraction. If a drug is given orally and has a high hepatic extraction ($E_h > 0.7$), then its bioavailability will be low ($F < 0.3$). On the other hand, if a drug has low hepatic extraction ($E_h < 0.3$), then the extent of bioavailability will be high, provided it is completely absorbed and not significantly metabolized by the intestine.

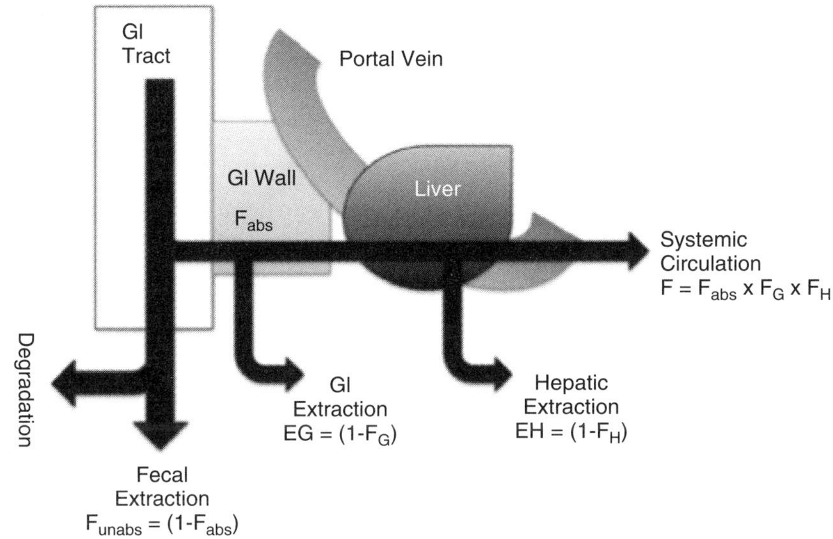

Figure 7.1 Oral bioavailability is a function of intestinal absorption (F_a), the fraction escaping intestinal metabolism (F_G), and the fraction escaping hepatic extraction (F_H).

7.4 GENOTOXICITY, ONCOGENICITY, REPRODUCTIVE TOXICITY VERSUS TOXICOGENOMICS AND BIOMARKERS IN PRECLINICAL SPECIES

7.4.1 Genotoxicity Studies

Systemic exposure in the test species and indicator tissue should be characterized as completely as possible to aid in subsequent risk assessments. Genotoxicity studies primarily focused on small molecule assessments are described here. Such studies aim to detect genetic damage following exposures to the chemical of interest. Genotoxicity studies following a standard battery utilize both in vitro and in vivo assays. Chemicals that test positive in these assays constitute possible human carcinogens or mutagens. Clear lines of evidence exist for chemical exposure and carcinogenesis and most assays are used to quantify this carcinogenic risk. However, mutations to the germ line may also be induced and heritable disease risk is as serious as carcinogenesis.

The standard battery of tests includes an assessment of mutagenicity using a bacterial reverse gene mutation assay and a test in mammalian systems. The mammalian test systems may be either in vitro or in vivo. The most common bacterial reverse gene mutation system is the Ames test [15]. Mammalian in vitro tests that have been well validated include the metaphase chromosome aberration assay [16], micronucleus assay [17], and mouse lymphoma L5178Y thymidine kinase gene mutation assay

[18]. All assays are generally interchangeable for compliance with industry standard testing protocols.

In vivo testing is maintained owing to historical data showing some chemicals that are not mutagenic in vitro are mutagenic in vivo [19]. Additionally, it may be desirable to account for variations in chemical disposition (absorption, metabolism, distribution, elimination). Blood or bone marrow micronucleated erythrocytes are typically chosen for analysis, although metaphase bone marrow cells or cultured lymphocytes may be used [20]. Chromosomal aberration in metaphase cells is indicative of a broad spectrum of alterations in chromosomal integrity.

7.4.2 Carcinogenicity (Oncogenicity) Studies

Oncogenicity study designs will ideally ensure that dose levels result in a range of systemic exposures that exceed maximum therapeutic, occupational, or environmental exposures in target populations of humans by given multiple-fold indices. This may be confounded by species-specific problems, but the aim of such studies is to highlight the comparative exposure for humans versus the animal model [21]. For practical reasons, the majority of oncogenicity studies are carried out in rodents.

Chronic toxicity and oncogenicity studies are often combined. The object of combining these studies is to determine the effects of a chemical in a mammalian species following prolonged and repeated exposure to identify the majority of dose-dependent chronic and oncogenic effects. Test animals are observed for a majority of their lifespan for the development of neoplasticism lesions during or after exposures to the chemical. Both malignant and benign tumors are generally reported. Statistically significant increases in tumor number, size, severity, and so forth above controls are indicative of carcinogenic potential for the chemical in question. Inclusion of sufficient controls, both concurrent and historical, can help identify potentially spurious and misleading false positives. Foreknowledge of the pharmacodynamic response will aid in selection of the appropriate strain to best reflect analogous outcomes in human populations. Oncogenicity studies may require the use of species or strains not included in earlier toxicity studies, as well as routes or methods of administration—in such cases, study profile monitoring is important for parameters being evaluated for the first time.

7.4.3 Reproductive Toxicity Studies

Alteration of development or reproduction by a chemical may be assessed by segment 1 tests. They examine general fertility that does not result in maternal toxicity over a range of two to three doses administered before mating, during gestation, and through lactation. Segment 2 tests test for teratogenicity (e.g., disruption of normal embryonic or fetal development), with exposures occurring during organogenesis in the first trimester. Segment 3 studies perinatal and postnatal toxicities examined following chemical exposures from gestational day 15 through delivery and lactation, with birth weights, survival, and growth through postnatal day 21. A multigenerational study may be conducted in place of a segment 3 study.

Rats are typically the test species of choice for reproductive studies, though rabbits are commonly chosen as a secondary species for teratogenicity study, owing to species-specific teratogenicities (for historical perspectives, see the description of lessons learned following thalidomide poisonings in the late 1950s by Ridings (2013) [22]). In all cases, two to three doses are chosen and numbers of viable pregnancies, stillborn versus live, birth weight, survival and growth through postnatal day 21 are assessed. Teratogenicity studies will also include fetal delivery by cesarean prior to parturition to assess uterine weights and numbers of live, dead, or resorbed fetuses and to examine living fetuses for skeletal or soft tissue malformations.

7.4.4 Toxicogenomics Studies

Toxicogenomics is the integration of the *-omics* technologies genomics, proteomics, and metabonomics, bioinformatics, and toxicology to better understand drug- or toxicant-induced alterations in biochemical networks (gene, protein, and metabolite) of drug candidate development in a pharmaceutical setting. Consequently, toxicogenomics data can be used as drug toxicity/exposure biomarkers or signatures that provide insights into the toxic mechanism of action of a drug candidate and support safety risk assessment. It can be conducted in high-throughput screening in discovery research, and gene expression is now used in the clinic to predict pathologic conditions (e.g., breast cancer), prognosis, and response to therapy. A good example of the application of toxicogenomics was its use to identify a transcriptional biomarker of the histopathological liver change of oval cell-mediated bile duct hyperplasia that occurs in all animal species and can progress to cholangio-carcinoma with low margins of safety, which can lead to costly, late-stage compound terminations and increased risk to patient safety. Multidisciplinary collaboration between toxicologic pathologists, toxicologists, biologists, personnel in toxicokinetics, and statisticians is needed for successful toxicogenomic efforts.

7.5 DRUG METABOLISM AND DRUG RELATED-TOXICITIES

In 2002, Baillie et al. [23], proposed a guideline for assessing the contributions of metabolites to toxicity named "metabolites in safety testing," or "MIST," which attempted to define those situations in which metabolites should be further studied to help define risk assessment for the parent drug. Briefly, the guideline defined the quantitative assessment of a metabolite present in humans at 25% or more of the total drug-related material in circulation. If this occurs, it merits further investigation as a potential contributor to safety findings. This assessment was based on the technical feasibility of using radiolabeled drugs as a method of quantitation that reliably reveals the metabolite quantities as a percentage of total, but not in absolute concentration terms. In 2005, Smith and Obach [24] proposed a set of criteria to be used to determine whether a metabolite should be more extensively studied above what the MIST guideline has indicated. The criteria focus less on relative abundance (proportion, percentage) and more on absolute abundance (concentration, mass),

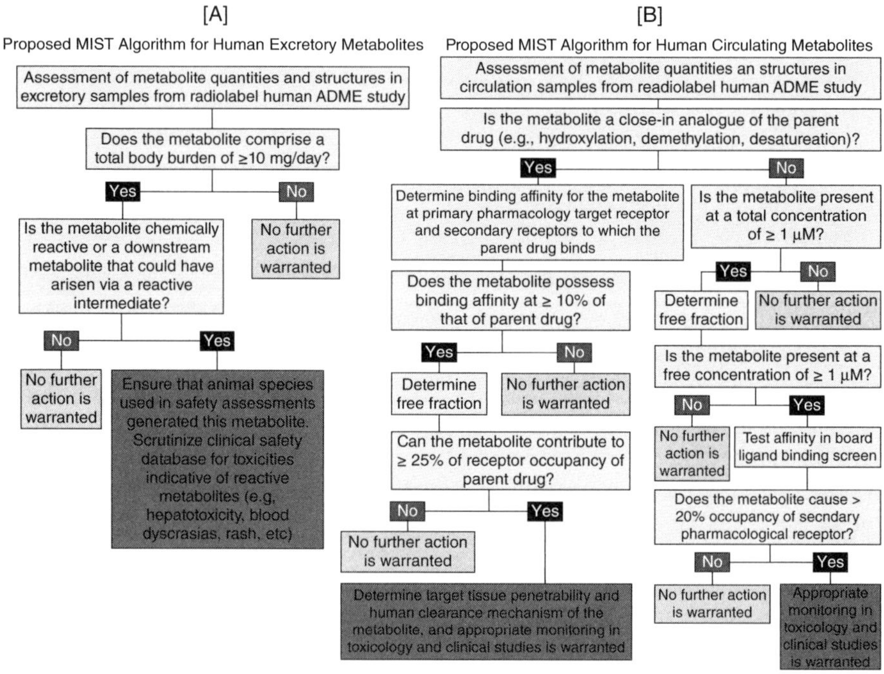

Figure 7.2 Schematic presentation of decisions for risk assessment and safety testing of human excretory (A) and circulating (B) metabolites. Ref [24] with permission.

and they also take into consideration the structure of the metabolite relative to the parent drug and the potential toxic mechanisms of metabolites. Figure 7.2 shows the schematic presentation on decisions for risk assessment and safety testing of human excretory [Figure 7.2(A)] and circulating [Figure 7.2(B)] metabolites.

To develop a strategy for metabolites in toxicity and safety testing, the types of mechanisms of toxicity that could be caused by metabolites and the chemical structure of the metabolite relative to parent drug need to be considered. Smith and Obach proposed that the toxicity might be categorized into four overall types, A, B, C, and D:

1. In *type A toxicity*, the toxic mechanism has a pharmacological basis. There are two subtypes: the target pharmacology (A1) and the nontarget pharmacology (A2). For type A1 toxicity, the parent drug is the most common culprit. This can arise by too much receptor occupancy or enzyme inhibition, or either factor occurring for too extended a period ("suprapharmacological" effects). It can also occur merely due to concurrent side effects of the pharmacological mechanism. Simple examples include gastrointestinal bleeding due to cyclooxygenase-1 inhibition by nonsteroidal anti-inflammatory drugs, gastrointestinal motility decreases due to opioid agonism, or extrapyramidal effects of dopamine antagonists. Metabolites in which structural modifications

are minor occur on substituents not critical for target receptor activity and do not substantially change the physicochemical properties of the parent drug; they are the ones most likely to contribute to pharmacological activity and, hence, any suprapharmacological toxic effects. Thus, any pharmacologically active metabolite can be important if observed toxicity is due to suprapharmacology. Type A2 toxicity is that elicited by binding to and altering the activity of a specific receptor or enzyme that is not the primary pharmacological target. In many cases, the binding affinity may be weaker (i.e., may have a higher K_d), but slight alterations in the receptor function can have profound physiological consequences. The most well-known example of this would be binding to the I_{Kr} channel that can cause QT interval prolongation and in rare cases result in fatal cardiac arrythmias. Partial block of this channel can have profound influence on cardiac function; therefore, the intrinsic potency of the drug or metabolite for the I_{Kr} channel does not have to be as high as that for the primary pharmacology to exert an undesirable side effect. Other examples of type A2 are frequent among neuroleptic agents in which drugs can bind to receptors closely related to the target pharmacological receptor but are responsible for other functions.

2. *Type B, C, and D* toxicities tend to be more related to mechanisms that are not for specific enzymes or receptors but, rather, for nonselective effects. In many cases, the structural elements of metabolites that are associated with type B, C, and D toxicity involve the introduction of reactive electrophilic groups, or structural entities that can cause oxidation (e.g., quinones), and in most cases, it is the observation of metabolites downstream from these reactive intermediates that arise via reaction with nucleophiles (e.g., mercapturic acids, diols, etc.) that are actually seen in vivo. *Type B* refers to idiosyncratic toxicities, such as drug-induced allergy, that do not necessarily exhibit classic dose-response relationships and are observed in very low numbers of patients. Mechanisms of type B toxicity are not well established, but the first pivotal event is considered to be activation of the drug to a reactive metabolite that, nonselectively, covalently bonds to proteins. Some of the haptenized proteins can trigger an immune response that could either target only haptenized proteins (resulting in toxicity only when the drug is administered) or could begin to also recognize native proteins (resulting in autoimmune toxicity that does not require continued drug administration). Type B toxicity can occur in a variety of tissues and can even occur in different tissues in different patients for the same drug. Normally, though, the three prime sites of toxicity are the liver, blood cells, and skin. Many of the drugs causing type B toxicity exhibit effects on all three. These organs and tissues may be uniquely sensitive due to their high intrinsic activity in terms of oxidizing systems (e.g., activated neutrophils and the release of hypochlorous acid) or the presence of a highly developed and active immunological defense system [25]. It should also be noted that toxicity in these organs and tissues is among that most easily detected.
3. In *type C toxicity*, the effect of the drug is due to a chemical reaction between drug or metabolite and tissue macromolecules resulting in a rapidly ensuing

response. It is rare that drugs themselves elicit this type of toxicity (an exception is direct alkylating agents used in cancer chemotherapy), but some drugs can be bioactivated to chemically reactive entities that can act directly by covalently binding to proteins. It is also possible that drugs can be bioactivated to metabolites that undergo redox cycling (e.g., quinones/hydroquinones), deplete intracellular stores of reduction potential, such as reduced glutathione, and cause oxidative stress. This can lead to cell death and tissue necrosis. Liver toxicity elicited by high doses of acetaminophen via generation of the N-acetyl paraquinoneimine metabolite is a good example of type C toxicity.

4. *Type D toxicity* is similar in underlying mechanism to types B and C; however, the response is delayed, even for years. Examples of type D toxicity include carcinogenesis and teratogenesis. For carcinogenesis, mechanisms could be due to genotoxins or could possess an endocrinological basis (which is actually more related mechanistically to type A toxicity). Genotoxins can also cause teratogenesis, but teratogenesis can have other underlying mechanisms often similar to type A1 and A2. It is important to note that for types B, C, and D, metabolites that could be responsible for toxicity can be reactive enough that they are not detected per se in circulation or excreta.

When assessing the potential toxicity of a drug, what is actually being examined is a complex mixture of chemicals: drug plus all metabolites and impurities. Developed policies have been described for assessing the safety of impurities in drugs [26]. These policies are not directly applicable for metabolites. Chemical structures of impurities can be substantially different from those of the parent drug, since they may derive from the chemical process used to synthesize the drug. As completely unrelated structures, impurities can elicit toxic responses very different from those of the parent drug; thus, the examination of the safety of impurities can have stringent criteria. Metabolites are generated from the parent drug; therefore, in most cases they bear structural similarity to the parent drug.

REFERENCES

1. MacGowan AP. Role of pharmacokinetics and pharmacodynamics: does the dose matter? Clin Infect Dis 2001;33 Suppl 3:S238–S239.
2. Varma MVS, Radi ZA, Rotter CJ, Litchfield J, El-Kattan AF. Pharmacokinetics and toxicokinetics in drug discovery and development. In: Encyclopedia of drug metabolism and interactions, ed. Lyubimov AV. Hoboken, NJ: Wiley, 2012, 3:1–18
3. Aimone LD, de Lannoy IA. Overview of pharmacokinetics. Current Protocols Pharmacology 2014;66:7111–7131.
4. Krewski D, Acosta D Jr., Andersen M, Anderson H, Bailar JC 3rd, Boekelheide K, Brent R, Charnley G, et al. Toxicity testing in the 21st century: a vision and a strategy. J Toxicol Environmental Health 2010;13:51–138.
5. Attene-Ramos MS, Miller N, Huang R, Michael S, Itkin M, Kavlock RJ, Austin CP, Shinn P, Simeonov A, Tice RR, Xia M. The Tox21 robotic platform for the assessment of environmental chemicals—from vision to reality. Drug Discovery Today 2013;18:716–723.

6. Baber N. International Conference on Harmonization of Technical Requirements for Registration of Pharmaceuticals for Human Use (ICH). British J Clin Pharmacol 1994;37:401–404.
7. Ledwith BJ, DeGeorge JJ. Changes to ICH Guideline M3: new and revised guidance on nonclinical safety studies to support human clinical trials and marketing authorization. Clin Pharmacol Ther 2011;89(2):295–299.
8. Peters SA, Ungell AL, Dolgos H. Physiologically based pharmacokinetic (PBPK) modeling and simulation: applications in lead optimization. Curr Opin Drug Discov Devel 2009;12(4):509–518.
9. Lorke D. A new approach to practical acute toxicity testing. Arch Toxicol 1983;54:275–287.
10. Haseman JK. Issues in carcinogenicity testing: dose selection. Fundamental Applied Toxicol 1985;5:66–78.
11. Borzelleca JF, Olson JW, Reno FE. Lifetime toxicity/carcinogenicity studies of FD & C red no. 40 (allura red) in mice. Food Chemical Toxicol 1991;29:313–319.
12. Kola I, Landis J. Can the pharmaceutical industry reduce attrition rates? Nat Rev Drug Discov 2004;3(8):711–715.
13. Hellriegel ET, Bjornsson TD, Hauck WW. Interpatient variability in bioavailability is related to the extent of absorption: implications for bioavailability and bioequivalence studies. Clin Pharmacol Ther 1996:60(6):601–607.
14. Williams JA, Hyland R, Jones BC, et al. Drug–drug interactions for UDP-glucuronosyl-transferase substrates: a pharmacokinetic explanation for typically observed low exposure (AUCi/AUC) ratios. Drug Metab Dispos 2004;32(11):1201–1208.
15. Ames BN, McCann J, Yamasaki E. Methods for detecting carcinogens and mutagens with the Salmonella/mammalian-microsome mutagenicity test. Mutation Research 1975;31:347–364.
16. Misenko SM, Bunting SF. Rapid analysis of chromosome aberrations in mouse B lymphocytes by PNA-FISH. J Vis Exp 2014;A19:90.
17. Maier P, Schmid W. Ten model mutagens evaluated by the micronucleus test. Mutation Research 1976;40:325–337.
18. Amacher DE, Paillet S, Ray VA. Point mutations at the thymidine kinase locus in L5178Y mouse lymphoma cells. I. Application to genetic toxicological testing. Mutation Research 1979;64:391–406.
19. Tennant RW. Relationships between in vitro genetic toxicity and carcinogenicity studies in animals. Annals New York Academy Sciences 1988;534:127–132.
20. Ishidate M Jr., Sofuni T, Nohmi T. Quantitative evaluation on the genotoxic potency of chemicals. J Toxicological Sciences 1991;16 Suppl 1:83–92.
21. Cook JL, Hibbs JB Jr., Lewis AM Jr. Resistance of simian virus 40-transformed hamster cells to the cytolytic effect of activated macrophages: a possible factor in species-specific viral oncogenicity. Proc National Academy Sciences U S A 1980;77:6773–6777.
22. Ridings JE. The thalidomide disaster, lessons from the past. Methods Mol Biol 2013;947:575–586.
23. Baillie TA, Cayen MN, Fouda H, Gerson RJ, Green JD, Grossman SJ, Klunk LJ, LeBlanc B, Perkins DG, Shipley LA. Drug metabolites in safety testing. Toxicol Appl Pharmacol 2002;182:188–196.

24. Smith DA, Obach RS. Seeing through the MIST: abundance versus percentage. Commentary on metabolites in safety testing. Drug Metab Dispos 2005;33:1409–1417.
25. Park BK, Kitteringham NR, Powell H, Pirmohamed M. Advances in molecular toxicology—towards understanding idiosyncratic drug toxicity. Toxicology 2000;153:39–60.
26. U.S. Department of Health and Human Services. Food and Drug Administration Guidance for Industry Q3B. Impurities in new drug products. FDA, 1997. Available: http://www.fda.gov.libproxy.lib.unc.edu/cder/guidance/1317fnl.htm#ATTACHMENT%20II (accessed March 28, 2005).

8

PBPK MODELING AND IN SILICO PREDICTION FOR ADME AND DRUG–DRUG INTERACTION

8.1 INTRODUCTION: COMPUTATIONAL ASSESSMENT OF ADME AND DRUG–DRUG INTERACTION (DDI) WITHIN PHARMACEUTICAL R&D PARADIGM

Over decades, properties related to absorption, distribution, metabolism, excretion, and toxicity (ADMET) have become one of the most important issues to assess the effects or risks of xenobiotics on the human body and to reduce the high rate of drug withdrawals and financial burden on pharmaceutical companies. To improve preclinical safety testing without the costly and laborious in vivo and in vitro evaluations, in silico techniques have been developed and widely used to estimate these properties. The recent advances of applying the in silico ADMET prediction and its challenges and limitations will be discussed in this chapter. Also, application domain of models, model validation techniques, global versus local models, computational systems toxicology (toxicogenomics), and data-integration and meta-decision-making systems that could be used for systemic in silico ADMET prediction in drug discovery and hazard risk assessment will be reviewed [1]. In addition, models including those for drug transport proteins will be included in this chapter. Models generally have very high throughput and can generate large quantities of data on large numbers of virtual compound structures. Consequently, the wider choice of models, coupled with their high throughput, indicate a growing need to integrate their output into collective analyses of molecules against preset criteria [2].

Translational ADMET for Drug Therapy: Principles, Methods, and Pharmaceutical Applications, First Edition. Souzan B. Yanni.

Mathematical descriptions that relay drug absorption, distribution, metabolism, and excretion (ADME) to drug concentration can be achieved by using pharmacokinetic (PK) models. One of the various modeling approaches that has been developed is classic PK compartmental modeling. This popular approach describes the body as rapidly or poorly equilibrating compartments for the distribution of drugs [3]. It is simple and most widely used in the clinical setting for determining the extents of drug distribution and elimination. The major pitfall of applying this approach is that the plasma drug concentrations do not always reflect tissue drug concentrations that relate to drug responses and toxicological effects. In addition, it possesses serious limitations by not describing the sequential handling of metabolites within organs where metabolites formed since the physiological processes can not assess function of key determinants of drug disposition, especially transporter function [4]. Therefore, PBPK approaches have been developed to correct for these deficits. Mathematical descriptions that relay ADME to drug concentration can be achieved by using PK models.

PBPK models are used to assess physiology, anatomy, and biochemistry-related parameters based on the assumption that compartments are homogeneous and well stirred. The PBPK model consists of a number of subcompartments that represent actual tissues and organs of discrete volumes. There are three parameters that construct the differential equations constructed for these models: physiological, thermodynamic/physical, and biochemical [5].

The physiological parameters include tissue volumes (V) and tissue blood flow rates (Q) (sum of cardiac output (Q_{CO}). The basis of PBPK models lies with the assumption that there is venous equilibration: crug concentration in tissue $AUC_{po}\{mi,\ P\}$ concentration in the emergent blood.

The thermodynamic parameters include protein binding (denoted as unbound fractions in plasma, blood, or tissue: f_P, f_B or f_u, f_T), the tissue to plasma/blood partition coefficient ($K_{P,T}$ or $K_{P,Tu}$) of the drug, and the transmembrane permeability.

The biochemical parameters exist as the result of metabolic and transport processes and are described by the Michaelis–Menten constants maximum velocity (V_{max}) and binding constant (K_m). Under first-order conditions, the CL_{int} is given by V_{max}/K_m. The secretory intrinsic clearance ($CL_{int,sec}$) is responsible for renal, biliary, or luminal excretion. For transport, the influx (CL_{in}) and efflux (CL_{ef}) clearances are included to represent both the transporter-mediated and passive diffusion processes [6–8].

In this chapter, the information needed for building a PBPK model and the general approaches used will be discussed. Also, a summary of a recent developed single-organ PBPK models for the intestine, liver, and kidney and whole-body PBPK models that are currently applied to gain a mechanistic understanding in drug disposition will included. The models to be reviewed are those that can predict drug and metabolite concentrations in tissues at any time and those that provide predictions of lipophilic compounds. To a lesser extent, the models that require transporter involvement for uptake and efflux or those that exhibit diffusion-limited transport will be reviewed. Furthermore, PBPK models used to determine DDIs will be reviewed. Integration of in vitro and in vivo data to establish human PK including PBPK models established for animals will all be discussed in the coming sections.

8.2 PBPK MODELS FOR ADMET AND DDI

8.2.1 General PBPK Model and Physiological Parameters that Affect Drug Disposition

A model that represents several major body organs and compartments where the drug is absorbed, distributed, metabolized, and excreted is shown in Figure 8.1 and

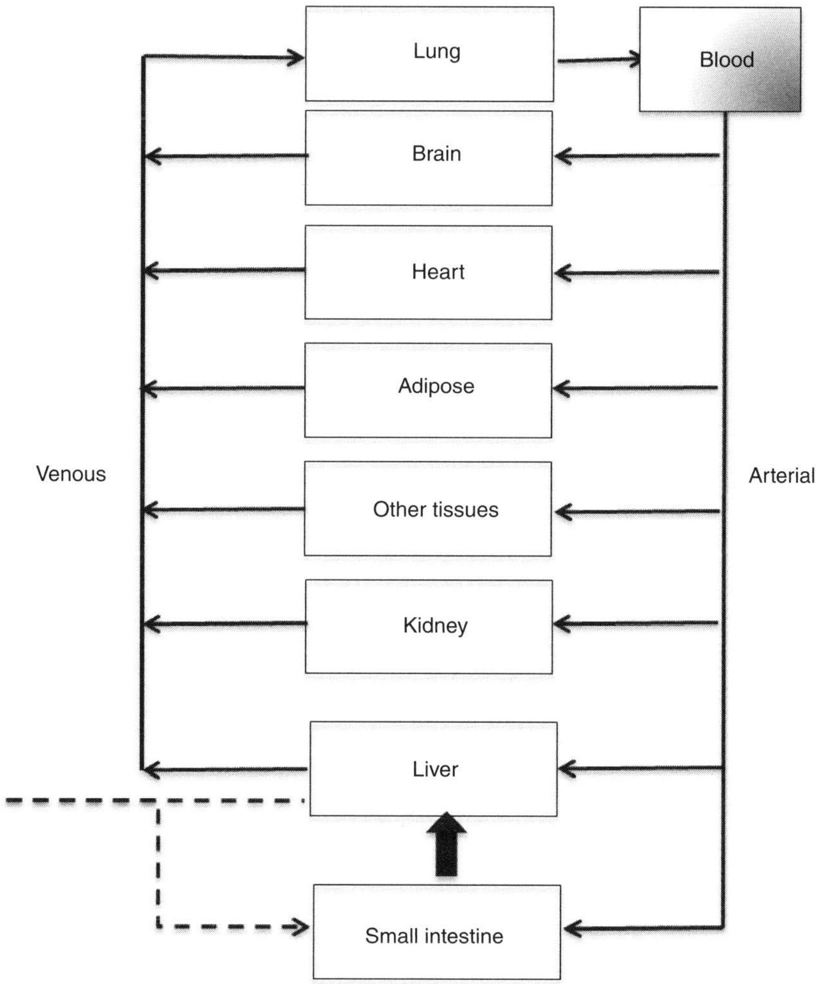

Figure 8.1 A PBPK model for the whole body.

according to Maeng et al. (2012) [5]. There is no general rule for the type of organ compartments to be considered in the model, though the model should include organs with parameters that contribute to the disposition of drugs, that is, exert pharmacological and toxicological effects, as indicated from in vitro studies, so that the model can predict the clinical human PK. It is worth mentioning that in developing a PBPK model, several assumptions need to be made before forming the mathematical relations that indicate the drug concentration in body, including blood and tissues, as well as that describe the physiological processes that lead to drug metabolism and secretion from body compartments considered in the model. In order for the mathematical relationship to assess these drug disposition factors, each organ drug concentration, physiological volume, blood flow rate, and tissue partition coefficient needs to be available.

As indicated in Table 8.1, the blood flow rates and tissue volumes for humans and rodents used in toxicity studies are tabulated according to values reported by other investigators [5,9,10]. Clearly, there are striking differences between the rate of blood flow across the species and across the body organs, which make significant differences in drug clearance, distribution, and tissue concentration as well as transport and metabolism activities among these tissues [11–13]. For example, in the intestine, the blood flow reaching the enterocytes where the absorptive uptake and efflux apical transporters are located represents a fraction of total intestinal blood flow [14].

As discussed in previous chapters, there are critical factors known to have a significant impact on the drug disposition and its clearance. These factors, such as plasma protein binding, tissue partition coefficient, and functions of drug mctabolizing enzymes and transporters, are included in PBPK models, and their accurate values are critical in the prediction of human (as well as animal) PK.

Protein binding, as discussed in previous chapters, affects the drug elimination, toxicity, and pharmacological response, and thus it is a key determinant in overall drug disposition. There are direct relationships between these processes and unbound drug fraction, as the unbound fraction is the drug portion that can metabolized, bound to transporters, and excreted in bile and urine. Physiochemical properties of drugs as well as extent of extraction can make an impact on the overall effect of protein binding in relation to drug disposition. For example:

1. Poor extracted drug can have a low excretion in kidney and metabolism in the liver with increase in binding to plasma proteins [15].
2. For high extracted drug, the elimination rate is dependent on blood flow, so plasma binding has a low influential role [16].
3. The change in plasma protein binding due to age, disease state, and DDI may alter the drug response and other clinical outcomes such as adverse events, and thus it requires dose adjustments in these particular conditions [17].

As indicated by several investigators, protein binding is a reversible process, and ratio between rate of association and rate of dissociation can indicate the extent of binding. When binding is high ($> M \times 10^{-4}$), that leads to negligible change in drug concentration with any change in extent of protein binding. Yang et al. (2006) [18]

TABLE 8.1 Physiological Parameters for Humans and Small Animal Toxicity Species used in Constructing PBPK Models.

Tissue and Compartment	Human (Based on 70 Kg BWT)		Rat (Based on 0.25 Kg BWT)		Mouse (Based on 0.02 Kg BWT)	
	Volume (mL)	Blood Flow Rate mL/min	Volume (mL)	Blood Flow Rate mL/min	Volume (mL)	Blood Flow Rate mL/min
Blood	5200	–	13.5		1.7	–
Brain capillary	1450	700	1.2	1.3	0.025	0.089
Brain tissue	–	–	–	–	0.226	–
Heart	310	240	1.2	3.9	0.095	0.28
Kidney	280	1240	3.7	9.2	0.34	1.3
Liver	1690	–	19.6	–	1.3	–
Hepatic artery	–	300	–	2.0	–	0.35
Intestine	1650	1100	11.3	7.5	1.5	1.5
Lung	1170	–	2.1	–	0.1	–
Cardiac output (mL/min)	–	5600	–	14.0	–	8.0
GFR (mL/min)	–	125	–	1.31	–	0.28
Urine flow (mL/day)	–	1400	–	50.0	–	1.0

demonstrated a mathematical equation to reveal the relationship between unbound plasma concentration, the plasma protein concentration, and number of the same class of binding sites. It is expressed as the difference between the total plasma concentration and unbound plasma concentrations. If the binding between a drug and plasma protein involves two classes of sites, one with low affinity and the other with higher affinity, such a binding involving two noncooperative affinities, then the relationship will include the binding constants of the two binding classes, their protein concentration, and number of binding site of each class. At low drug concentration, protein binding is not saturated and only the high affinity site is involved in the binding. When excess drug is present, the low affinity sites will be further recruited for binding. In such a case, the unbound fraction becomes a function of the binding to both the high and low affinity binding sites.

For *drug metabolizing enzymes* and *transporters*, as almost all chapters have emphasized, it is essential when building a PBPK model to consider the role of these two key determinants. For both processes, the Michaelis–Menten equation is often used as a mathematical equation that presents their function rate at maximum saturable rate V_{max}, the binding constant K_m, and unbound substrate concentration, S:

$$\nu = \frac{V_{max}}{K_m + [s]}\ [s] = Cl_{int}\ [s],$$

where the rate, v, is expressed in terms of the intrinsic clearance for transport, metabolism, and/or secretion. The V_{max} and K_m values are usually obtained from in vitro studies, commonly estimated from the rate plot (v vs. [S]), the Lineweaver–Burk plot, or the Eadie–Hofstee diagram [19]. Usually, the assumption is that elimination is first order, and CL_{int} is the ratio of V_{max}/K_m. The in vitro intrinsic metabolic clearance under linear conditions is presented by the following equation:

$$CL_{int\ in\ vitro} = \frac{V_{max}}{K_m}.$$

Metabolic capacity of enzymes is estimated from incubations with human liver microsomes, S-9 fraction, hepatocyte suspension, and/or recombinant expression system expressed by mg protein, then by using the appropriate scaling factors to convert to per gram liver. By using a scaling factor, the conversion (g liver/kg) is made to whole-body clearance, as shown in chapter 3 [20,21].

For the transporter-mediated processes, normally transporter function may be estimated in cell expression systems, combined with human transporter uptake studies. The influx transport intrinsic clearance (CL_{in}) is usually estimated from in vitro hepatic uptake studies. The localization of the transporter, whether within the cytosol or nucleus or on the membrane surface, would greatly affect the transport rate. For quantitative assessment of transporters, LC-MS/MS technologies can provide the relative proportions of transporters in the liver, kidney, and blood-brain barrier [22]. In the absence of these data, hepatocyte uptake information for extrapolation of in vitro parameters to in vivo, cell number per unit tissue weight and tissue weight per unit body weight, in addition to the physiological scaling factors, have been used [23,24]. At a qualitative level, in vitro uptake and biliary excretion data using the rat and human sandwich-cultured hepatocyte systems may be used to assess whether there is species similarity and correlation to the in vivo plasma and biliary clearances as presented in chapter 4 [23] or DDI in transport. It is worth mentioning that careful interpretation of results must be considered with regard to the differences in transporter expression between in vitro models and between in vitro and in vivo, as well as between human and animal species. In vitro data have identified marked species differences for the multidrug resistance-associated proteins (MRPs) and breast cancer resistance protein (BCRP) in rat and human hepatocytes; these significant species differences in efflux transporter activities could be integrated into animal data for the prediction of human transport [25]. When there is no significant interspecies difference between rat and man, the scaling factor determined in rats has been used to extrapolate human in vitro parameters to those in vivo [23]. The transporter-mediated transport clearance should then be considered, in addition to the passive clearance, for the prediction of in vivo intrinsic influx and efflux clearances [23].

8.2.2 Simple Organ-Based PBPK Models

The development of simple organ-based PBPK models to predict areas under the curve (AUCs) and clearances for a drug and its formed metabolite are achieved from equations of rate for each subcompartment, providing the essential coefficients for matrix inversion and solutions that yield the steady-state concentrations/AUCs of the drug and metabolite (AUC{mi, P}) [6]. The method of matrix inversion, appropriate only under linear kinetic conditions, is applied to examine whether AUC{mi, P} is suitable for estimation of the systemic bioavailability [26]. The following sections describe the specific PBPK models for liver, intestine, kidney, and the whole-body organ.

8.2.2.1 PBPK for Liver PBPK for liver describes membrane-limited transport, with either fast or slow transmembrane transport clearances for summed diffusion-limited and transporter-mediated transport cases [27,28]. According to Liu and Pang (2005) [28], the model comprises four compartments: the reservoir (R, or blood compartment), liver blood (LB), liver tissue (L), and bile compartment (bile) [6]. The hepatic flow rate and unbound fraction in blood and liver are denoted by Q_H, f_B, and f_L, respectively. The transport processes combined the carrier-mediated basolateral transporter activity and passive diffusion designated by the influx and efflux intrinsic clearances ($CL^H{}_{in}$ and $CL^H{}_{ef}$). Added to the $CL^H{}_{in}$ and $CL^H{}_{ef}$ is the biliary secretion at the canalicular membrane (biliary intrinsic clearance [$CL_{int,\ sec,\ H}$]) and the metabolism within hepatocytes (metabolic intrinsic clearance [$CL_{int,\ met,\ H}$]) of phase I and phase II enzymes. This model is also used to describe hepatic sequential metabolism [6,27], in which the metabolite formed within the liver tissue is prone to sequential metabolism to form other metabolites ($CL_{int,\ met,\ H}$ {mi}) or for excretion ($CL_{int,\ sec,\ H}$ {mi}).

As reported by Sun and Pang (2010) [26], for the equations of AUC and total hepatic clearance determined by either the membrane-limited case or the flow-limited case for the drug and the metabolite, the AUC of the drug is a complex equation that relates to the flow rate (Q_H), f_B, $CL^H{}_{in}$, $CL^H{}_{ef}$, and the total hepatic intrinsic clearance ($CL_{int,\ H}$). In the case of a drug of flow-limited distribution, the ratio of hepatic metabolic and biliary clearances, $CL_{H,\ met}/CL_{H,\ sec}$ equals the ratio of the intrinsic clearances, $CL_{int,\ met,\ H}/CL_{int,\ sec,\ H}$. From this relation, the unknown parameter, $CL_{int,\ sec}$, may be obtained with the data from liver perfusion studies or in vivo studies [28] when $CL_{int,\ met}$ is obtained from in vitro metabolism studies using S-9 fractions or liver microsomes. The equations reported by Sun and Pang are found to be useful in allowing inferences on mechanisms of DDIs in the liver.

In another report by Pang (2009) [29], the author describes the PK of formed metabolites after iv injection using PBPK equations. After iv dosing of parent drug, Pang stated that the AUC for the formed metabolite, AUC_{iv} {mi, P}, is dependent on the metabolic intrinsic clearance, $CL_{int,\ met\ 1,\ H}$, and individual components of the $CL_{int,\ H}$ (or $CL_{int,\ met\ 1,\ H} + CL_{int,\ met\ 2,\ H} + CL_{int,\ sec,\ H}$), as well as binding, transport clearances ($CL^H{}_{in}${mi} and $CL^H{}_{ef}$ {mi}), and intrinsic clearance, $CL_{int,\ H}$ {mi}, of the metabolite [sum of ($CL_{int,\ met,\ H}${mi}) and ($CL_{int,\ sec,\ H}${mi})]. Pang also mentioned that in contrast to these findings, the AUC obtained from administration of its preformed counterpart that yielded AUC_{po} {pmi} and AUC_{iv}{pmi} is independent of drug parameters. Formed metabolites after oral and intravenous drug dosing (AUC_{po}{mi, P}and AUC_{iv} {mi, P}) have revealed interesting differences from AUC_{po} {pmi} and AUC_{iv} {pmi} that result from the administration of the preformed counterpart [15,26].

The heterogeneity of drug-metabolizing enzymes (DMEs) and drug transporters contributes to differences between AUC of oral and iv dosing of parent drug (AUC_{po}{mi, P} and AUC_{iv}{mi, P}) and AUC oral and iv dosing of metabolite (AUC_{po}{pmi} and AUC_{iv}{pmi}). As an improvement, zonal PBPK liver models consisting of three zonal regions—zones 1 (periportal zones), 2 (midzonal zones), and 3 (perivenous zones)—or models similar to compartments in series have been developed [30,31].

The presence of a basolateral membrane barrier for the metabolite, hindering the entry or efflux of preformed metabolite into and out of cells and trapping the formed metabolite within the tissue, makes an impact on changing the AUC after parent dosing and AUC after metabolite dosing by the oral and iv routes [32]. There has been success applying zonal PBPK modeling administered with several drugs in a rat liver perfusion model, including enalapril morphine, estrone sulfate, estradiol−17β-glucuronide, and digoxin [30,33–35]. The efflux transporter Mrp2 for excretion of both the parent drug (estradiol−17β-glucuronide) and its sulfated metabolite (3-sulfate−17β-glucuronide) was elucidated comprehensively within the framework of the liver PBPK model [35,36]. The role of transporters was also elucidated using the PBPK model developed by Pang and her collaborators, as shown in a review article by Maeng at al. (2012) [5]. In the same review article, Pang and her colleagues described the small intestine, traditional model (TM), as well as kidney PBPK modeling.

8.2.2.2 Whole-Body PBPK Models Whole-body PBPK models provide an understanding of the behaviors of drugs and metabolites based on physiological parameters. Usually, when the major drug-eliminating organs/tissues—intestine, liver, and kidney—are made up of highly perfused, poorly perfused, and adipose tissues, that reveals the mechanisms of elimination by metabolism via DMEs, influx by transporters, and secretion via apical transporters as well as the rate-limiting step. Renal clearance may be embedded in the blood compartment to describe the parallel loss of drug from the kidney. The PBPK models provide an account of how the processes of transport and metabolism modulate the kinetics of the drug and its metabolites, and how drug behavior further affects the kinetics of the metabolites as described by Sun and Pang (2010) [26].

The PBPK model solutions reveal that the effect of intestinal secretion is diminished when avid absorption takes place. It is also determined that the ratio of $AUC_{po}\{mi, P\}/AUC_{iv}\{mi, P\}$ to drug oral and intravenous administration, corrected for dose, yields the fraction of dose absorbed, F_{abs}, into the portal circulation when either the intestine or the liver is the only drug-metabolizing organ, and the dose-corrected AUC_{po}/AUC_{iv} yields the systemic availability, F_{sys}. In conclusion, Sun and Pang stated that by dividing AUC_{po}/AUC_{iv} by $AUC_{po}\{mi, P\}/AUC_{iv}\{mi, P\}$, that resulted in obtaining the intestinal or hepatic availability [26].

In addition to the model built by Pang and her colleagues, a review article by Maeda and Sugiyama (2012) [37] reported that the Food and Drug Administration clearly recommends the use of dynamic models for the accurate prediction of DDIs in the process of drug development. Based on the review, the PBPK model developed by Watanabe et al. (2009) [38] of pravastatin, which is a substrate of organic anion transporting polypeptides (OATPs) and MRP2 in liver, was described. In this model, the liver compartment was divided into five compartments to mimic the dispersion model because the hepatic clearance of pravastatin is very close to the hepatic blood flow rate. The same intrinsic clearances of hepatic influx, back flux, biliary excretion, and metabolic clearance were set up in each segment. Then, all the kinetic parameters were set up only from the results of in vitro experiments using a simple scaling-up method. The intrinsic clearances of influx and passive diffusion of pravastatin were determined from uptake studies using isolated hepatocytes, intrinsic clearance of biliary excretion was determined from ATP-dependent transport using canalicular membrane vesicles (CMVs), and metabolic intrinsic clearance was determined from a metabolism assay using the liver S9 fraction. They confirmed the time profiles of the plasma concentration and cumulative biliary excreted amount of pravastatin at several doses in rats, and the plasma concentrations after intravenous and oral administration of pravastatin in humans were well reproduced by simulation in this PBPK model. One of the advantages of a PBPK model is in performing sensitivity analysis, which is a good method to identify critical processes dominating the phenotype (e.g., plasma and tissue concentrations) by monitoring any alteration of the target phenotype when a specific process is changed in silico. Watanabe et al. (2009) [38] performed the same type of simulation to understand the effects of changes in each intrinsic clearance on the plasma and hepatic concentrations of pravastatin. As data show in his report, changes in the influx intrinsic clearance largely altered the plasma concentration but not the hepatic concentration. However, changes in the biliary excretion intrinsic clearance greatly altered the hepatic concentration but not the plasma concentration, which is very similar to the results obtained from the simple model. A PBPK model was capable of simulating the severity of DDIs accurately by connecting the PBPK models of substrate drugs and inhibitors to their target enzymes/transporters, as will be discussed in the next section. As a whole, by utilizing a PBPK model, the authors [37] were able to simulate the plasma and tissue concentrations of drugs in any virtual situation easily by changing the parameters in the model without any additional clinical studies. This greatly helps pharmaceutical researchers to optimize the pharmacokinetic properties of drugs, because it is very difficult to estimate the target outcome intuitively from the complicated whole-body drug detoxification system.

8.2.3 PBPK Model for DDI

In addition to determination of PK, PBPK can be used to assess potential DDI. In vitro studies, as indicated in chapter 5, are conducted to generate hepatic enzyme inhibition parameters such as K_i or K_I that are used to predict which inhibitor concentration to use in vivo. PBPK modeling will not only provide a profile of substrates altered by a concomitantly given drug but also reveal the plasma concentration–time curve of the inhibitor instead of the fixed inhibitor concentrations, as in vitro studies reveal.

For *reversible inhibition*, estimation of in vivo relevant K_i value allowed quantitative prediction of CYP-mediated DDI [39]. In vivo K_i values were determined by fitting the plasma concentration–time profiles of a substrate and an inhibitor to a simple PBPK model. Increased ratios of AUCs (in presence to absence of inhibition) predicted inhibition. The results were in agreement with in vivo corresponding data and were more accurate relative to the prediction by the static model generated by in vitro [40]. When differences occur between in vivo K_i and in vitro K_i values, that may possibly be attributed to nonspecific binding to liver microsomes defined by lipophilicity.

For *irreversible inhibition*, the prediction using a PBPK model has not been as simple as in the case of reversible inhibition. For example, for DDI between 5-fluorouracil and sorivudine [41], triazolam and erythromycin [42], and midazolam and macrolides was challenging [40]. CYP turnover rate ($k_{deg,\ CYP}$) as indicated in chapter 5, is a key parameter that has an impact on the predictability. There has been difficulty in estimating the $k_{deg,\ CYP}$ of each CYP enzyme in the human liver in biological experiments, and hence rat CYP data of 0.0005/min was employed as human $k_{deg,\ CYP}$ in those studies.

Several assumptions were made to predict DDI [40,42]:

1. Substrate and inhibitor are orally administered according to the first-order rate constant and only undergo hepatic metabolism.

2. Unbound concentration in the hepatic vein and liver are equal due to a rapid equilibrium according to the well-stirred model.
3. Only unbound drug in the liver is subject to elimination.
4. Disposition and metabolism of inhibitor are not altered by the inhibition of CYP3A4.

In case of the interaction between triazolam and erythromycin, K_I and k_{inact} values were estimated by using human liver microsomes or recombinant CYP3A4, and the results were incorporated into a PBPK model. Repeated administration of erythromycin at 333 mg three times a day for 3 days resulted in simulated reduction in hepatic CYP3A4 enzyme to 50–60% of the initial level [42]. Subsequently, the predicted 2.0- and 2.6-fold increase in triazolam AUC was derived from human liver microsomes and recombinant CYP3A4, respectively. These values were similar to in vivo clinical data with a 2.1-fold increase.

Similarly, in case of interaction between midazolam and macrolide, the simulation using the K_I and k_{inact} of macrolides estimated in human liver microsomes yielded a 2.9- to 3.0-fold increase in midazolam AUC after pretreatment with erythromycin, and a 2.1- to 2.5-fold increase after pretreatment with clarithromycin, but little effect for azithromycin [40]. To increase the DDI predictability of drugs that are substrate for CYP3A4, first-pass metabolism in the small intestine after oral administration is important because the intestinal metabolism has a great influence on the bioavailability of CYP3A4 substrates and potential DDI [39], and the capability of the PBPK model beyond the static model in the DDI prediction.

In addition, the PBPK model in single and multiple ascending pharmacodynamics dose studies may provide the capability to estimate any change in the plasma concentration–time curves, including C_{max} and $t_{1/2}$. With the recent advancement in modeling combined with simulation tools, in vitro data have increasingly been utilized to predict in vivo profile to optimize the dose regimen in a clinical trial to ensure that any interaction is appropriately measured and avoided. This can be seen by using Simcyp (Simcyp Ltd., Sheffield, UK), which would give confidence to CL and DDI prediction [43,44], as would Gastroplus [45]. A case study was conducted by Sinha and colleagues (2012) [45] for one of the drug candidates under development by Janssen Pharmaceutical in which the PBPK modeling approach has been used to model the absorption of a lipophilic Biopharmaceutics Classification System (BCS) Class II compound (predominantly metabolized by CYP3A4 and not P-glycoprotein [P-gp] substrate) and to assess the interplay of absorption-related parameters with the DDI potential. The approach was based on building a PBPK model in the rat using Gastroplus to study the absorption characteristics of the compound. Subsequently, relevant model parameters were used to predict the nonlinear human PK observed during first in human study after optimizing the absorption model for colonic absorption, bile micelle solubilization, and unbound fraction in gut enterocytes (fu_{gut}) using Simcyp simulator. The model-fitted absorption parameters were then used to assess the DDI potential of the test compound when administered along with multiple doses of a potent CYP3A4 inhibitor, ketoconazole. The impact of fu_{gut} in the extent of DDI was assessed using parameter sensitivity analysis. After optimizing the preclinical model and taking into consideration bile micelle solubilization and colonic absorption, the nonlinear PK of the test compound was satisfactorily predicted in man. Sensitivity analysis performed with the absorption parameter fu_{gut} indicated that it could be an important parameter in predicting oral absorption. In addition, DDI simulations using Simcyp suggest that C_{max} and AUC ratios may also be sensitive to the fu_{gut} input in the model. Since fu_{gut} cannot be measured experimentally, sensitivity analysis may help in assessing the importance of fu_{gut} in human PK and DDI prediction using Simcyp.

8.2.4 PBPK and Genetic Polymorphism

Very recently, Djebli and his collaborators have reported the use of PBPK modeling in assessing the role of genetic polymorphism in PK of drugs when the PK varies among human populations due to the genetic polymorphism of P450 enzymes mediated the metabolism and disposition of these drugs. The model drug used by the investigators is clopidogrel, which is a prodrug that needs to be converted to its active metabolite (clopi-H4) in two sequential cytochrome P450 (CYP)-dependent steps [46].

In their study, a dynamic PBPK model was developed in Simcyp for clopidogrel and clopi-H4, using a specific sequential metabolite module in four populations with phenotypically different CYP2C19 activity: poor (PM), intermediate (IM), extensive (EM), and ultrarapid metabolizers (UM). The four groups received a loading dose of 300 mg followed by a maintenance dose of 75 mg. After validating the model using three approaches, first by comparing predicted to observed AUC_{0-24} obtained from a randomized crossover study conducted in four balanced CYP2C19-phenotype metabolizer groups using a visual predictive check method; second by using the interindividual and intertrial variability from the AUC_{0-24} comparisons between the predicted trials and the observed trial of individuals for each phenotypic group; and third by assessing DDI prediction from the comparison with observed values of clopidogrel and clopi-H4 with or without moderate CYP3A4 inhibitor (dronedarone) that was coadministered using a previously developed and validated PBPK dronedarone model. The PBPK model was well validated for both clopidogrel and its active metabolite clopi-H4, in each CYP2C19-phenotypic group, whatever the treatment period (300 mg loading dose and 75 mg last maintenance dose). The investigation, as stated by the authors, is the first study proposing a full dynamic PBPK model able to accurately predict simultaneously the PK of the parent drug, and its primary and secondary metabolite, in populations with genetically different activity for DMEs [46].

8.3 IN SILICO PREDICTION OF ADMET

Since in vivo and in vitro evaluations are costly and laborious, in silico techniques have been widely used to estimate these properties. Unlike PBPK modeling based on physiological parameters and organ functions, in silico ADMET prediction emphasizes a substructure pattern recognition method [47,48].

In this section, a brief description of the latest advances in recently developed models as well as challenges and limitations in the area of in silico ADMET prediction will be discussed. Application domain of models, model validation techniques, and data integration that could be used for systemic in silico ADMET prediction in drug discovery and hazard risk assessment will be reviewed.

8.3.1 Significance of Using In Silico Modeling: In Silico versus PBPK Modeling

In silico modeling approaches will be discussed here, but unlike the PBPK modeling discussed in previous sections, information about the existence or extent of the interaction between a compound and a protein will be provided. These data may be useful in early stage drug development, providing an indication of the likelihood of such an interaction, though they provide no indication of the clinical significance of this interaction. However, for PBPK, as mentioned, the data provide information about the clinical significance of a compound-protein interaction, which is necessary to integrate interaction data with PK data.

As mentioned in previous chapters, all ADME properties can be determined by appropriate in vitro and/or in vivo assays, but they are costly with the chance of failing the clinical investigation due to poor in vitro–in vivo correlation. This generates competition for resources in lead identification and lead optimization that requires careful prioritization for successful outcomes at affordable costs. Improvements in assay miniaturization and throughput have helped offset these costs, but there is still a potentially crucial role for in silico modeling to inform prioritization of assays, and hence to ease the competition for resources throughout drug discovery. A variety of in silico techniques are used across the entire range of ADME, toxicity, pharmacology, and physicochemical properties [49–53] and applications to drug metabolism, DDI, and toxicity. However, in silico modeling for ADME prediction can't be specific in addressing the mechanism underlying the metabolism, transporter-mediated drug disposition, or the mechanism underlying DDI. The assessment can be based on total rate of metabolism or total metabolic clearance, but not on the isozyme responsible for metabolism. It can predict the overall rate of drug transporters but not by a given transporter and the rate of such transport. Similarly, in silico prediction can't provide information on interactions with a given transporter or DME.

8.3.2 Methods for In Silico ADMET Prediction

There are two types of methods for in silico ADMET prediction: molecular modeling and data-derived modeling (data modeling). Molecular modeling methods, such as molecular docking, molecular dynamics simulation, and QM/MM calculation, are based on the three-dimensional structures of proteins involved in the ADMET process, such as P450s.

8.3.2.1 Data Modeling For data modeling, refer to review articles that explicitly cover the data modeling method [47,48]. Briefly, it is a quantitative structure activity relationship (QSAR) based on a mathematical equation between molecular descriptors and properties using statistic methods, such as multilinear regression and partial least squares (PLS), in which molecular descriptors are used to represent the molecules. Recently, Tang and collaborators developed a novel in silico method for ADMET prediction, namely substructure pattern recognition [47,48]. In this method, each molecule is first represented as a fingerprint according to the predefined substructure dictionary, then machine learning methods are applied to build the prediction model. Therefore direct connections between molecular structures and properties are built, and molecular descriptors are skipped. Furthermore, information gain (IG) analysis is adopted to identify the essential substructures or structural alerts that contribute mostly to the corresponding properties, which is helpful to interpret the "black box" models.

8.3.2.2 Molecular Modeling Molecular modeling will be discussed in detail, including different in silico approaches to drug metabolism and DDI, and the compound interaction with varying promiscuous proteins, whether transporters, DMEs, or nuclear receptors.

Protein structure methods: These methods require three-dimensional structure of the protein and limit its use to protein homology models [54]. Because of the massive molecular structure of proteins, these methods are computationally expensive and are not high throughput but may reveal important information such as the exact nature of binding into the active site of an enzyme.

Molecular dynamic simulation: This simulation reviews the binding of enzyme-substrate over time from the atomistic level. It can grasp the different orientations and conformations of the compound in the active site. If included in the simulation, the role of solvent molecules can be determined [55]. The method can't be used as a general screening method in drug discovery, as its accuracy depends on the model of interactions between compound and enzyme and computationally intensive molecular dynamics simulation.

Docking: This method reviews the binding of compound as it is docked into the binding site of the enzyme. Each atom pair between the ligand and the enzyme is considered. The results indicate the separation between the two atoms is converted into an atom-pair potential according to a scoring function. The total binding potential is the sum of the potential energies across all atom pairs. The solvation effects have been incorporated into the scoring function and the free energy of binding is obtained [56].

8.4 APPLICATIONS OF IN SILICO MODELS IN ADME, DDI, AND DRUG TOXICITY

In silico methods have been used to characterize the compounds in drug discovery as substrates of P450, UDP-glucuronosyltransferase (UGT), transporters, and nuclear receptors.

8.4.1 Prediction of the Rate of Metabolism

Prediction of overall rate of metabolism using in silico modeling due to the significance of the parameter of candidate progression in drug discovery can reduce the cost, time, and laboratory screening efforts to determine the parameters by in vitro methods for hundreds of drugs. Despite the efforts by many investigators in the field of in silico model development, prediction of the properties of individual DMEs is the route that offers greatest potential in drug discovery. The work of Chang and his colleagues (2009) [57] on developing models for microsomal CL_{int} implies that a large "global" model for CL_{int} using data generated across many areas of chemistry has potential benefit in screening new chemistry and could thus be used to virtually screen libraries or hits for lead identification.

8.4.2 DDI of Metabolism

Studying drug metabolism and potential DDI in a simpler system, such as individual enzyme isoforms, rather than more complex systems, such as hepatocytes, or subcellular fractions, such as microsomes, cytosol, or S-9, increases both the toolset available for model development and the number of properties that can potentially be predicted. By using additional tools such as docking and molecular dynamics methods with in silico models, the metabolism of a particular enzyme isoform of a given compound or the potential of a given compound to inhibit the metabolism of other compounds by a particular isoform can be determined.

Several investigators used the in silico approach to predict specific P450 enzyme inhibitors and substrates like the approach employed by Yap and Chen (2005) [58], who predicted CYP3A4, 2D6, and 2C9 inhibitors and substrates using a support vector machine approach to decision trees. Others, such as Kriegl et al. (2005) [59], used multivariate modeling to predict CYP3A4 inhibition. With high-confidence predictions of DDI, affinity for CYP2C9 was achieved using multiple computational methods, as reported by Hudelson et al. (2008) [60], who built four models to predict compounds that inhibit CYP2C9. Vasanthanathan et al. (2009) [61] tested a variety of models on a large test set of 7,000 compounds, resulting in predictive accuracies in the range of 67–76%, and since the test set is so large, these values can be considered to be a reliable measure of the predictive accuracy of enzyme inhibition, specifically for CYP1A2 by machine learning methods.

8.4.3 Identifying Substrates for Transporters

To fully understand drug metabolism and DDIs, one must consider active transport into and out of cells, as it may be significant compared with passive diffusion for certain drugs, and there has been corresponding interest in developing in silico prediction models to predict transporter activity [62]. Selected studies have developed QSAR models to predict the affinities for a range of transport proteins—P-gp, organic cation transporters, and nucleotide transporters—and were summarized by Dimelow et al. (2012) [54]. For example, Osterberg and Norinder (2000) [63] have reported mathematical calculations and prediction of P-gp-interacting drugs using MolSurf parametrization and PLS statistics. The authors concluded that variables associated with hydrogen bonding and high polarizability correlate with high P-gp ATPase activity. Other studies also highlight the importance of hydrophobic features, employing receptor surface analysis [64] and three-dimensional QSARs of P-gp inhibitors and substrates [65], which may represent aliphatic or aromatic hydrophobes [65].

REFERENCES

1. Cheng F, Li W, Liu G, Tang Y. In silico ADMET prediction: recent advances, current challenges and future trends. Curr Top Med Chem 2013;13(11):1273–1289.
2. Beresford AP, Segall M, Tarbit MH. In silico prediction of ADME properties: are we making progress? Curr Opin Drug Discov Devel 2004;7(1):36–42.
3. Gibaldi M, Perrier D. Pharmacokinetics, 2nd ed. New York: Marcel Dekker, 1982.
4. Pang KS, Durk MR. Physiologically-based pharmacokinetic (PBPK) modeling of metabolites. J Pharmacokinet Pharmacodyn 2010;37:591–615.
5. Maeng H-J, Chow ECY, Fan J, Pang SK. Physiologically based pharmacokinetic (PBPK) modeling: usefulness and applications. In: Encyclopedia of drug metabolism and interactions, ed. Lyubimov AV. Hoboken, NJ: Wiley, 2012, 3:1–48.
6. Pang KS, Morris ME, Sun H. Formed and preformed metabolites: facts and comparisons. J Pharm Pharmacol 2008;60:1247–1275.
7. Pang KS, Maeng HJ, Fan J. Interplay of transporters and enzymes in drug and metabolite processing. Mol Pharm 2009;6:1734–1755.
8. Pang KS, Rodrigues AD, Peter RM. Enzyme- and transporter-based drug–drug interactions: progress and future challenges. New York: Springer, 2010.
9. Davies B, Morris T. Physiological parameters in laboratory animals and humans. Pharm Res 1993;10:1093–1095.
10. Kawahara M, Sakata A, Miyashita T, et al. Physiologically based pharmacokinetics of digoxin in mdr1a knockout mice. J Pharm Sci 1999;88:1281–1287.
11. Pang KS, Rowland M. Hepatic clearance of drugs. I. Theoretical considerations of a "well-stirred" model and a "parallel tube" model. Influence of hepatic blood flow, plasma and blood cell binding, and the hepatocellular enzymatic activity on hepatic drug clearance. J Pharmacokinet Biopharm 1977;5:625–653.
12. Perl W, Chinard FP. A convection-diffusion model of indicator transport through an organ. Circ Res 1968;22:273–298.
13. Roberts MS, Rowland M. Hepatic elimination-dispersion model. J Pharm Sci 1985;74:585–587.
14. Cong D, Doherty M, Pang KS. A new physiologically based, segregated-flow model to explain route-dependent intestinal metabolism. Drug Metab Dispos 2000;28:224–235.
15. Perl W, Chinard FP. A convection-diffusion model of indicator transport through an organ. Circ Res 1968;22:273–298.
16. Roberts MS, Rowland M. Hepatic elimination-dispersion model. J Pharm Sci 1985;74:585–587.

17. Yanni SB, Smith PB, Benjamin DK Jr, Augustijns PF, Thakker DR, Annaert PP. Higher clearance of micafungin in neonates compared with adults: role of age-dependent micafungin serum binding. Biopharm Drug Dispos. 2011;32(4):222–232.
18. Yang F, Tong X, McCarver DG, et al. Population-based analysis of methadone distribution and metabolism using an age-dependent physiologically based pharmacokinetic model. J Pharmacokinet Pharmacodyn 2006;33:485–518.
19. Gillette JR. Factors affecting drug metabolism. Ann N Y Acad Sci 1971;179:43–66.
20. Rostami-Hodjegan A, Tucker GT. Simulation and prediction of in vivo drug metabolism in human populations from in vitro data. Nat Rev Drug Discov 2007;6:140–148.
21. Stringer R, Nicklin PL, Houston JB. Reliability of human cryopreserved hepatocytes and liver microsomes as in vitro systems to predict metabolic clearance. Xenobiotica 2008;38:1313–1329.
22. Kamiie J, Ohtsuki S, Iwase R, et al. Quantitative atlas of membrane transporter proteins: development and application of a highly sensitive simultaneous LC/MS/MS method combined with novel in-silico peptide selection criteria. Pharm Res 2008;25:1469–1483.
23. Yanni SB, Augustijns PF, Benjamin DK Jr, Brouwer KL, Thakker DR, Annaert PP. In vitro investigation of the hepatobiliary disposition mechanisms of the antifungal agent micafungin in humans and rats. Drug Metab Dispos 2010;38(10):1848–1856.
24. Kusuhara H, Sugiyama Y. In vitro–in vivo extrapolation of transporter-mediated clearance in the liver and kidney. Drug Metab Pharmacokinet 2009;24:37–52.
25. Li M, Yuan H, Li N, et al. Identification of interspecies difference in efflux transporters of hepatocytes from dog, rat, monkey and human. Eur J Pharm Sci 2008;35:114–126.
26. Sun H, Pang KS. Physiological modeling to understand the impact of enzymes and transporters on drug and metabolite data and bioavailability estimates. Pharm Res 2010;27:1237–1254.
27. de Lannoy IA, Barker F 3rd, Pang KS. Formed and preformed metabolite excretion clearances in liver, a metabolite formation organ: studies on enalapril and enalaprilat in the single-pass and recirculating perfused rat liver. J Pharmacokinet Biopharm 1993;21:395–422.
28. Liu L, Pang KS. The roles of transporters and enzymes in hepatic drug processing. Drug Metab Dispos 2005;33:1–7.
29. Pang KS. Safety testing of metabolites: Expectations and outcomes. Chem Biol Interact 2009;179:45–59.
30. Abu-Zahra TN, Pang KS. Effect of zonal transport and metabolism on hepatic removal: enalapril hydrolysis in zonal, isolated rat hepatocytes in vitro and correlation with perfusion data. Drug Metab Dispos 2000;28:807–813.
31. Liu L, Mak E, Tirona RG, et al. Vascular binding, blood flow, transporter, and enzyme interactions on the processing of digoxin in rat liver. J Pharmacol Exp Ther 2005;315:433–448.
32. Schwab AJ, Barker F III, Goresky CA, et al. Transfer of enalaprilat across rat liver cell membranes is barrier-limited. Am J Physiol 1990;258:G461–G475.
33. Doherty MM, Poon K, Tsang C, et al. Transport is not rate-limiting in morphine glucuronidation in the single-pass perfused rat liver preparation. J Pharmacol Exp Ther 2006;317:890–900.
34. Tan E, Lu T, Pang KS. Futile cycling of estrone sulfate and estrone in the recirculating perfused rat liver preparation. J Pharmacol Exp Ther 2001;297:423–436.

35. Sun H, Liu L, Pang KS. Increased estrogen sulfation of estradiol 17beta-D-glucuronide in metastatic tumor rat livers. J Pharmacol Exp Ther 2006;319:818–831.
36. Sun H, Zeng Y-Y, Pang KS. Interplay of Phase II enzymes and transporters in futile cycling: influence of multidrug resistance-associated protein 2-mediated excretion of estradiol 17β-D-glucuronide and its sulfate metabolite on net sulfation in perfused TR^- and wistar rat liver preparations. Drug Metab Dispos 2010;38:769–780.
37. Maeda K, Sugiyama Y. Excretion systems. In: Encyclopedia of drug metabolism and interactions, ed. Lyubimov AV. Hoboken, NJ: Wiley, 2012, 3:1–32.
38. Watanabe T, Kusuhara H, Maeda K, et al. Physiologically based pharmacokinetic modeling to predict transporter-mediated clearance and distribution of pravastatin in humans. J Pharmacol Exp Ther 2009;328:652–662.
39. Kato M, Shitara Y, Sato H, et al. The quantitative prediction of CYP-mediated drug interaction by physiologically based pharmacokinetic modeling. Pharm Res 2008;25:1891–1901.
40. Ito K, Ogihara K, Kanamitsu S, et al. Prediction of the in vivo interaction between midazolam and macrolides based on in vitro studies using human liver microsomes. Drug Metab Dispos 2003;31:945–954.
41. Kanamitsu S, Ito K, Okuda H, et al. Prediction of in vivo drug–drug interactions based on mechanism-based inhibition from in vitro data: inhibition of 5-fluorouracil metabolism by (E)-5-(2-bromovinyl)uracil. Drug Metab Dispos 2000;28:467–474.
42. Kanamitsu S, Ito K, Green CE, et al. Prediction of in vivo interaction between triazolam and erythromycin based on in vitro studies using human liver microsomes and recombinant human CYP3A4. Pharm Res 2000;17:419–426.
43. Einolf HJ. Comparison of different approaches to predict metabolic drug–drug interactions. Xenobiotica 2007;37:1257–1294.
44. Youdim KA, Zayed A, Dickins M, et al. Application of CYP3A4 in vitro data to predict clinical drug–drug interactions; predictions of compounds as objects of interaction. Br J Clin Pharmacol 2008;65:680–692.
45. Sinha VK, Snoeys J, Osselaer NV, Peer AV, Mackie C, Heald D. From preclinical to human—prediction of oral absorption and drug–drug interaction potential using physiologically based pharmacokinetic (PBPK) modeling approach in an industrial setting: a workflow by using case example. Biopharm Drug Dispos 2012;33(2):111–121.
46. Djebli N, Fabre D, Boulenc X, Fabre G, Sultan E, Hurbin F. Physiologically-based pharmacokinetic modeling for sequential metabolism: effect of CYP2C19 genetic polymorphism on clopidogrel and clopidogrel active metabolite pharmacokinetics. Drug Metab Dispos 2015;43(4):510–522.
47. Shen J, Cheng F, Xu Y, Li W, Tang Y. Estimation of ADME properties with substructure pattern recognition. J Chem Inf Model 2010;50(6):1034–1041.
48. Cheng F, Li W, Liu G, Tang Y. In silico ADMET prediction—recent advances, current challenges and future trends. Curr Top in Med Chem 2013;13:1273–1289.
49. Ekins S, Mestres J, Testa B. In silico pharmacology for drug discovery: applications to targets and beyond. Br J Pharmacol 2007;152:21–37.
50. Ekins S, Mestres J, Testa B. In silico pharmacology for drug discovery: methods for virtual ligand screening and profiling. Br J Pharmacol 2007;152:9–20.
51. Lagorce D, Reynes C, Camproux AC, et al. Silico ADME/tox predictions. In: ADMET for medicinal chemists: a practical guide, ed. Tsaioun K, Kates SA. Hoboken, NJ: Wiley, 2011, 29–124.

52. Feenstra KA, De Graaf C, Vermeulen NPE. Cytochrome P450 protein modelling and ligand docking. In: Drug–drug interactions, ed. Rodrigues AD. New York: Informa Healthcare, 2008, 435–470.
53. Marechal JD, Yu J, Brown S, et al. In silico and in vitro screening for inhibition of cytochrome P450 CYP3A4 by comedications commonly used by patients with cancer. Drug Metab Dispos 2006;34:534–538.
54. Dimelow RJ, Metcalfe PD, Thomas S. In Silico models of drug metabolism and drug interactions. In: Encyclopedia of drug metabolism and interactions, ed. Lyubimov AV. Hoboken, NJ: Wiley, 2012, 6:1–55.
55. Shan Y, Kim ET, Eastwood MP, et al. How does a drug molecule find its target binding site? J Am Chem Soc 2011;133:9181–9183.
56. Sousa SF, Fernandes PA, Ramos MJ. Protein-ligand docking: current status and future challenges. Proteins 2006;65:15–26.
57. Chang C, Duignan DB, Johnson KD, et al. The development and validation of a computational model to predict rat liver microsomal clearance. J Pharm Sci 2009;98:2857–2867.
58. Yap CW, Chen YZ. Prediction of cytochrome P450 3A4, 2D6, and 2C9 inhibitors and substrates by using support vector machines. J Chem Inf Model 2005;45:982–992.
59. Kriegl JM, Eriksson L, Arnhold T, et al. Multivariate modeling of cytochrome P450 3A4 inhibition. Eur J Pharm Sci 2005;24:451–463.
60. Hudelson MG, Ketkar NS, Holder LB, et al. High confidence predictions of drug–drug interactions: predicting affinities for cytochrome P450 2C9 with multiple computational methods. J Med Chem 2008;51:648–654.
61. Vasanthanathan P, Taboureau O, Oostenbrink C, et al. Classification of cytochrome P450 1A2 inhibitors and noninhibitors by machine learning techniques. Drug Metab Dispos 2009;37:658–664.
62. Chang C, Pang KS, Swaan PW, et al. Comparative pharmacophore modeling of organic anion transporting polypeptides: a meta-analysis of rat Oatp1a1 and human OATP1B1. J Pharmacol Exp Ther 2005;314:533–541.
63. Osterberg T, Norinder U. Theoretical calculation and prediction of P-glycoprotein-interacting drugs using MolSurf parametrization and PLS statistics. Eur J Pharm Sci 2000;10:295–303.
64. Zalloum HM, Taha MO. Development of predictive in silico model for cyclosporine- and aureobasidin-based P-glycoprotein inhibitors employing receptor surface analysis. J Mol Graph Model 2008;27:439–451.
65. Ekins S, Kim RB, Leake BF, et al. Application of three-dimensional quantitative structure-activity relationships of P-glycoprotein inhibitors and substrates. Mol Pharmacol 2002;61:974–981.

9

TRANSLATIONAL TOOLS TOWARD BETTER DRUG THERAPY IN HUMAN POPULATIONS

9.1 INTRODUCTION: TRANSLATIONAL ADMET AND ITS THERAPEUTIC VALUE

Drug discovery and development is viewed as an outstanding advancement of biomedical technologies and innovative approaches to bring to the market therapeutic agents that can be used to treat patients with broad illness and diseases. Despite these extensive efforts that involve an array of scientific disciplines and tools, there has been little success in increasing the approvals of new drugs reaching the market each year.

Several sequential steps and processes are involved with discovery and development of a new drug molecule, ranging from target identification to clinical studies and final drug registration. These efforts involve computational modeling, medicinal chemistry, biological screening, lead generation and optimization, preclinical, and selection of pharmaceutics for delivery of a drug. With these activities, cutting-edge technologies that have been developed in the past two decades are applied, including high-throughput screening assays, genetically modified animal models, ultrasensitive LC-MS/MS systems enabled with structure elucidation tools, advanced imaging systems, nano-technology, siRNA technology, -omics (genomics, proteomics, and metabonomics), and many more. However, these technological advances have not been used for more meaningful information that translates predictive data generated from in vitro and in vivo systems to an appropriate dosing regimen in humans that ensures efficacious and safe therapy, and thus success of new drugs reaching the market. The number of new drugs approved now is no greater than 50 years ago. However,

Translational ADMET for Drug Therapy: Principles, Methods, and Pharmaceutical Applications, First Edition. Souzan B. Yanni.

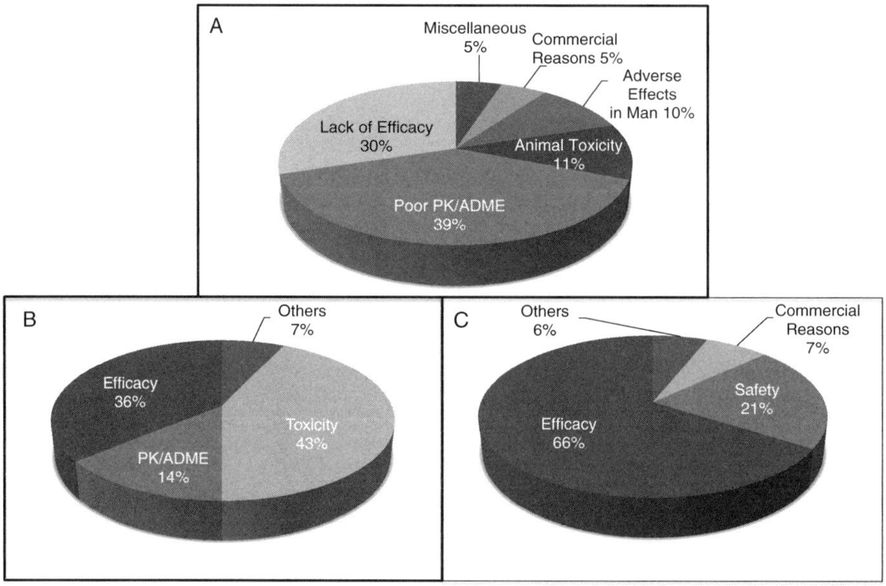

Figure 9.1 Why is drug development still failing? (A) Due to poor ADME/PK, (B) failure reduced due to advancement of predicting human ADME/PK, and (C) failure in Phase 3 due to poor efficacy and safety reasons. Adapted from Ref [2] with permission.

the number has increased more during the last 20 years than the previous 30 years. Concurrently, the cost of new drug development has been rising by ~15% annually, making the recent price of developing a new drug an estimated \$1.3 billion–\$1.7 billion [1]. These points underline the need for a critical review of factors that hamper the success of new drugs reaching the market in a more effective manner with regards to time and cost.

In order to improve the process, case studies of several drug candidates, or even drugs that reach the market, were conducted to reveal the reasons for failure of drug approval for human use in the past three decades. This investigation indicates that the major reason for failures of drugs developed between the 1960s and 1980s was unfavorable pharmacokinetics (PK) (39%) in Phase 1 clinical trials [2], while failures due to lack of efficacy and safety were estimated to be 30% and 10%, respectively. As shown in Figure 9.1, early assessment of PK liabilities in drug discovery using in vitro and in vivo approaches led to a reduced number of failures (to 14%) in Phase 1 clinical trials, but with a minor change in failure in the efficacy [3]. But, during the 1990s to early 2000, failures due to safety (toxicity) issues rose to 43%, which can be attributed to an increase in drug exposures due to improved PK profiles of the drug candidates. In 2010, failure in Phase 3 trials due to lack of efficacy was 66%, but improvement to meet safety was observed (by 22%). In the analysis, 28% of the drugs that failed in Phase 3 clinical investigation were for oncology and 18% for neurological disorders. It has been shown that success in developing a drug for one type of cancer can be translated to success for another type of cancer. For

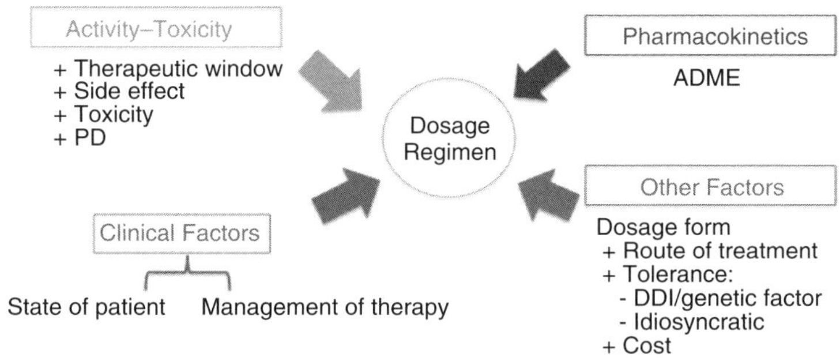

Figure 9.2 Factors implicated in selection of dosing regimen (exposure) and to ensure therapeutic response.

example, sunitinib (Sutent, Pfizer) was approved in 2006 for renal cell carcinoma and imanitib-resistant gastrointestinal (GI) stromal cell carcinoma and subsequently for rare pancreatic neuroendocrine cancers [4]. In addition to translation from one target to other, success can also be achieved by translation from one population to another (adult vs. children), or a more commonly applied translational approach is from animal species to humans, although proof-of-concept (POC) trials in animal models in certain cases may not translate to the desired clinical outcome [5].

The most significant and promising role that translational research offers is closing the gaps between efficacy and safety outcomes through rational drug design [6]. Rational drug discovery requires an early assessment of all factors affecting the likely success of a drug candidate in the subsequent preclinical and clinical phases of drug development, including drug-related issues and subject-related issues, as indicated in Figure 9.2. The use of mechanistic pharmacokinetics/pharmacodynamics (PK/PD) models early in the drug discovery process along with diagnostic biomarker monitoring could potentially reduce attritions during late-stage drug development. The advancements in bioanalytical technologies have dramatically improved the absorption, distribution, metabolism, excretion, and toxicology (ADMET) data, thus the development of powerful in silico predictive models and screening tools toward more robust drug design. From the pharmacology side, models (cell-based model or diseased animal model) have also considerably shifted from traditional approaches to molecular biology approaches in dissecting the cell signaling pathways, humanized animal models, and imaging tools to select appropriate and selective biomarkers that play a direct or indirect role in assessing drug efficacy and safety. The increasing use of these technologies has led to employing mechanistic PK/PD studies during early drug discovery to select promising candidates for drug development. Last but not least, Internet accessibility and computing power that allow data sharing have enabled researchers to interrogate thousands of publicly available databases containing molecular, clinical, and epidemiology data [7]. Translational bioinformatics, along with the increase in new approaches, can now provide solutions to unsolved problems ranging from potential adverse drug reactions in specific populations to clinical implication of genetic fingerprints. Although there are some challenges, the positive promises are beyond the expectations.

This chapter provides an overview of the translational research in drug discovery with special emphasis on integrating clinical pharmacology, genomics, DMPK, toxicology, and bioinformatics to individualize medicine. Key PK/PD concepts combined with applications of recent advances in DM, drug transport (DT), exploratory toxicology, and biomarker tools allow accurate prediction of clinical efficacy and safety during preclinical evaluation. In the following sections, the strategic use of translational PK/PD approaches starting from early drug discovery through late-stage clinical development for successful recognition of drug product with minimal failures will be discussed. Furthermore, the attempts to achieve the promise of personalized or individualized medicine will be illustrated.

9.2 TRANSLATIONAL BIOINFORMATICS AND BIOMARKERS: UTILIZATION FOR BETTER DRUG THERAPY

Translational medicine, or in other words "personalized medicine," is mostly based on the evaluation of genetic variation or genetic fingerprints of patients or healthy volunteers as a way to better diagnose, evaluate, or effectively treat an individual patient [8]. The analysis of biomarkers can provide an indication of biological action and is essential to understanding complex disease states. In many areas of medicine, biomarkers aid in early diagnosis, prognosis, patient recruitment, and treatment selection. There are several areas of medicine, such as oncology, cardiovascular, central nervous system (CNS), and metabolic disease, where the biomarkers are considered as a critical step in accelerating clinical drug development and bringing new efficacious and safe drugs to target patient populations. The utilization of genetic data becomes a new standard of patient care and in a much earlier stage, during clinical study design, which has brought about remarkable improvement in the quality of pharmacotherapy.

As already mentioned, the utilization of translational informatics and applying genetic data to patient care has a variety of benefits, ranging from improved treatment efficacy and increased patient safety to reducing the cost of R&D and drug prices. As expected, the utilization of translational informatics potentially accelerates drug development by reducing the occurrence of unexpected safety concerns and by simply reaching the target efficacious dose, hence achieving successful clinical study protocol and design [9]. These interventions could benefit patients with increases in the quality of life, overall survival, and achievement of economical therapeutic outcome. However, the journey from the discovery of a biomarker to its implementation in clinical practice is long and complex, and it can face many obstacles, including lack of consistency in evidence and results, inadequate validation of the biomarker, inadequate of evidence of clinical utility, operation barriers in clinical implementation of its utility and its effectiveness or impact in clinical care. Though there have been improvements in the pitfalls in implementation, most important is bringing many disciplines to work together. Their input allows the progression of clinical practice from the previous paradigm of "one size fits all" to more personalized treatment pathways that are more predictable toward health gain and improvement of the quality of pharmacotherapy to all patient populations, regardless of age, gender disease stage, or ethnicity.

9.2.1 In Cancer

The personalized medicine using genomics approach is demonstrated in the treatment of multiple myeloma (MM) whereby its practice may surpass the application of traditional clinical practices. In a recent review article, Fall et al. (2014) [8] illustrated the application of translational bioinformatics and genomic biomarker utilization to improve treatment strategies in MM. More specifically, in the review, the authors addressed how gene expression profiling (GEP) can be applied to report the clinical challenge of bortezomib (Bz) resistance that frequently develops when patients undergo treatment for MM. MM is an incurable malignant neoplasm hallmarked by a clonal expansion of plasma cells, the presence of a monoclonal protein in the serum and/or urine (M-spike), lytic bone lesions, and end organ damage. Patients with MM can clinically improve as a result of treatment with drugs such as thalidomide derivatives, as well as with novel chemotherapeutic agents including first- and second-generation proteasome inhibitors Bz and carfilzomib; unfortunately, the majority of patients relapse following treatment. Although Bz is currently the most commonly and successfully used proteasome inhibitor in clinical practice, some patients with MM later develop drug resistance following treatment. The lack of a reliable clinical test that would predict which MM patients have or will later develop Bz resistance is a significant deficit in the MM pharmacotherapy, and development of Bz resistance remains a significant challenge. Thus research efforts are needed to identify novel biomarkers of early Bz resistance, particularly when an early therapeutic intervention can be initiated. With the recent advances in MM research, it has become clear that genomic data can be utilized to identify novel biomarkers toward more effective, personalized treatment protocols for individual patients. By integrating large patient databases with data from whole transcriptome profiling and laboratory-based models, better understanding of the mechanisms leading to MM disease can be reached. This approach has provided rational therapeutic targets by revealing the identity of novel biomarkers, thus the risks in drug treatment and/or response and resistance to Bz in MM are avoided. In this review, the authors discuss the use of gene expression profiling to reach their research target.

9.2.2 In Chronic Kidney Disease (CKD)

The use of biomarker has been successful to assess the cardiovascular mortality in CKD patients, which is a critical clinical challenge due to poor clinical outcome and increasing prevalence. Clinicians such as nephrologists and transplant specialists need suitable biomarkers to predict the occurrence of cardiovascular events and/or mortality in practice. Biomarkers offer a platform for clinicians to dynamically monitor the alteration of patient condition. In a recent review article, Sun et al. (2014) [10], looked at studies that identified potential biomarkers to analyze their predictive value and significance. The study revealed that the identified potential biomarkers are involved in magnesium and phosphate metabolism, hormone dysregulation, proinflammatory processes, and cardiovascular pathogenesis. By combining the use of those biomarkers, it might allow early identification of subclinical cardiovascular system organ damage, thus effectively predicting cardiovascular mortality. The authors also stated that by using those biomarkers, it significantly intensifies the mechanistic understanding of the occurrence of cardiovascular events and mortality, and thus will help to develop preventive measures.

In other clinical venues, post-transplant care medicine and immune suppressants have been greatly improved due to the appearance of various influencing factors. Prediction of graft dysfunction and development of noninvasive biomarkers are becoming the highlight of research topics in the field of renal transplantation due to the fact that they can be convenient approaches to the physicians tasked to give preemptive rescue treatment. Diagnostic technique is used to identify applicable early biomarkers in clinic to dynamically monitor renal graft status and accurately predict transplant outcome independent of various confounders. In a recent review article by Ge et al. (2013) [11], a summary of recent studies on the predictive value of biomarkers and methods to quantify for monitoring kidney transplant was reported. The analysis and comparison of different biomarkers for living and nonliving donors were conducted to identify clinically relevant biomarkers, thus advancing the understanding of distinct molecular and cellular mechanisms of transplantation. This approach has provided insight into the development of novel potential approaches to induce transplant tolerance.

In this review, it has been reported that high expression of microRNA (miR)-142-5p, miR-155, and miR-223 in the peripheral blood mononuclear cells and renal allograft biopsy of living donor transplant recipients were predictive of an acute rejection event, and that can be considered as a biomarker during selection of live donors. With respect to a single factor of deceased-donor graft, three critical genes (CXCR4, CCL5, and ITGB2) were identified from 112 specimens by using microarray-profiling examination. Based on a random forest analysis, these were suggested to be suitable biomarkers to assess organ quality and predict kidney graft function.

9.2.3 Role of Biomarkers in CNS

One of the CNS diseases that benefits most from the biomarker diagnosis is Alzheimer's disease (AD). Insoluble proteinaceous aggregates made up of the microtubule-associated protein tau constitute pathological hallmarks of several neurodegenerative diseases, including AD [12]. In AD, the most common cause of dementia that primarily affects the elderly, senile plaques (SPs) made of Aβ peptides are also abundant in brain regions affected by tau pathology. Unambiguous diagnosis of AD calls for the postmortem detection of both plaques and tangles in the brain. However, the deposition of tau and Aβ in tangles and plaques, respectively, is believed to take place at different time points or stages of AD as well as during normal aging, while tau deposits are thought to occur very early, and the most dramatic presymptomatic AD pathology appears to be an accumulation of Aβ aggregates. The development of increased tau aggregate burden is believed to occur after substantial Aβ deposition has occurred. Tau burden in the AD brain correlates much better with cognitive decline than does SP pathology. Thus, as indicated by Ballatore et al.

(2014) [13], the noninvasive detection and quantification of tau aggregates could be useful to enable (1) accurate diagnosis of AD during life; (2) staging of AD, especially early in the disease course; and (3) monitoring the effectiveness of therapies or target engagement of a potential therapy targeting tau pathology. Consequently, noninvasive brain imaging positron emission tomography (PET) and single-photon emission computed tomography (SPECT) have been recognized as the most promising translational tools to detect preferential binding to Aβ plaques relative to tau aggregates, and that could be utilized to visualize pathology in the brains of living patients [14]. Representatives of these ligands are listed in Figure 9.3. In the field of tau-selective imaging agents, a number of promising compounds have been reported, including arylquinoline derivatives such as THK-5105, pyridoindole derivatives such as T-807, and several phenyl- or pyridinyl-butadienyl-benzothiazoles such as PBB-3. Heterocyclic compounds as imaging probes of tau have been recently discovered and reported. The report discloses the identification of a novel class of tau imaging agents, the aminothienopyridazines (ATPZs), as shown in Figure 9.3.

9.2.4 Biomarkers in Diabetes and Their Role in AD

Diabetes is a major cause of illness and death in our time. While 285 millions battle diabetes in this aging society [15], more than 30 million patients suffer from dementia [16]. Numerous epidemiological studies have also demonstrated that patients with diabetes have a significantly higher risk of developing AD [17, 18]. While genetic and nongenetic risk factors contribute to sporadic AD [19], APOEε4 allele may be associated with an earlier onset of memory loss, and it is the strongest genetic risk factor for sporadic AD and is believed to promote the development of SPs.

- Approximately 25% of the general population, including type 2 diabetic patients, carry at least one copy of APOE4 (1–3% homozygotes)
- Approximately 40–50% of the AD population carry APOE4 (10–12% homozygotes)

Glucose hypometabolism has been detected in brain regions commonly affected in AD in asymptomatic APOE4 allele carriers compared to non-APOE4 allele carriers [20, 21].

A recent review by Sato and Morishita (2014) [22] summarized the brain alterations and clinical symptoms in dementia associated with diabetes and its Aβ/τ -dependent and independent mechanisms. Clinically, diabetic patients show decreased executive function, information processing, planning, visuospatial construction, and visual memory. Therefore, in comparison with the characteristics of AD brain structure and cognition, diabetes seems to affect cognitive function through not only simple AD pathological feature-dependent mechanisms but also independent mechanisms. As an Aβ/τ -independent mechanism, diabetes compromises cerebrovascular function, increases subcortical infarction, and might alter the blood-brain barrier (BBB). Diabetes also affects glucose metabolism, insulin signaling, and mitochondrial function in the brain. It also modifies metabolism of Aβ and tau and causes Aβ/τ -dependent pathological changes.

Animal models of AD or insulin resistance or both demonstrate that AD pathology and impaired insulin signaling form a reciprocal relationship [23]. Of note are

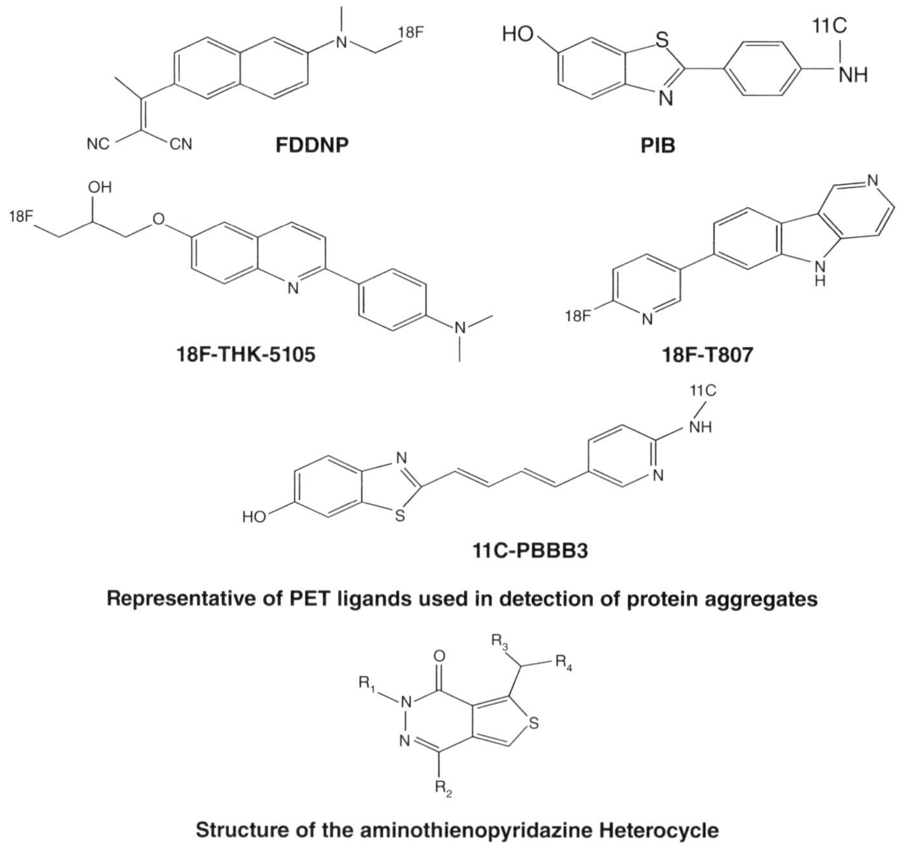

Figure 9.3 Tau-selective imaging agents, including a number of promising compounds (arylquinoline derivatives [THK-5105]; pyridoindole derivatives [T-807]; phenyl- or pyridinyl-butadienyl-benzothiazoles [PBB-3], the new heterocyclic compounds as imaging probes: aminothienopyridazines (ATPZs).

human and animal model studies geared toward improving insulin resistance that have led to the identification of the nuclear receptor and transcription factor peroxisome proliferator-activated receptor gamma (PPARγ) as an intervention tool for early AD. Insulin enhances memory in diabetic patients as well as AD. Glucose uptake by GLUT1 transporters at the BBB can be altered, and when it decreases it leads to AD. By treatment with rosiglitazone, significant effect on both memory and attention versus placebo was observed in moderate AD patients.

In clinical study with rosiglitazone in mild to moderate AD patients who genotyped with respect to APOE4 allele [24], the results suggested that better response to rosiglitazone correlates with patients that are APOE4 negative, and those patients without APOE4 improved while APOE4 carriers did not improve and continued to deteriorate.

To directly treat type 2 diabetes (T2DM), selective inhibition of sodium-glucose cotransporter 2 (SGLT2) has recently become a focus of potential therapeutics. Targeting renal glucose reabsorption as a mechanism to reduce the hyperglycemia, Pfister and his colleagues (2011) [25] describes the mechanism of SGLT2-induced urinary glucose excretion (UGE) and its effects on parameters of glycemic control in animal models, healthy humans, and patients with T2DM. The authors reviewed the development of various SGLT2 inhibitors, listed as an insulin-independent mechanism of treatment for T2DM under various treatment strategies (found under the PD parameters of SGLT2 inhibitors). The authors concluded that this therapeutic approach holds promising outcomes in a wide spectrum of patients.

9.3 GENOMICS AND PHARMACOGENOMICS IN TRANSLATIONAL ADMET

The field of pharmacogenomics ranges from the research conducted on gene discovery to studies in clinical trials to individualize drug therapy in various patient populations. With advancement in this field, there is significant potential to improve the clinical outcome of individual patients. The translational tools and strategy from chemical structure to clinical investigation and its outcome are depicted in Figure 9.4, which indicates the path of data and decisions that should be considered as a reversible process, "from bench to bedside and from bedside to bench." The use of genetic finger prints has facilitate the flow of data, although the translation of pharmacogenomics research findings into clinical practice has faced many barriers and consequently, drugs are prescribed to patients without identify the genotype and without customize the drug dose accordingly.

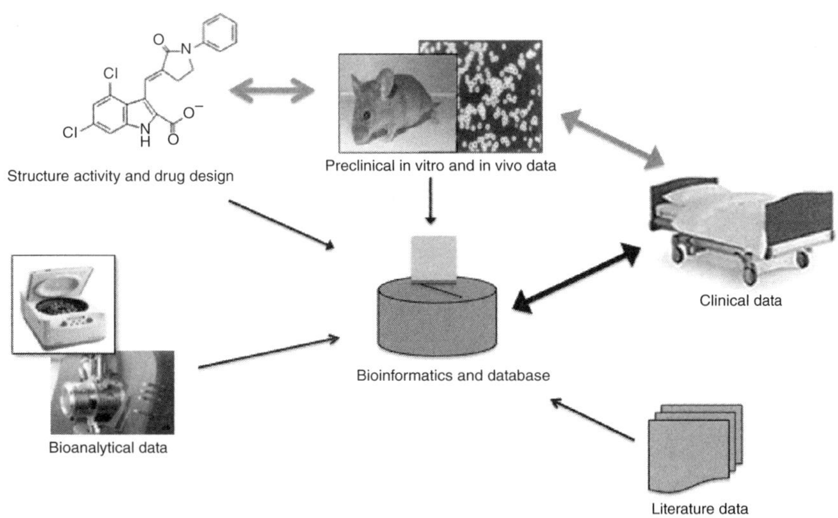

Figure 9.4 Translational tools and strategy for better design of a clinical study and its outcome.

9.3.1 Influence of Pharmacogenomics on Drug Metabolism-Mediated Drug Development

Several P450 genes are associated with polymorphism, and that causes a significant impact on the metabolic functions of the enzyme as well as the PK/PD of drugs involved. Hepatic P450 CYP2D6, which catalyzes the metabolism of many drugs, such as codeine, is one of the P450 enzymes that is strongly associated with genetic polymorphism and consequently causes significant variability of the therapy by these drugs among patient populations [26]. Codeine (see Figure 9.5) is a prodrug

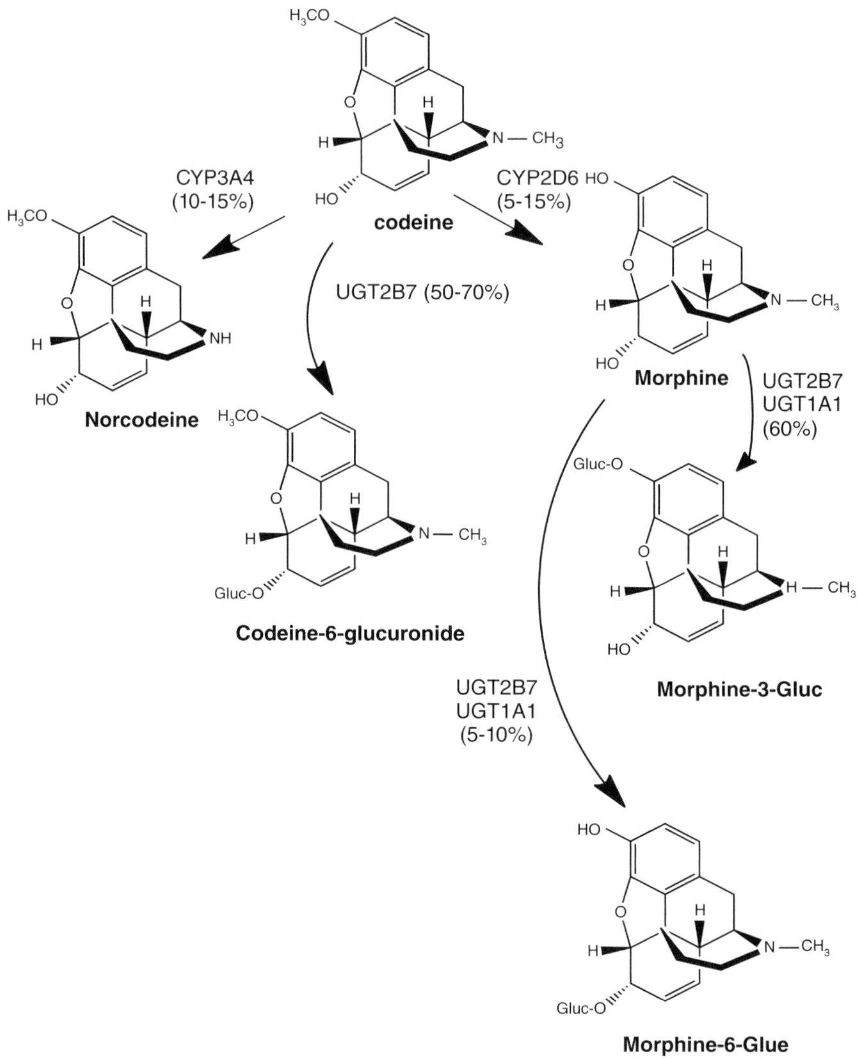

Figure 9.5 Metabolism of codeine.

that metabolizes by CYP2D6 to morphine. Morphine, not codeine, is therapeutically active with a strong opioid agonist function. Therefore, the efficacy and safety of codeine are influenced by CYP2D6 polymorphisms [26–28]. CYP2D6, as shown in Tables 9.1, 9.2, and 9.3, is subjected to genetic polymorphism with several identified alleles that vary functions of CYP2D6, causing a variable extent metabolism of CYP2D6 substrates among individuals and patient populations. For example, individuals carrying more than two copies of functional alleles (e.g., $*1/*1 \times N$; $*1/*2 \times N$) (see Tables 9.1 and 9.2) are ultrametabolizers (UMs), while individuals carrying no function alleles (e.g., combination of *4/*4, *4/*5, *5/*5, *4/*6) are poor metabolizers (PMs). Individuals that carry two alleles encoding full or reduced function or one full function with either one nonfunctional or one reduced function allele are extensive metabolizers (EMs) and those that carry one reduced allele function and other with nonfunction are intermediate metabolizers (IMs). The frequency of various CYP2D6 diplotypes among different ethnic groups is listed in Table 9.3. Clinical phenotype data are available for common alleles; however, many alleles have not been evaluated in clinical trials. The clinical phenotypes of unevaluated alleles are predicted based on the expected functional impact or phenotype of the defining genetic variation or extrapolated based on in vitro functions of probe substrates.

As of today, codeine is still prescribed to patients without defining their CYP2D6 genotype, although patients who are CYP2D6 PMs may have no need for the drug treatment because it is unlikely the PM patients experience analgesia. In contrast, patients that are UMs are at increased risk of toxicity. Therefore, it is advisable to note a patient's CYP2D6 diplotype and to calculate an activity score before making therapeutic decisions about codeine therapy.

TABLE 9.1 Allele Combination of Diplotypes in Prediction of CYP2D6 Phenotypes.

	Predicted Metabolizer Phenotype (Range Multiethnic Frequency)									
Allele	**1*	**2*	**1 × N or *2 × N*	**3*	**4 or *4 × N*	**5*	**6*	**10*	**17*	**41*
**1*	EM	EM	UM	EM	EM	EM	EM	EM	EM	EM
**2*		EM	UM	EM	EM	EM	EM	EM	EM	EM
**1 × N or *2 × N*			UM	EM or UM	EM or UM	EM or UM	EM or UM	UM	UM	UM
**3*				PM	PM	PM	PM	IM	IM	IM
**4*					PM	PM	PM	IM	IM	IM
**5*						PM	PM	IM	IM	IM
**6*							PM	IM	IM	IM
**10*								EM	EM	EM
**17*									EM	EM
**41*										EM

Notes: EM = extensive metabolizer; IM = intermediate metabolizer; PM = poor metabolizer; UM = ultrarapid metabolizer.

TABLE 9.2 CYP2D6 Phenotypes, Frequency, and Corresponding Genotypes and Diplotypes.

Phenotype	% Frequency*	Genotype	Example of Possible Diplotype
Ultrametabolizer	~1 – 2	An individual carrying more than two copies of functional alleles	$*1/*1\times N$, $*1/*2\times N$
Extensive metabolizer	~77 – 92	An individual carrying two alleles encoding full or reduced function; or one full function with either one nonfunctional or one reduced-function allele	*1/*1, *1/*2, *2/*2, *1/*41, *1/*4, *2/*5, *1/*10
Intermediate metabolizer	~2 – 11	An individual carrying one reduced allele function and other with nonfunction	*4/*10, *5/*41
Poor metabolizer	~5 – 10	An individual carrying no function alleles	*4/*4, *4/*5, *5/*5, *4/*6

Notes: *Based on Caucasians patients. The activity scores for advisable treatment are > 2.0 for the ultrametabolizer, 1.0–2.0 for extensive metabolizer, 0.5 for intermediate metabolizer, and 0 for poor metabolizer.

TABLE 9.3 Frequencies of CYP2D6 Alleles among Various Ethnic Groups.

Allele	African	African American	Caucasian (European + North American)	Middle Eastern	East Asian	South/ Central Asian	Americas
**1*	39.23	40.60	53.63	58.04	34.17	53.70	64.28
**2*	20.12	14.15	26.91	21.72	12.82	31.90	23.48
**3*	0.03	0.31	1.32	0.10	0.00	0.00	0.73
**4*	3.36	6.23	18.50	7.80	0.42	6.56	11.28
**5*	6.07	6.14	2.69	2.34	5.61	2.54	1.88
**6*	3.05	0.24	0.95	0.72	0.02	0.00	0.43
**7*	0.00	0.00	0.11	0.00	0.00	*ND*	0.00
**8*	0.00	0.00	0.02	0.00	0.00	*ND*	0.07
**9*	0.10	0.48	2.14	0.00	0.07	1.43	1.32
**10*	6.77	4.18	3.16	3.49	42.31	19.76	3.37
**14*	0.26	0.00	0.00	0.00	0.86	0.00	0.33
**17*	19.98	18.22	0.32	1.58	0.01	0.38	3.0
**36*	0.00	0.56	0.00	0.00	1.58	*ND*	0.25
**41*	10.94	9.41	8.56	20.37	1.97	10.50	5.93
$*1\times N$	1.47	0.44	0.80	3.07	0.28	0.50	0.73
$*2\times N$	1.56	1.61	1.27	3.87	0.38	0.5	2.38
$*4\times N$	1.40	2.07	0.25	0.00	0.00	0.00	0.60

When reviewing the metabolism of codeine, as shown in Figure 9.5 the CYP2D6 catalyzes the O-Demethylation of codeine into morphine. This reaction is a minor pathway in EMs, accounting for 5–10% of codeine clearance (CL) in such individuals, but it is essential for its opioid activity. As expected, the formation of morphine from codeine is much higher in UMs than EMs, and that reflects on the further glucuronidation of morphine to morphine-3-glucuronide and morphine-6-glucuronide, which is known to have analgesic activity in humans. In conclusion, association of CYP2D6 polymorphism with the formation of morphine from codeine and the observed PK/PD is well defined: in PMs, morphine level decreases and in analgesia in PMs it also decreases compared with EMs. Furthermore, incidence of gastrointestinal side effects was less in PMs versus EMs. As mentioned earlier, the conversion of codeine to morphine in CYP2D6 UMs is higher than EMs, and that results in higher systemic toxicity due to higher concentrations of morphine even at low codeine doses [30]. Despite the distinct CYP2D6 genotype and corresponding phenotype in relation to the observed PK/PD and toxicity of codeine therapy, great variability within the EMs and some of those patients may develop symptoms similar to UMs. The variation among individuals with the same diplotype is great, and the occurrence of severe or life-threatening side effects following standard doses of codeine in UMs was observed [29, 30].

Despite these unknown causes of variability among the same diplotype, there is a considerable level of benefit in CYP2D6 genotype testing. The potential benefit is that patients with genotypes that confer a higher risk of ineffective analgesia or of an adverse event may be identified and an alternative analgesic may be administered. It is worth mentioning that CYP2D6 genotyping is reliable when performed in qualified laboratories. However, as with any laboratory test, a possible risk to the patient is an error in genotyping that could have long-term adverse health implications for the patient.

In addition to codeine drug therapy that was affected by CYP2D6 genetic polymorphism and variability among patient populations, there are several other drug therapies affected by it, such as tamoxifen treatment for breast cancer [31]. In common clinical practice, all patients treated with tamoxifen receive a standard dose. Because of the implication of CYP2D6 genotype on metabolism of tamoxifen to active metabolite endoxifen and its outcome on PK/PD, genotyping before tamoxifen therapy is encouraged (Figure 9.6). Recent attention is now on the adequate exposure to endoxifen. Data from several studies indicated that genotype only partially explains interindividual differences in endoxifen concentrations. Phenotyping approaches, as well as therapeutic drug monitoring strategies, are now being tested to individualize tamoxifen treatment.

Recent revised labels of several medicines such as abacavir, carbamazepine, cetuximab, imatinib mesylate, irinotecan, mercaptopurine, and traztuzumab indicated that these drugs are associated with genetic polymorphism, and their labels include pharmacogenomics information The influence of pharmacogenomics on the development of these drugs has been well documented [31]. The influence of pharmacogenomics on Plavix (clopidogrel) PK/PD is also well investigated, and the new findings demonstrated that CYP2C19*2 (IM) carriers treated with clopidogrel were at higher risk of

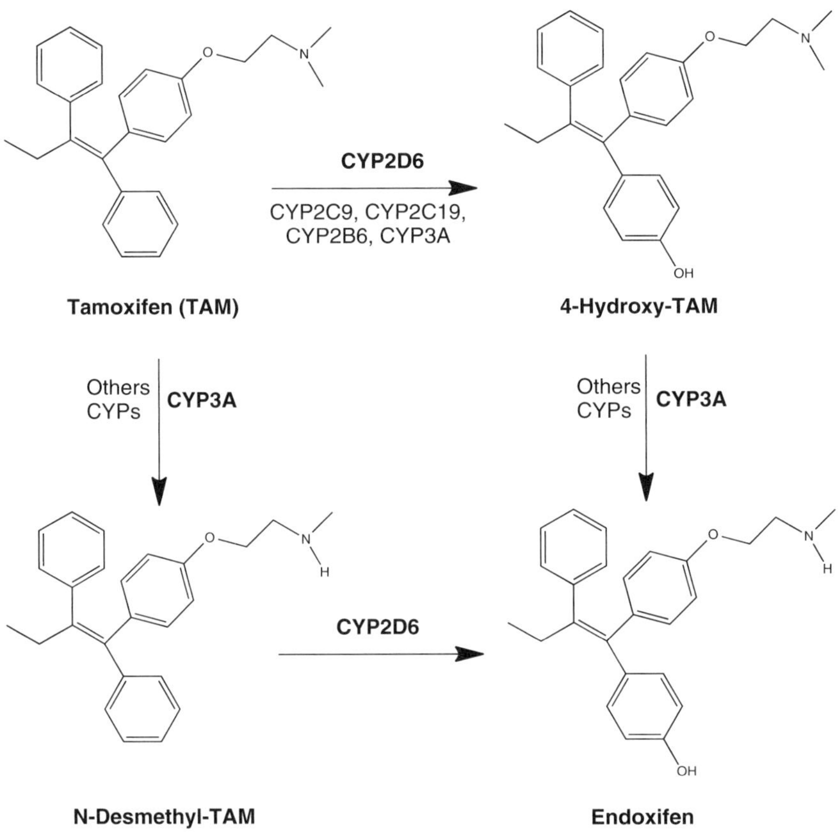

Figure 9.6 Metabolism of tamoxifen by CYP2D6 and formation of active metabolite endoxifen.

major adverse cardiovascular events, particularly stent thrombosis, than were noncarriers [32], and this information was recently included to the drug label information. A similar action based on evidence of a pharmacogenomic relationship has been taken for warfarin [33]. Variability in warfarin dose requirements is significant and can be explained on the basis of genotypes of both CYP2C9 and VKORC1 (the gene that encodes for the enzyme that is associated with the formation of blood clotting). Warfarin is an anticoagulant characterized by a wide interindividual variation in dose requirements. Furthermore, warfarin treatment has a narrow therapeutic index; consequently, accurate dosing is critical for the patient's safety. Nongenetic caveats, such as body size and age, are poor predictors of an individual's dose requirement. However, several investigations have revealed that genetic factors significantly influence warfarin dose requirements. Warfarin is metabolized primarily via oxidation in the liver by CYP2C9, and it exerts its anticoagulant effect by inhibiting the protein vitamin K epoxide reductase complex, subunit 1 (VKORC1). Three single-nucleotide polymorphisms (SNPs), two in the CYP2C9 gene and one in the VKORC1 gene, have been found to play key roles in determining the effect of warfarin therapy on coagulation. The CYP2C9 diplotypes are *1/*1, *2/*2, and *3/*3 (combination of alleles such as *1/*2, *2/*3 also occurs) and the phenotypes of these corresponding diplotypes are EM, IM (30% reduced metabolism function compared to EM), and PM (90% reduced metabolism function compare to EM), respectively. The frequency of these genetic polymorphisms varies among different ethnic populations; 10% and 6% of Caucasians carry the *2 and *3 variants, respectively, but both variants are rare (<2%) in those of African or Asian descent [34]. Because warfarin given to patients with *2 or *3 variants will be metabolized less efficiently, the drug will remain in circulation longer, so lower warfarin doses will be needed to achieve anticoagulation. In the VKORC1, lower warfarin doses are needed to inhibit VKORC1 and to produce an anticoagulant effect in carriers of the A allele. The prevalence of these variants also varies by race, with 37% of Caucasians and 14% of Africans carrying the A allele [35].

9.3.2 Influence of Pharmacogenomics on Drug Transporter-Mediated Drug Development

Drug efficacy and dose requirement are not influenced solely by variation in drug-metabolism genes but also by polymorphisms in genes that encode receptors (as seen in the case of VKORC1 associated with warfarin), transporters, drug targets, and, most important, those associated with drug transporters [29]. Several transporters have been shown to have pharmacogenomic relationships with drug PK or effects. For example, a synonymous SNP in the MDR1 (ABCB1) gene has been associated with the maximally achievable digoxin concentration [36]. The well-reported case of genetic polymorphism and its clinical adverse drug reaction (ADR) is associated with the polymorphisms in the transporter OATP1B1 (SLCO1B1*)* and statin drug therapy. OARP1B1 has been associated with several phenotypes, including an increased risk of simvastatin-induced myopathy [37], methotrexate-related gastrointestinal toxicity [38], and disposition of kinase inhibitor flavopiridol [39].

To relate the genetic polymorphism of drug transporters to clinical outcome sometimes is challenging translational research due to the fact that most of the drugs depend on metabolism and transporters for their disposition and observed PK/PD. Accordingly, translational investigations are required to elucidate the mechanism underlying the disposition, efficacy, and safety of tested drugs and whether the genetic polymorphism of either transporters or drug-metabolizing enzymes (DMEs) is the key factor that needs to be considered; for example, the reports by Herrlinger et al. (2011) [40] that describe the limitation of nonresponsive and unpredictable adverse events from tacrolimus (Tac) therapy for the treatment of steroid-refractory ulcerative colitis (UC) in specific patient populations. Similarly, a recent report by Markova et al. (2013) [41], who tested 12 functional polymorphisms in five genes involved with bosentan disposition used in treatment of pulmonary arterial hypertension, found its treatment is limited by drug-induced liver injury (DILI) in selected patients.

Previous reports have indicated that the dispositions of both Tac [42] and bosentan [43] are associated with the combined functions of DMEs and drug transporters.

For Tac, response in patients who have undergone solid-organ transplantation has been associated with the presence of variants in CYP3A and ABCB1. The authors elucidated the contributions of variants of CYP3A: CYP3A4*1B and CYP3A5*3, as well the variants of ABCB1: 1236C > T, 2677G > T, A, and 3435C > T polymorphisms to Tac response in 89 patients with UC. In a linear logistic regression model, patients with homozygous variants for one of the three ABCB1 alleles showed significantly higher short-term remission rates as compared with those of other genotypes. The effects held true after multivariate analysis including multiple comparisons and were more pronounced after correction for dose-adjusted Tac blood trough levels. The authors concluded that ABCB1, but not CYP3A5, may predict short-term remission of Tac in steroid-refractory UC. In the case of bosentan, Markova and his colleagues in 2013 [41] reported the results generated from in vitro and clinical data to assess the factor that associated with DILI—mediated by the treatment with bosentan in pulmonary arterial hypertension (PAH) patients. To identify genetic markers of DILI, association analyses were performed on 56 Caucasian PAH patients receiving bosentan. Twelve functional polymorphisms in five genes (ABCB11, ABCC2, CYP2C9, SLCO1B1, and SLCO1B3) implicated in bosentan PK were tested for associations with alanine aminotransferase (ALT), aspartate aminotransferase (AST), and DILI. After adjusting for body mass index, CYP2C9*2 was the only polymorphism associated with ALT, AST, and DILI ($\beta = 2.16$, $P = 0.024$; $\beta = 1.92$, $P = 0.016$; odds ratio 95% CI = $2.29 - \infty$, $P = 0.003$, respectively). Bosentan metabolism by CYP2C9*2 in vitro was significantly reduced compared with CYP2C9*1 and was comparable to that by CYP2C9*3. These results suggest that CYP2C9*2 is a potential genetic marker for prediction of bosentan-induced liver injury and warrants investigation for the optimization of bosentan treatment. However, the same authors in 2014 [44] comment on data by other investigators that indicated that CYP2C9-dependent intrahepatic bosentan concentrations may be more relevant and may not be directly related to bosentan plasma concentrations. In the 2014 report [44], Markova et al. commented on 2013 results that although hepatic transporter polymorphisms were not associated with bosentan hepatotoxicity, the possibility exists that transporter function is altered in PAH patients as a result of disease or drug treatment. Furthermore, in the 2014 paper, the authors commented that there is a possibility of CYP2C9*2 not to be causative SNP but that it tags as a regulatory SNP. In addition, difference in study populations from healthy volunteers versus PAH patients may also have influenced the reported results regarding the role of CYP2C9 with DILI in bosentan treatment. The pharmacokinetic study by Markert et al. (2014) [45] was carried out in healthy volunteers, whereas our genetic association study was performed in PAH patients receiving steady-state bosentan for clinical indications. As also noted by Markert et al., differences in PK between CYP2C9 metabolizer groups might be more apparent in PAH patients, especially because they may have a longer duration of exposure. In previously published bosentan PK data for PAH patients [41], Markova et al. showed that bosentan exposure is not proportional to dose and is approximately 2-fold higher than that in healthy subjects. In addition, because PAH affects expression of endothelin receptors, it may affect the tissue-to-plasma ratios of bosentan in these patients. Finally, the possibility that concurrent medications in PAH patients affect bosentan pharmacological and toxicological properties cannot be excluded.

9.4 TRANSLATIONAL ADMET, APPROACHES AND TOOLS

9.4.1 From Bedside to Bench to Bedside: POC Investigations

As mentioned earlier, translational research is defined as the biomedical and pharmaceutical discipline that integrates with public health research aiming to individualize toward improving the health of patients by "translating" findings into diagnostic tools, medicines, procedures, and policies (see Figure 9.7). In this chapter, the translation of pharmaceutical research and results from the bedside to bench to bedside will be discussed. Although those few successes have been notable, the translation of results from one patient population to another, from healthy to diseased, or from animal model to human is currently extremely low. Here, several specific case studies involving a pediatric antifungal drug therapy for juvenile myelomonocytic leukemia, as well as others, will be discussed.

9.4.1.1 Individualized Antifungal Drug Therapy in Pediatric Patients Children are typically not included in most product submission studies, and therefore the data on dosing, efficacy, and safety that are available for adults are usually not available for pediatric patients. Consequently, 75% of drug products have insufficient label information for pediatric dosing, safety, or efficacy [46], and often dosing and safety information has been based on simple allometric scaling of adult body weight or body surface area. Because children physiologically and anatomically are different than adults, drug disposition can be altered as a result of these differences. A child's maturation process can affect the absorption, distribution, metabolism, and excretion (ADME) of drugs, and hence the disposition and CL of drugs change with age and are often different compared to adults. In the past, therapeutic or toxicological assessment in pediatric populations was done only on the basis of an approximate linear relationship between postnatal age and body weight or surface area. Hence, dose was adjusted on the basis of age-associated differences in body weight (BW) or

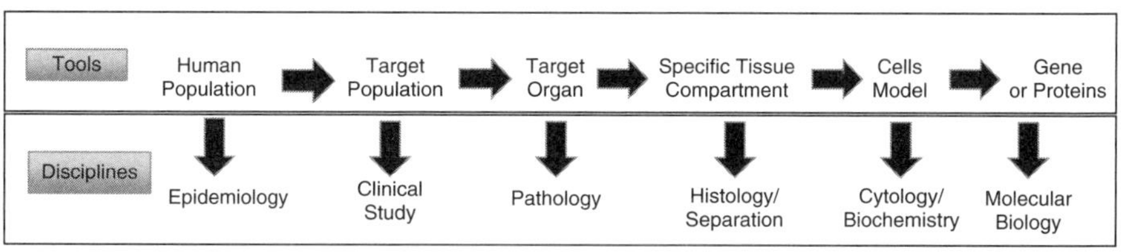

Figure 9.7 Translational path: from bedside to bench.

body surface area (BSA) [18] according to Equations (9.1) and (9.2):

$$\text{Dose}_{\text{child}} = \text{Dose}_{\text{adult}} \times \frac{\text{BW}_{\text{child}}}{\text{BW}_{\text{adult}}}, \tag{9.1}$$

$$\text{Dose}_{\text{child}} = \text{Dose}_{\text{adult}} \times \frac{\text{BSA}_{\text{child}}}{\text{BSA}_{\text{adult}}}. \tag{9.2}$$

Dose-level adjustments based on simple allometric scaling of adult body weight or body surface area and not based on maturation can result in substantial over- or underexposure of children to drugs, thus placing them at a risk of adverse events and denying them potential therapeutic benefits [47–50]. The several incidents of inadequate drug therapy in the 1960s and 1970s occurred due to the lack of our knowledge on the changes in ADME with maturation. Therefore, the National Institute of Child Health and Human Development (NICHD) funded 13 different "Pediatric Pharmacology Research Units" (PPRUs) in the United States to facilitate and promote adequate labeling of new drugs or drugs already in the market for safe and effective pediatric use. However, ethical considerations can sometimes prevent clinical studies from being conducted in pediatric populations. Hence, alternative translational approaches to predict potential differences in the PK behavior of drugs in children compared with adults are promoted, and several pediatric clinical research sites have been set up along with related translational research in pediatric drug therapy. Consequently, well-designed comprehensive pediatric clinical studies have been conducted both in the United States and Europe, causing the off-label therapeutic products to decline as adequate drug exposures in pediatric populations have been recommended and delivered. Currently drug CL, a key PK parameter and determinant of disposition of drugs, is being used for scaling of dosage and for determining the potential of adverse drug reactions in children.

To support drug therapy in children, a translational approach from bedside to bench to bedside and from adult to pediatric population was conducted. First, in vitro approaches were applied to elucidate the factors that alter the mechanisms of drug disposition of selected therapeutic agents and how these factors are influenced by the ontogeny when comparisons between children and adults were made. Because fungal infection is one of the most common and sometimes life-threatening medical issues in pediatric populations, selected agents used for fungal infection were chosen for this investigation. Furthermore, the dispositions of antifungals were known to be age dependent [51]. The two antifungal agents selected in the POC translational investigations were voriconazole (VORI) and micafungin (MIC), known to be cleared predominantly via the oxidative metabolism in liver in the case of VORI [51] and predominantly via biliary excretion in the case of MIC [52].

As seen in Figure 9.7, the translational tools and the corresponding disciplines applied in diversified investigations, depending on the specific goals, were also applied in this translational investigation of antifungals therapy in children. The target goal in this investigation was to select the adequate drug dose that ensured efficacy without causing any overdosing drug-induced toxicity. This goal can be achieved by identifying the relationship between drug dose and exposure and drug CL according to the following equation:

$$AUC_0^{\infty} = \frac{F \times Dose}{Cl_s}. \tag{9.3}$$

Assessing accurate drug CL is the foundation of selecting the adequate dose in children as in adults. The initial results that were used were the finding generated from clinical investigation in adults (in the case of VORI) or in general pediatric study (in the case of MIC), and it was found that the CL in children is greater than adults. Based on the in vitro and in vivo investigations of these drugs in human adults, the mechanism underlying the disposition and drug CL of both drugs targets the liver and its specific target tissue compartment or cell models. In vitro investigations using biological samples generated from children and adults livers were found not only to provide a rationale for the difference in CL between both age groups but also generated CL values that are similar to those generated from the clinical studies in both age groups. Most importantly, these translational investigation were conducted to provide a better estimation for drug dose based on exposure and the CL values as shown in Equation (9.3) and not based on the scaling value from adult dose based on body weight ratio between adults and pediatrics as shown in Equation (9.1). Figure 9.8 shows the comparison between the dose estimation from these two approaches for VORI and MIC in pediatric populations under investigation. Clearly, the doses estimated based on body weight difference were 2- to 5-fold lower than dose estimated

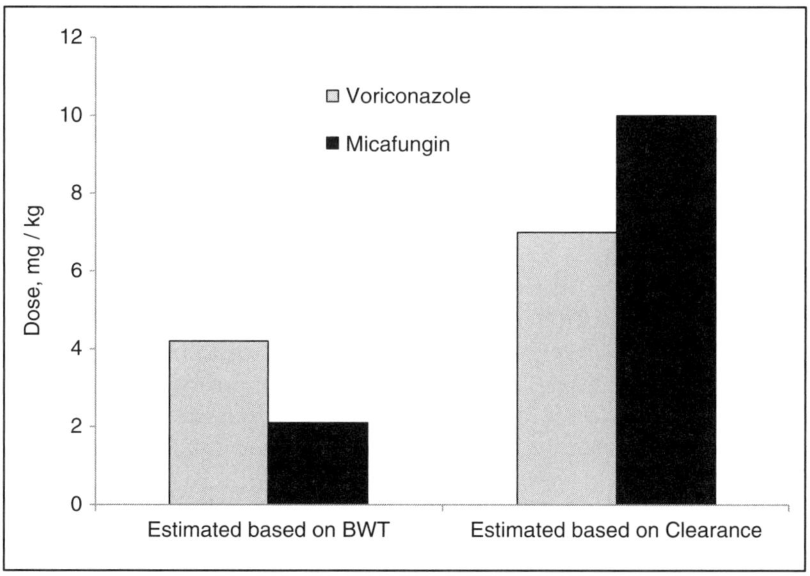

Figure 9.8 Pediatric estimated dose of antifungal drugs based on body weight ratio of pediatrics and adults compared to dose estimated from actual clearance value in pediatrics.

from actual drug clearance. Consequently, pediatric patients had not been treated with the effective therapeutic dose. However, when dose adjustment was made based on actual clearance, antifungal drug therapy was remarkably improved [53, 54].

Because our ultimate goal is to predict the PK behavior of drugs in pediatric populations, this was a retrospective research program that used PK data of selected model drugs in pediatrics and adults along with the in vitro kinetic data generated by using reagents prepared from adults and pediatrics. Beside appropriately designed in vitro studies to mimic the clinical investigations (e.g., in vitro drug concentration to match C_{max}, the age of pediatric populations, and the use of in vitro model corresponding to the organ of CL), the selection of model drugs was a critical element in this research program. The two antifugal drugs, VORI and MIC, were selected according to the following criteria: (1) these drugs have a significant therapeutic value against life-threatening infection in pediatrics; (2) known mechanism of CL in clinic in adults and pediatrics; (3) CL in pediatrics is different than adults; and (4) CL is predominantly dependent on one route of elimination (either metabolism or biliary excretion).

VORI, which is eliminated mainly by hepatic metabolism, exhibits a 3-fold higher CL in children (age 2–8 years) compared to adults. The in vitro and in vivo metabolism studies indicated that the oxidative metabolism generated the N-oxide of VORI as a predominant circulating metabolite in humans [51]. The in vitro investigation revealed that, in addition to P450 enzymes. the N-oxide metabolite is formed by flavin-containing monooxygenase (FMO) enzymes. In vitro metabolism of VORI by human liver microsomes from adults and from children indicated that oxidative metabolism of VORI in children was higher than in adults. Enzyme kinetics analysis for the formation of N-oxide indicated that children have higher maximal metabolic rate of VORI (V_{max}), while the binding affinity (K_m) was similar to adults. In vitro metabolism studies using selective inhibition of CYP3A4, CYP2C19, or FMO demonstrated that the latter two enzymes contributed much more to N-oxidation of VORI in children compared to adults. Although the expression of the enzymes responsible for the formation of N-oxide was only slightly higher (CYP2C19 and FMO but not CYP3A4) in children compared to adults, CYP2C19 and FMO appear responsible for the age-dependent CL of VORI. Prediction of in vivo CL from the in vitro studies of N-oxide formation normalized to individualized scaling factors was in agreement with the observed CL for both populations (predicted value was 80% of the observed). On the other hand, MIC has been used to treat life-threatening fungal infections of the CNS in neonates and young infants and is mainly cleared into the bile. CL in neonates (age < 28 days) is reported to be 3-fold higher than adults [52]. The role of hepatobiliary transporters has been investigated in the CL of MIC not only in adults but also in neonates. In vitro characterization of micafungin disposition using freshly isolated sandwich-cultured hepatocytes from human and rat indicated that the cellular uptake and excretion to the bile were comprised of carrier-mediated processes, with biliary secretion as the rate-limiting step in humans. The uptake was found to be mediated by NTCP >> OATP, while the excretion was found to be mediated by BSEP > BCRP. The relative protein expression of these transporters in liver from neonates was similar to that in adults, thus we assumed that the in vitro plasma intrinsic CL and biliary intrinsic CL of neonates could be similar to values generated by sandwich-cultured hepatocytes from human adult donors. This indirect approach to estimate the hepatobiliary CL in neonates was employed to accommodate the shortcomings of the ethical complications involved in obtaining neonatal hepatocytes isolated from fresh tissue. To calculate the in vivo CL, the in vitro MIC fraction unbound in serum from neonates was measured and compared to that in adults. Interestingly, unbound fraction of MIC in neonate serum was higher compared to adults. In addition, the hepatic blood flow in neonates is known to be higher than in adults. As these parameters (intrinsic CL, fraction unbound, hepatic blood flow) were used to calculate in vivo CL, good in vitro-in vivo relationship of MIC plasma CL in neonates and adults was found. These results also revealed that differences in protein binding and/or physiological parameters (hepatic blood flow) between neonates and adults might be key factors that contribute to the higher CL of MIC observed in neonates compared to adults, while the role of hepatobiliary drug transporters seems to be a noncontributing factor.

These results indicated that, similar to VORI, MIC in vitro studies can not only be used to predict in vivo PK behavior in pediatrics and in adults but can also be used to elucidate the key factors that contribute to the PK differences between the two groups. The validation of the methods that were used in generating the in vitro data of VORI and MIC disposition in human population arises from the good agreement between in vitro and in vivo data for humans.

We anticipate that similar mechanistic studies with similar drugs based on employing tissue samples from children and adults and known PK behavior in adults might predict the CL of these drugs, prospectively. The findings from this translational research can benefit the antifungal and other drug therapies in pediatric populations.

9.4.1.2 "From Bedside to Bench" in Rare Pediatric Leukemia One of the rare leukemias in pediatric patients is juvenile myelomonocytic leukemia (JMML). It is a typically aggressive myeloid neoplasm of childhood that is clinically characterized by overproduction of monocytic cells that can infiltrate organs, including the spleen, liver, GI tract, and lung. JMML shares some clinical and molecular features with chronic myelomonocytic leukemia, a similar disease in adults. Currently, relapse in treatment that relies on allogeneic hematopoietic stem cell transplant as a standard care for patients with JMML is the most frequent cause of treatment failure. Defining the genomic landscape of JMML has improved the care of pediatric patients. Insights from cancer predisposition syndromes have led to the discovery of nearly 90% of driver mutations in JMML, all of which thus far converge on the Ras signaling pathway. In a recent review article by Chang et al. (2014) [55], the authors summarized the efforts to improve the ability to accurately diagnose patients, develop molecular markers to measure disease burden, and choose therapeutic agents to test in clinical trials. Furthermore, this review emphasizes recent advances in the field, including mapping of the genomic and epigenome landscape, insights from new and existing disease models, targeted therapeutics, and future directions.

9.4.2 From Juvenile Animal Model to Human Adult

The mouse genetic models were used to achieve new discoveries in neurobiology of Down syndrome (DS). A summary of this research is included in a review article by Kleschevnikov et al. (2012) [56], where the authors provided an outline of experimental approaches aimed at development of pharmacological restoration of cognitive function in people with this developmental disorder. Changes in structure and function of synaptic connections within the hippocampal formation of DS model mice, as well as alterations in innervations of the hippocampus by noradrenergic and cholinergic neuromodulatory systems, have provided important evidence for potential pharmacological treatments of cognitive disabilities in DS. By using the mouse models, possible molecular and cellular mechanisms underlying this genetic disorder have been addressed. Furthermore, this review article described novel mechanisms engaging amyloid precursor protein (App) and other proteins through their effect on axonal transport and endosomal dysfunction to "Alzheimer-type" neurodegenerative processes that affect cognition later in life. Finally, the authors defined a number of therapeutic strategies that may restore cognitive function in mouse models of DS in both juvenile and young animals. These strategies focus on restoration of synaptic plasticity, rate of adult neurogenesis, and functions of the neuromodulatory subcortical systems. That may lead to identifying an increasing number of potential targets for pharmacotherapy of cognitive deficits in DS and potential initiation and completion of clinical trials. When designing a translational study in pediatric populations using animal models, Table 9.4 can provide some guidance on selecting the corresponding age of juvenile animal model.

It has been mentioned that improvements in human health must be translated into practical applications, typically beginning at "the bench" with basic research. This can then progress to the clinical level, which has been obviously revealed in the fields of interventional cardiology, CNS, diabetes, and others. Preclinical studies with new therapy using animal models play a very important role in the evaluation of efficacy and safety before their use in human clinical studies.

Although animal models were used in the preclinical stage of drug development prior to Phase 1 clinical trials, there has been a considerable and ongoing debate as to how, or even if, results from animal studies can be extrapolated to determine the potential toxicity in man [57]. Furthermore, the advancement in science of risk assessment has simply indicated that it is impossible to set a direct translation of results generated from laboratory animal studies conducted under highly controlled

TABLE 9.4 Translational of Growth Classification from Animal Species to Human.

Classification	Rat(Days)	Dog(Weeks)	Primate (Month)	Min-pig(Weeks)	Man
Neonate	<10	<3	<0.5	<2	≤ 1 month
Infant	10–12	3–6	0.5–6	2–4	1–23 months
Child	21–35	6–20	6–36	4–14	2–12 years
Adolescent	35–60	2–28	36–48	14–26	12–18 years

conditions to the diversified cultural and genetic human populations. However, to permit the safe use of chemicals/drugs, they must be tested in laboratory animal studies, sometimes using more than one animal species, to protect the general human population and patients. A number of actual measurements and theoretical extrapolations (such as allometric scaling) have been developed to understand the ADMET, PK/PD, and biochemical effects of drugs, as well as the critical differences that occur in these parameters between different animal species and man [58].

9.4.3 Use of Chimeric Rodents with Humanized Liver as a Translation Model in Bridging the Gap between Preclinical and Clinical Trials in ADMET

The standard species used in safety evaluation studies are rodents, normally the rat and/or mouse, and a nonrodent species, for example, the dog or nonhuman primates, such as the cynomolgus monkey [59, 60]. The ability to assess the safety and PK/PD of drug candidates is an important part of the drug discovery process, and translation of data from preclinical animal models or from various in vitro approaches has been used to estimate the safety margin in relation to exposure of a subsequent drug dose to man prior to clinical trials. The differences between in vitro and in vivo systems, and between the biology of the different species used to assess PK and those in human, have been comprehensibly described in a review article by Tang and Prueksaritanont (2010) [61] in relation to DMEs and drug transporter-mediated ADME. Some concerns and recommendations have been raised in establishing a new translational in vivo model that can be used to predict ADME and toxicity in humans prior to Phase 1 clinical investigation. Modified animal models for assessing impact of drug and metabolite exposure early can be critical in making the appropriate decision toward selection of drug candidate and /or drug dose before the human study. Because primary human hepatocyte is the best predictor in vitro model to assess in vivo ADMET (as discussed in previous chapters), "humanizing" laboratory animal species, mostly mice, by replacing their own hepatocytes with human hepatocytes generates an animal model that more closely mirrors human disposition and toxicity of drugs. A high level of replacement of host liver cells with human hepatocytes would clearly be of great potential for studying human diseases. The science of developing and characterizing such "chimeric liver" mice has advanced significantly in the past few decades. Human metabolism for compounds such as quinidine and cefmetazole was reasonably estimated using the model. The chimeric mouse model not only retains CYP activity, as was the case for CYP2D6, but also responds to classical CYP-inducing chemicals by increasing expression levels of CYPs 3A4 and 1A2 [62]. The disposition of other drugs, such as troglitazone [63], warfarin [64], and debrisoquin [65], was also evaluated. Additional studies on phase II enzymes, including UDP-glucuronosyltransferase (UGT), sulfonyl transferase (SULT), N-acetyltransferase (NAT), and glutathione S-transferase (GST) and drug transporters such as bile salt export pump (BSEP) and multidrug resistance protein 2 (MRP2), revealed that a chimeric mouse model with human liver is indeed capable of assessing ADME and safety of drugs in humans.

In a 2014 volume of the journal of *Xenobiotica*, several review articles and study investigations were listed to demonstrate the validity of the model and its application as a valuable translational tool to investigate the safety and efficacy of drug candidates prior to human use [66–69]. In one of these review articles, Foster et al. (2014) [66] summarized all the applications of the chimeric mouse with humanized liver. The authors described that in addition to the assessment of ADMET, another effective application of the model is to investigate therapeutic strategies against malaria, the greatest infectious disease killer of man that has an obligatory hepatic stage in its life cycle. *Plasmodium falciparum* is the deadliest form of malaria, responsible for millions of deaths annually, especially in children in developing countries. The ability of chimeric mice with chimeric humanized livers to support *P. falciparum* malaria infection through its liver stage [70] and to then infect human erythrocytes, completing its life cycle, was assessed. The model effectively supported the studies involved in developing a new malaria vaccine utilizing parasites with alteration of genes required for maturation through the liver stage as immunogens [66, 71].

9.5 SCALING OF PK IN PREDICTION OF HUMAN PK AND DOSING

9.5.1 From Adult PK to Pediatric: Calculation of In Vivo CL in Children

Some of the ontogeny models that predict hepatic metabolic CL were built by using in vitro enzyme ontogeny information from the literature and/or by using in vivo CL studies in children. However, to develop a CL versus age prediction model, the age dependence of the involved enzyme had to be determined. In vitro literature values of the age dependence of enzyme activity of probe substrates in hepatic microsomes or hepatocytes can initially be used to describe the enzyme-specific ontogeny. When the relationship between a child to adult metabolic CL by specific drug metabolizing enzyme is determined retrospectively, unknown child CL in metabolic study can be extrapolated from the adult value when the same enzyme is involved. This approach can be used when there is sufficient data that can help in making an accurate assessment on the difference in drug metabolism between children and adults.

As mentioned in this chapter, PK of drug in children was determined from adult PK allometric scaled with body weight ratio of both populations, as shown in Equation (9.4):

$$\text{PK parameter}_{\text{child}} = \text{PK parameter}_{\text{adult}}(\text{BW}_{\text{child}} / \text{BW}_{\text{adult}})^{\text{b}}. \tag{9.4}$$

However, this approach is not accurate, as growth from birth to adulthood is not a linear relationship.

Based on the age of the child and the specific pathway of CL (metabolism, renal excretion, or biliary CL), the adult intrinsic CL can be modified by the percentage of change based on child value, ratio of change in unbound fraction (f_u), and hepatic blood flow (Q_H) of the organ responsible for CL. Intrinsic CL in children can be used to calculate in vivo plasma CL using a well-stirred model of drug disposition. The age-dependent changes in these scaling factors that are being used to extrapolate the in vitro intrinsic CL to in vivo intrinsic CL and in vivo calculated whole-body CL are described below. For example, to calculate intrinsic CL for a particular enzyme pathway, Michaelis-Menten constants V_{max} and K_m can be determined experimentally. The V_{max}/K_m ratio can be scaled to whole-body intrinsic CL according to Equations (9.5) and (9.6):

$$CL_{int} = V_{max}/K_m \ (\text{ml}/\text{min}/\text{mg}) \times \text{Scaling Factor} \ (\text{mg}/\text{kg}), \tag{9.5}$$

$$\text{Scaling Factor} = [\text{MPPG} \ (\text{mg}/\text{g}) \times \text{liver wt} \ (\text{g})]/\text{body wt} \ (\text{kg}), \tag{9.6}$$

where MPPG is milligram microsomal protein per gram liver; liver weight (wt) and body wt can be obtained from published data [72]. In vivo CL of drug can then be calculated based on in vitro data from children and adults using the well-stirred model of hepatic disposition [51], as shown in Equation (9.7):

$$\text{Predicted CL} = Q_H \times f_u \times Cl_{int}/(Q_H + f_u \times Cl_{int}). \tag{9.7}$$

The use of the well-stirred model to calculate CL requires a consideration of the age-related changes in hepatic blood flow (Q_H) and f_u. Therefore, the values of Q_H and f_u, specific for the age group, must be used to obtain estimates of in vivo drug CL in specific pediatric groups. It is known that plasma protein levels are lower in the newborn than in the adult and gradually increase with age. At birth, human serum albumin (HSA) concentrations are close to adult levels (75–80%), while α_1-acid glycoprotein (AAG) is initially half the adult concentration. McNamara and Alcorn (2002) [73] have reported that age-dependent changes in f_u can be estimated for a particular pediatric age group on the basis of known pediatric plasma protein-binding ontogeny data. This approach scales the unbound fraction from adult to pediatric subjects on the basis of a correlation between unbound fraction and the relative levels of the major plasma-binding proteins (albumin and a1-acid glycoprotein), assuming a constant value for binding affinity between adults and pediatric subjects according to Equation (9.8):

$$f_{u,\,infant} = \frac{1}{1 + \frac{P_{infant}\,(1-f_{u,\,adult})}{P_{adult}\ \ f_{u,\,adult}}}, \tag{9.8}$$

where P_{infant} is the molar plasma protein concentration in infants, and P_{adult} is the molar plasma protein concentration in adults. The authors showed that the extent of drug binding to HSA in children is close to that in the adult; however, the extent of protein binding for drugs largely bound to AAG increased with age. McNamara and Alcorn tabulated the observed versus predicted f_u values in infants and compared the results to those in adults for several drugs. However, Yanni et al. (2011) [52] have observed, from experimental data, that in unbound fraction of highly lipophilic drug (mostly to HSA), there are 8-fold differences in protein binding between neonates and adults; accordingly, it is recommended, particularly with young infants, to determine the f_u values prior to calculation of clearance.

For age-associated changes in Q_H, it is one factor that may influence hepatic clearance when hepatic elimination is limited by the rate of hepatic drug transport and not enzyme activity [50]. With blood flow rate-limited drugs, developmental changes in metabolic enzymes will impact hepatic clearance only when the enzyme activity is exceedingly underdeveloped such that enzyme activity becomes the rate-limiting factor for overall drug elimination. For blood flow rate-limited drugs, the use of an adult parameter or allometric scaling model for Q_H may not be appropriate for newborns, as a significant portion of hepatic portal blood flow will be shunted away from the functioning hepatocytes in the first week or two of life. The Q_H values for the various pediatric populations have been reported by Björkman (2004) [72].

When biliary CL is the major CL pathway, most physiologically based CL scaling models preclude biliary excretion from consideration in their estimation of pediatric systemic CL. Edginton et al. (2006) [74] proposed that in order to predict CL (ml/min) from adult to child, the following equation can be used:

$$Cl_{\text{biliary child}} = Cl_{\text{biliary adult}} \times (\text{Liver wt}_{\text{child}} / \text{Liver wt}_{\text{adult}})^{0.75}. \tag{9.9}$$

For each compound, adult CL is the starting point for the scaling procedure. Alcorn and McNamara (2008) [50] reported that the allometric exponent, b, typically assumes a value of 1 when the parameter is directly proportional to age-dependent changes in body weight, 3/4 when the PK parameter is related to age-related changes in physiological processes, or 2/3 when the PK parameter is approximately proportional to age-dependent changes in body surface area:

$$\text{PK parameter}_{\text{child}} = \text{PK parameter}_{\text{adult}} \; (\text{BW}_{\text{child}} / \text{BW}_{\text{adult}})^{b}. \tag{9.10}$$

It is unlikely that a correlation exists between body size and CL, as we have discussed before.

Mechanistic studies using well-established polarized cell models, such as sandwich-cultured hepatocytes (SCH), that form bile canalicular networks suitable for measuring active excretion of compounds in a vectorial manner can be used. This model has been widely used for assessing the transporter-related complexity of ADME/tox issues [75] and, most importantly, to calculate biliary and plasma CL. In vitro biliary intrinsic CL (Cl_{biliary}) determined in SCH correlates well with in vivo Cl_{biliary}, suggesting that this system is useful for predicting the in vivo biliary excretion and plasma CL of drug candidates. The accumulation (pmol/mg protein), biliary excretion index (% BEI), and in vitro intrinsic Cl_{biliary} (ml/ min/kg) were calculated in hepatocytes using B-CLEAR technology (Qualyst Transporter Solutions, Durham, NC) as explained by Equations (9.11)–(9.13):

$$\text{BEI} = \frac{\text{Accumulation}_{\text{cells+bile}} - \text{Accumulation}_{\text{cells}}}{\text{Accumulation}_{\text{cells+bile}}} \times 100, \tag{9.11}$$

$$\text{Intrinsic } Cl_{\text{biliary}} = \frac{\text{Accumulation}_{\text{cells+bile}} - \text{Accumulation}_{\text{cells}}}{\text{AUC}_{\text{medium}}}, \tag{9.12}$$

The intrinsic accumulation clearance is determined as follows:

$$\text{Intrinsic Cl}_{\text{Accumulation}} = \text{saturable} + \text{nonsaturable clearance}, \tag{9.13}$$

where AUC is the product of the incubation time and the medium concentration of drug in the medium (the initial concentration in the incubation); saturable CL is the ratio of Michaelis-Menten kinetics of drug uptake (V_{max}/K_m) normalized to drug concentration in the medium, while the nonsaturable CL is the estimated K_d parameters of nonsaturable uptake kinetic component using a mixed kinetic model, as shown in Equation (9.14):

$$v = \frac{V_{max} \times C}{K_m + C} + K_d \times C. \tag{9.14}$$

The in vivo $Cl_{biliary}$ or Cl_{plasma} is calculated according to the well-stirred model shown in Equation (9.7) and by using scaling in vitro intrinsic CL as reported by Yanni et al. (2011, 2010) [52, 75].

For renal clearance, Edginton et al. (2006) [74] described that the term can be calculated by comparing glomerular filtration rate (GFR) between both populations. In summary, the following is recommended: prediction of CL in pediatric population can be based on available PK data in adult populations and comparative data in a relevant in vitro assay using tissue from adults and children can be employed to provide an insight for the potential PK in clinic in pediatric patients. Several steps should be considered:

1. The design of in vitro studies should be guided by the known mechanism of CL in adults, such as metabolism, biliary excretion, or renal excretion.
2. In vitro CL of a drug in biological samples from adults should be experimentally determined using in vitro concentrations comparable to plasma drug concentrations, then intrinsic CL calculated.
3. The well-stirred model can be used to calculate plasma CL by including f_u and Q_H values corresponding to age group. For neonates, f_u has to be measured in neonatal serum. For older children, f_u values were not statistically different than adults, thus the adult value of f_u can be used for older children.
4. For a metabolically cleared drug, V_{max} for the formation of major metabolites by pediatric liver samples was determined in vitro. Intrinsic CL in children was calculated by multiplying the ratio of ($V_{max\ child}/V_{max\ adult}$) by the ratio of scaling factor for child/scaling factor for adult.
5. The good relationship of adult in vitro intrinsic CL-in vivo CL will ensure the prediction of the in vivo CL in children based on in vitro studies as indicated above. Factors that might contribute to the observed disposition of drugs in children might also be revealed from the resulting in vitro–in vivo relationships.

9.5.2 From Animal PK to Human Dose

As mentioned in all previous chapters as well as in this chapter, PK is defined as the parameter that describes the time course of drug concentration in the body, resulting from administration of a certain drug dose, while PD is the term that describes the relationship of the time course of drug concentration and the drug effects in the body [76]. The PK/PD of a drug is usually accompanied with another important therapeutic-related term, the toxicokinetics (TK), which reveals the toxicity effects as a function of drug exposure as a function of time. Therefore, PK and TK can be considered biomarkers of drug exposure as well as markers of efficacy and safety. Key determinants of the PK/TK of a drug include ADME [77]. The development of any drug candidate requires accurate determination of PK and TK at the time of a preclinical animal model to determine the first in human (FIH) studies. In addition, the determination is used to select dose for preclinical toxicity testing and dosage route and drug exposure.

PK studies are usually conducted in rodents to evaluate the extent of drug exposure in vivo followed by dosing in nonrodents, such as dogs or monkeys, to better characterize the PK profile of the compound and to support safety risk assessment studies. Allometry studies or PK scaling follow to predict human PK profile using preclinical data. These studies predict the drug's half-life, dose, and extent of absorption in humans. They are based on empirical observations that various physiological parameters are a function of body size and assume that the same metabolic and disposition processes in the species evaluated are correlated with those observed in humans.

A power equation,

$$Y = aW^{b}, \tag{9.15}$$

is often used in allometric scaling, where Y is the physiologic variable of interest, W is body weight, and log a and b are the y intercept and slope obtained from the plot of log Y versus log W, respectively.

The pitfalls here are, for example, the P450 enzymes in the rat are not the same as those in humans and thus may exhibit altered disposition of the compound or even produce different metabolite patterns. In addition, as shown in previous sections, the uptake and efflux transporters in the animal species may differ in substrate specificity or rate compared to humans [61], thus may mislead the prediction of human PK, safety, and efficacy under the predicted dose.

In other words, animals commonly used in preclinical tests (i.e., mouse, rat, rabbit, monkey, and dog) may not eliminate drugs with the same rate and pathway as humans, and small mammals usually eliminate chemicals faster than large mammals. Since the elimination of many drugs is associated with physiologic properties well described among species (e.g., size and blood flow for the elimination organs), it is assumed that drug elimination may be scaled among mammals, and many successful case studies have been reported. However, the majority of the small-molecule drugs are eliminated via oxidative hepatic metabolism. If there are significant cross-species differences in substrate specificity and/or in enzyme activity, the animal to human extrapolation may fail. To reduce the error in the prediction of metabolic CL in humans, several correction factors have been introduced and reported in a review article by Feng et al. (2012) [78].

Briefly, from Equation (9.15), the coefficients a and b can be obtained from a plot of the body weight from several species against the PK parameter of interest on a log-log scale; log(a) is the y intercept and b is the slope obtained from the plot of log(Y) versus log(W). The allometric exponents for most drugs are in a similar range to those from the corresponding biological processes, with exponents of 0.60–0.80 for CL and 0.8–1.0 for volume of distribution (V_d). If the PK data are only available in a single animal species, the CL and V_d in the human are often estimated using a fixed exponent of 0.75 and 1.0, respectively. Feng et al. 2012 [78] tested the single species (mouse, rat, dog, or monkey) scaling of CL using a fixed b value ranging from 0.60 to 0.85, and the prediction errors from this analysis of 62 small molecules and 32 macromolecule drugs were reported [78]. The analysis may suggest that an optimal value of 0.70–0.80 for the exponent b can be applied when using large animals (e.g., monkey or dog) for single-species scaling [41]. In general, the error of human CL prediction is within 2-fold using monkey PK data and body weight. With the allometry exponent b fixed at 0.70, 0.75, or 0.80, the average error was 1.88-, 1.87-, and 1.86-fold for small-molecule and 1.97-, 1.75-, and 1.61-fold, respectively, for macromolecule drugs. The average error of human CL prediction is higher for the small-molecule drugs using dog as a single species and ranged from 2.4- to 2.9-fold. For macromolecule drugs, the predicted human CL is generally within 2-fold of the observed value using dog data, with average error ranging from 1.6- to 1.8-fold.

More accurate prediction of human PK profiles can be determined from in silico methodologies and/or PBPK modeling and simulation as well as by using Simcyp PK prediction, as discussed in more detail in chapter 8 and in many excellent reviews [79]. An in-depth discussion of various PK concepts and their application can be found in various references [80].

For TK analysis, several toxicology studies are conducted during early drug discovery and all the way to the late stages of drug development before a new drug application (NDA) filing is made. It is common that clinical investigations are terminated due to toxicity issues in spite of the comprehensive toxicity assessment in preclinical investigations. This can be due to the disconnection in predictions of safety margin in humans based on preclinical data obtained from animal models. However, proper preclinical study design and appropriate scaling and modeling tools will improve predictions and minimize % failure in clinical drug development. In general, the correlation between human and animal toxicities is good for conditions such as cardiovascular, hematological, and gastrointestinal diseases and the poorest correlation for adverse drug reactions such as idiosyncratic reactions, hypersensitivity, and hepatotoxicity. The origin of any toxicological or safety outcome is multifactorial and complex and thus demands use of sophisticated systems for definitive assessment; hence introducing in vivo toxicology studies in animal models as early as possible becomes recommended, quite often in the lead optimization (LO) stage. Extensive and appropriate toxicology studies of varying duration ranging from acute, single dose to chronic, repeat dose in rodent and nonrodent species are needed to establish safe human clinical trials, as discussed in chapters 6 and 7. Acute toxicity (single-dose ranging) studies in preclinical species (usually rodents) are performed to support selection of a drug candidate for potential advancement to repeat-dose toxicology studies and ultimately to enable initial FIH clinical trials. The objective of such studies is to identify a dose at which the major adverse effects are observed. The information obtained may be translated to select the dose levels for the FIH studies and also to give an indication of potential effects of acute overdose in humans.

Repeat-dose toxicity studies in early development (for 7–14 days) in both rodent and nonrodent species are used to better refine safety margins and PK/PD modeling, as well as to set appropriate dosages for good laboratory practice (GLP) one-month general toxicology and safety pharmacology, usually in dogs for cardiovascular testing and in rodents for CNS and respiratory function tests. These studies will be required and filled in on an investigational new drug (IND) application before starting FIH clinical trials. TK and PK assessment are critical in defining a margin of safety between the no observed adverse effect level (NOAEL) and the projected plasma concentrations achieved in humans. It is generally considered that a 100-fold safety factor (rodent-to-human exposure ratio) from the most sensitive species NOAEL provides good safety margin in clinical studies. The toxicology assessment profiles usually examined and considered are the maximum tolerated dose (MTD), safety margins and therapeutic index, target organ toxicities, most sensitive preclinical species, and reversibility of an effect/toxicity. Furthermore, investigating biomarkers characterization and preclinical to clinical translation can be conducted in these GLP toxicology studies. Several repeat-dose toxicology studies (general toxicology, embryo-fetal and developmental, fertility, juvenile, carcinogenicity) of longer duration (3 months and up to 2 years) in both rodent and nonrodent species are conducted to support clinical trials of longer duration in patients.

9.5.2.1 CL and PK/TK Modeling in Predicting Clinical Dose CL as discussed earlier in this chapter is the primary parameter that describes the process of irreversible elimination of a drug from the systemic circulation. It is defined as the volume of blood or plasma that is totally cleared of its content of drug per unit time. And it is the basis of calculating drug dose as shown in Equation (9.3). For PK/TK, it is the term that deals with the exposure of test compound and metabolites, which is determined by the kinetics of exposure and ADME parameters. Generally, the extent and duration of exposure are related to the pharmacological or toxicological effects. Clearly, alteration of dose or exposure time can change the extent of exposure or the occurrence of the observed effects (efficacy or toxicity).

As reported by Varma et al. (2012) [81], NOAEL or no observed effect level (NOEL) or low observed effect level is the basis for toxicity assessment. NOAEL represents the highest dose at and below which no significant adverse effects are seen. Owing to ethical and practical reasons, assessment of the NOAEL is derived only from animal toxicity data and extrapolated to identify a clinical dose that is significantly lower than the NOAEL. The same strain and species used in the toxicology studies should be used in the TK studies. Initial studies may involve one sex of each species; however, the use of multiple species is relevant to build confidence in models predicting human effects. By using the dose-response relationship, the incidence or frequency of a toxic effect can be established. While it is important to understand whether the parent molecule or its metabolite is responsible for toxicity, an estimate of whether humans are of less, equal, or greater sensitivity in comparison to the test species is needed for translating the preclinical data to predict human effects. To establish a relationship between exposure and dosimetry, a PK/TK model should incorporate the rate and extent of absorption, distribution of compound and/or metabolites in the body, metabolism and kinetics of metabolites, elimination rate and elimination route(s), and the influence of dose on all the above processes.

REFERENCES

1. Collier R. Drug development cost estimates hard to swallow. Can Med Assoc J 2009;180:279–280.
2. Vangala S, Jakir Pinjari J, Patole P, Ravindran S, Gangal R, Wangikar P, Basu S, Ahmed T, Rastogi H. Translational drug discovery research: integration of medicinal chemistry, computational modeling, pharmacology, ADME, and toxicology. In: Encyclopedia of drug metabolism and interactions, ed. Lyubimov AV.. Hoboken, NJ: Wiley, 2012, 3:1–54.
3. Schuster D, Laggner C, Langer T. Why drugs fail? A study on side effects in new chemical entities. Curr Pharm Des 2005;11:3545–3359.
4. Shikawa T, Kanda T, Kosugi S, et al. Sunitinib as a second-line therapy for imatinib-resistant gastrointestinal stromal tumors. Gan To Kagaku Ryoho 2011;38:916–921.
5. Schäfer S, Kolkhof P. Failure is an option: learning from unsuccessful proof-of-concept trials. Drug Discov Today 2008;13:913–916.
6. Hurko O, Ryan JL. Translational research in central nervous system drug discovery. NeuoRx 2005;2:671–682.
7. Butte AJ, Ito S. Translational bioinformatics: data-driven drug discovery and development. Clin Pharmacol Ther 2012;91(6):949–952.
8. Fall DJ, Stessman H, Patel SS, Sachs Z, Van Ness BG, Baughn LB, Linden MA. Utilization of translational bioinformatics to identify novel biomarkers of bortezomib resistance in multiple myeloma. J Cancer 2014;5(9):720–727.
9. Kelloff GJ, Sigman CC. New science-based endpoints to accelerate oncology drug development. Eur J Cancer 2005;41(4):491–501.
10. Sun W, Liu D, Gong P, Shi X, Wang Y, Wang P, Gong W. Predicting cardiovascular mortality in chronic kidney disease (CKD) patients. Ann Transplant 2014;19:513–518.
11. Ge F, Dai Q, Gong W. Biomarkers for renal transplantation: where are we? Int J Nephrol Renovasc Dis 2013;6:187–191.
12. Lee VMY, Goedert M, Trojanowski JQ. Neurodegenerative tauopathies. Annu Rev Neurosci 2001;24:1121–1159.
13. Ballatore C, Smith AB 3rd , Lee VM, Trojanowski JQ, Brunden KR. Aminothienopyridazines as imaging probes of tau pathology: a patent evaluation of WO2013090497 Expert Opin Ther Pat 2014;24(3):355–360.
14. Furumoto S, Okamura N, Iwata R, et al. Recent advances in the development of amyloid imaging agents. Curr Top Med Chem 2007;7(18):1773–1789.
15. Type 2 diabetes epidemic: a global education. Lancet 2009;374(9702):1654.

16. Dartigues JF. Alzheimer's disease: a global challenge for the 21st century. Lancet Neurol 2009;8(12):1082–1083.
17. Maher PA, Schubert DR. Metabolic links between diabetes and Alzheimer's disease. Expert Rev Neurother 2009;9(5):617–630.
18. Kopf D, Frolich L. Risk of incident Alzheimer's disease in diabetic patients: a systematic review of prospective trials. J Alzheimers Dis 2009;16(4):677–685.
19. Fotuhi M, Hachinski V, Whitehouse PJ. Changing perspectives regarding late-life dementia. Nat Rev Neurol 2009;5(12):649–658.
20. Roses AD. On the discovery of the genetic association of Apolipoprotein E genotypes and common late-onset Alzheimer disease. J Alzheimers Dis 2006;9(3 Suppl):361–366.
21. Combarros O, Rodríguez-Rodríguez E, Mateo I, Vázquez-Higuera JL, Infante J, Berciano J, Sánchez-Juan P. APOE dependent-association of PPAR-γ genetic variants with Alzheimer's disease risk. Neurobiol Aging 2011;32(3):547.
22. Sato N, Morishita R. Brain alterations and clinical symptoms of dementia in diabetes: aβ/tau-dependent and independent mechanisms. Front Endocrinol (Lausanne) 2014;5:143.
23. Dineley KT, Jahrling JB, Denner L. Insulin resistance in Alzheimer's disease. Neurobiol Dis 2014;72:92–103.
24. Brodbeck J, Balestra ME, Saunders AM, Roses AD, Mahley RW, Huang Y. Rosiglitazone increases dendritic spine density and rescues spine loss caused by apolipoprotein E4 in primary cortical neurons. Proc Natl Acad Sci USA 2008;105(4):1343–1346.
25. Pfister M, Whaley JM, Zhang L, List JF. Inhibition of SGLT2: a novel strategy for treatment of type 2 diabetes mellitus.Clin Pharmacol Ther 2011;89(4):621–625.
26. Droll K, Bruce-Mensah K, Otton SV, Gaedigk A, Sellers EM, Tyndale RF. Comparison of three CYP2D6 probe substrates and genotype in Ghanaians, Chinese and Caucasians. Pharmacogenetics 1998;8(4):325–333.
27. Ciszkowski C, Madadi P, Phillips MS, Lauwers AE, Koren G. Codeine, ultrarapid-metabolism genotype, and postoperative death. N. Engl. J. Med 2009;361:827–828.
28. Stamer UM, Musshoff F, Kobilay M, Madea B, Hoeft A, Stuber F. Concentrations of tramadol and O-desmethyltramadol enantiomers in different CYP2D6 genotypes. Clin Pharmacol Ther 2007;82:41–47.
29. Crews KR, Gaedigk A, Dunnenberger HM, Leeder JS, Klein TE, Caudle KE, Haidar CE, Shen DD, Callaghan JT, Sadhasivam S, Prows CA, Kharasch ED, Skaar TC. Clinical Pharmacogenetics Implementation Consortium guidelines for cytochrome P450 2D6 genotype and codeine therapy ypdate. Clin Pharmacol Ther 2014;95:376–382.
30. Gasche Y, et al. Codeine intoxication associated with ultrarapid CYP2D6 metabolism. N Engl J Med 2004;351:2827–2831.
31. Binkhorst L, van Gelder T, Mathijssen RH. Individualization of tamoxifen treatment for breast carcinoma. Clin Pharmacol Ther 2012;92(4):431–433.
32. Sofi F, et al. Cytochrome P450 2C19*2 polymorphism and cardiovascular recurrences in patients taking clopidogrel: a meta-analysis. Pharmacogenomics J 2011;11:199–206.
33. Avery PJ, Jorgensen A, Hamberg AK, Wadelius M, Pirmohamed M, Kamali F; EU-PACT Study Group. A proposal for an individualized pharmacogenetics-based warfarin initiation dose regimen for patients commencing anticoagulation therapy. Clin Pharmacol Ther 2011;90(5):701–706.

34. Au N, Rettie AE. Pharmacogenomics of 4-hydroxycoumarin anticoagulants. Drug Metab Rev 2008;40(2):355–375.

35. Rieder MJ, Reiner AP, Gage BF, et al. Effect of VKORC1 haplotypes on transcriptional regulation and warfarin dose. N Engl J Med 2005;352(22):2285–2293.

36. Hoffmeyer S, Burk O, von Richter O,Arnold HP, Brockmüller J, Johne A, Cascorbi I, Gerloff T, Roots I, Eichelbaum M, Brinkmann U. Functional polymorphisms of the human multidrug-resistance gene: Multiple sequence variations and correlation of one allele with P-glycoprotein expression and activity in vivo Proc Natl Acad Sci USA 2000;97(7):3473–3478.

37. Link E, et al. SLCO1B1 variants and statin-induced myopathy-a genomewide study. N Engl J Med 2008;359:789–799.

38. Trevino LR, et al. Germline genetic variation in an organic anion transporter polypeptide associated with methotrexate pharmacokinetics and clinical effects. J Clin Oncol 2009;27:5972–5978.

39. Ni W, et al. Flavopiridol pharmacogenetics: clinical and functional evidence for the role of SLCO1B1/OATP1B1 in flavopiridol disposition. PLoS One 2010;5:e13792.

40. Herrlinger KR, Koc H, Winter S, Teml A, Stange EF, Fellermann K, Fritz P, Schwab M, Schaeffeler E. ABCB1 single-nucleotide polymorphisms determine tacrolimus response in patients with ulcerative colitis. Clin Pharmacol Ther 2011;89(3):422–428.

41. Markova SM, De Marco T, Bendjilali N, Kobashigawa EA, Mefford J, Sodhi J, Le H, Zhang C, Halladay J, Rettie AE, Khojasteh C, McGlothlin D, Wu AHB, Hsueh W-C, Witte J S, Schwartz J B, Kroetz D L. Association of CYP2C9*2 with bosentan-induced liver injury. Clin Pharmacol Ther 2013;94:678–686.

42. Saeki T, Ueda K, Tanigawara Y, Hori R, Komano T. Human P-glycoprotein transports cyclosporin A and FK506. J Biol Chem 1993;268(9):6077–6080.

43. Treiber A, Schneiter R, Häusler S, Stieger B. Bosentan is a substrate of human OATP1B1 and OATP1B3: inhibition of hepatic uptake as the common mechanism of its interactions with cyclosporin A, rifampicin, and sildenafil. Drug Metab Dispos 2007;35:1400–1407.

44. Markova SM, Schwartz JB, Kroetz DL. Response to "CYP2C9 polymorphism is not a major determinant of bosentan exposure in healthy volunteers." Clin Pharmacol Ther 2014;95(3):252

45. Markert C, Burhenne J, Weiss J, Mikus G, Haefeli WE. CYP2C9 polymorphism is not a major determinant of bosentan exposure in healthy volunteers. Clin Pharmacol Ther 2014;95:250–251.

46. Benjamin DK Jr , Smith PB, Murphy MD, Roberts R, Mathis L, Avant D, Califf RM, Li JS. Peer-reviewed publication of clinical trials completed for pediatric exclusivity. JAMA 2006;296:1266–1273.

47. Johnson TN. The problems in scaling adult drug doses to children. Arch Dis Childhood 2008;93;207–211.

48. Anderson BJ, McKee AD, Holford NH. Size, myths and the clinical pharmacokinetics of analgesia in pediatric patients. Clin Pharmacok 1997;33;313–327.

49. Anderson GD, Lynn AM. Optimizing pediatric dosing: A developmental pharmacologic approach. Pharmacotherapy 2009;29;680–690.

50. Alcorn J, McNamara PJ. Using ontogeny information to build predictive models for drug elimination. Drug Discovery Today 2008;13:507–512.

51. Yanni SB, Annaert PP, Augustijns P, Ibrahim JG, Benjamin DK, Thakker DR. In vitro hepatic metabolism explains higher clearance of voriconazole in children versus adults: role of CYP2C19 and FMO. Drug Metab Dispos 2010;38;45–51.
52. Yanni SB, Smith PB, Benjamin DK Jr, Augustijns PF, Thakker DR, Annaert PP. Higher clearance of micafungin in neonates compared with adults: role of age-dependent micafungin serum binding. Biopharm Drug Dispos 2011;32(4):222–232.
53. Karlsson MO, Lutsar I, Milligan PA. Population pharmacokinetic analysis of voriconazole plasma concentration data from pediatric studies. Antimicrob Agents Chemother 2009;53:935–944.
54. Hope WW, Smith PB, Arrieta A, et al. Population pharmacokinetics of micafungin in neonates and young infants. Antimicrob Agents Chemother 2010;54:2633–2637.
55. Chang TY, Dvorak CC, Loh ML. Bedside to bench in juvenile myelomonocytic leukemia: insights into leukemogenesis from a rare pediatric leukemia. Blood 2014;124(16):2487–2497.
56. Kleschevnikov AM, Belichenko PV, Salehi A, Wu C. Discoveries in Down syndrome: moving basic science to clinical care. Prog Brain Res 2012;197:199–221.
57. Shanks N, Greek R, Greek J. Are animal models predictive for humans? Philos Ethics Humanit Med 2009;4:2.
58. Faustman EM, Omenn GS. Risk assessment. In: Casarett and Doull's toxicology: the basic science of poisons 2008, ed. Klaassen CD. New York: McGraw-Hill, 107–128.
59. Bailey J. Nonhuman primates in medical research and drug development: a critical review. Biogenic Amines 2005;19:235–55.
60. Smith D, Combes R, Depelchin O, et al. Optimising the design of preliminary toxicity studies for pharmaceutical safety testing in the dog. Toxicol Pharmacol 2005;41:95–101.
61. Tang C, Prueksaritanont T. Use of in vivo animal models to assess pharmacokinetic drug-drug interactions. Pharm Res 2010;27(9):1772–1787.
62. Katoh M, Yokoi T. Application of chimeric mice with humanized liver for predictive ADME. Drug Metab Rev 2007;39:145–57.
63. Schulz-Utermoehl T, Sarda S, Foster JR, et al. Evaluation of the pharmacokinetics, biotransformation and hepatic transporter effects of troglitazone in mice with humanized livers. Xenobiotica 2012;42:503–17.
64. Inoue T, Sugihara K, Ohshita H, et al. Prediction of human disposition toward S-3H-Warfarin using chimeric mice with humanized liver. Drug Metab Pharmacokinet 2009;24:153–60.
65. Katoh M, Sawada T, Soeno Y, et al. In vivo drug metabolism model for human cytochrome P450 enzyme using chimeric mice with humanized liver. J Pharm Sci 2007;96:428–37.
66. Foster JR, Lund G, Sapelnikova S, Tyrrell DL, Kneteman NM. Chimeric rodents with humanized liver: bridging the preclinical/clinical trial gap in ADME/toxicity studies. Xenobioteca 2014;44(2):109–122.
67. Scheer N, Wolf CR. Genetically humanized mouse models of drug metabolizing enzymes and transporters and their applications. Xenobiotica 2014;44(2):96–108.
68. Kitamura S, Sugihara K. Current status of prediction of drug disposition and toxicity in humans using chimeric mice with humanized liver. Xenobiotica 2014;44(2):123–134.
69. Samuelsson K, Pickup K, Sarda S, Foster JR, Randall K, Abrahamsson A, Jacobsen M, Weidolf L, Wilson I. Troglitazone metabolism and transporter effects in chimeric mice: a comparison between chimeric humanized and chimeric murinized FRG mice. Xenobiotica 2014;44(2):186–195.

70. Sacci Jr JB, Alam U, Douglas D, et al. Plasmodium falciparum infection and exoerythrocytic development in mice with chimeric human livers. Int J Parasitol 2006;36:353–60.
71. Vaughan AM, Mikolajczak SA, Wilson EM, et al. Complete Plasmodium falciparum liver-stage development in liver-chimeric mice. J Clin Invest 2012;122:3618–28.
72. Björkman S. Prediction of drug disposition in infants and children by means of physiologically based pharmacokinetic (PBPK) modelling: theophylline and midazolam as model drugs. Br J Clin Pharmacol 2004;59:691–704.
73. McNamara PJ, Alcorn J. Protein binding predictions in infants. Pharm Sci 2002;4:E4.
74. Edginton AN, Schmitt W, Voith B, Willmann S. A mechanistic approach for the scaling of clearance in children. Clin Pharmacokinet 2006;45:683–704.
75. Yanni SB, Augustijns PF, Benjamin DK Jr, Brouwer KL, Thakker DR, Annaert PP. In vitro investigation of the hepatobiliary disposition mechanisms of the antifungal agent micafungin in humans and rats. Drug Metab Dispos 2010;38(10):1848–1856.
76. MacGowan AP. Role of pharmacokinetics and pharmacodynamics: does the dose matter? Clin Infect Dis 2001;33 Suppl 3:S238–S239.
77. Singh SS. Preclinical pharmacokinetics: an approach towards safer and efficacious drugs. Curr Drug Metab 2006;7(2):165–182.
78. Feng MR, Zheng N, Zhang X, Huh YM, Yu JU. Interspecies pharmacokinetic scaling and the prediction of human pharmacokinetics using animal and cell models. In: Encyclopedia of drug metabolism and interactions, ed. Lyubimov AV. Hoboken, NJ: Wiley, 2012, 3:1–36.
79. Poggesi I. Predicting human pharmacokinetics from preclinical data. Curr Opin Drug Discov Devel 2004;7(1):100–111.
80. Gibaldi M, Perrier D. Pharmacokinetics. New York: Marcel Dekker, 1982.
81. Varma MVS, Radi ZA, Rotter CJ, Litchfield J, El-Kattan AF. Pharmacokinetics and toxicokinetics in drug discovery and development. In: Encyclopedia of drug metabolism and interactions, ed. Lyubimov AV. Hoboken, NJ: Wiley, 2012, 3:1–18.

10

PHASE 1–PHASE 3 CLINICAL STUDIES, PROCEDURES, RESPONSIBILITIES, AND DOCUMENTATION

10.1 INTRODUCTION: WHAT IS CLINICAL INVESTIGATION? GOALS, UTILITY, AND PROCESSES OF FOUR PHASES IN CLINICAL DRUG DEVELOPMENT

Clinical investigations are prospective biomedical research studies on human subjects or patients with specific medical conditions. The investigations are designed to answer specific questions about biomedical interventions, which can be novel vaccines, new drugs, drug treatments, new food products, dietary supplements, or biomedical devices, or they can be new ways of using known interventions. These investigations are generating safety and efficacy data and are conducted only after health authorities have reviewed and approved the information gathered about the specific therapeutic product in the country where approval of the therapy is being pursued. The health authorities responsible for review and approval of therapeutic products are the Food and Drug Administration (FDA) and/or the European Medicines Agency (EMA), as well as ethics committees, such as institutional review boards (IRBs), also known as independent ethics committees (IECs), ethical review boards (ERBs), or research ethics boards (REBs). The product type and its development stage direct the investigators toward the enrollment, whether for healthy volunteers and/or patients, first into small pilot studies, followed by larger-scale

Translational ADMET for Drug Therapy: Principles, Methods, and Pharmaceutical Applications, First Edition. Souzan B. Yanni.

comparative studies. When these studies generate positive safety and efficacy data, subsequently greater numbers of patient will be enrolled. Though there are no rules on specific size of clinical studies, it varies and can involve a single research site in one country or multiple sites in multiple countries. In general, depending on the number of patients and number of investigation sites, a complete set of clinical investigations may cost hundreds of millions of dollars; clinical investigations are considered to be the most costly process in the discovery and development of a new biomedical product. The sponsor, which may be a governmental organization or a pharmaceutical, biotechnology, or medical device company, has the full financial responsibility of the clinical investigation, while the trial may be managed by an outsourced partner, such as a contract research organization or an academic clinical trials unit.

The concepts behind clinical trials are ancient: beginning in 1025 [1], when advice about determining the efficacy of medical drugs and substances was revealed, then in 1700 and 1800, medical experimentation was performed but the use of control groups with the treated group was lacking. After several efforts, physicians during these early times began to learn what we already know about the modern development of clinical trials. In the 1920s, experimental design was developed with an idea of randomization, which meant randomly assigning individuals to different treatment groups in the specific treatment. The first randomized trial was carried out 1946–1947 and aimed to test the efficacy of streptomycin for curing pulmonary tuberculosis [2]. This trial was a double-blind and placebo-controlled study, which will be discussed later in this chapter. The methodology of clinical trials was further developed by applying statistics to medicine and collaboration, resulting in the study of correlation between smoking and lung cancer. It was carried out in a case-control study, which compared lung cancer patients with matched controls and also began a sustained long-term prospective study into the broader issue of smoking and health, which involved studying the smoking habits and health of over 30,000 doctors over a period of several years.

In an article by Zweig and Robertson (1982) [3] that described why there is a need for better test evaluation in clinical studies, the authors stated that a laboratory test is clinically useful only if it successfully answers a question of consequence to patient management. Unfortunately, the results and conclusions of many published test evaluations are misleading or of uncertain validity because commonsense principles of study design are overlooked. This is illustrated by examples from literature in the period 1950–1980. The authors suggested that tests should be evaluated with prospective studies of patients representative of the population for which the test will ultimately be used and the clinical question to be addressed by the test should be clearly stated, thus providing an answer for each patient by means independent of the test being evaluated. When comparing tests with each other, decision levels should be chosen to give either the same sensitivity or specificity for each. The use of soundly designed protocols for the clinical evaluation of tests provides the information needed to select the most effective tests. With this article, the authors seem to have introduced the concept of "individualized medicine" that became the golden practice of clinical investigations.

A review article by Farrington and Miller (1996) [4] confirmed the recommendation of Zweig and Robertson [3]. In their article, Farrington and Miller indicated that the transition from laboratory testing to evaluation in humans marks an important stage in therapeutic development (specified the use of vaccine). They addressed the issues of therapeutic safety versus benefit not only to the patient but also to the community at large. They also raised the issue of ethics, with the inherent variability of human responses to therapy. Due to the variability of human responses, the authors stressed the necessity of why a coherent statistical framework for the design, analysis, and interpretation of clinical trials is regarded as essential today. Furthermore, the authors described the features of clinical trial research, based on its broad collaborative nature and involving laboratory, epidemiological, clinical, and statistical skills.

Those disciplines considered the foundation of any medical research were mentioned in another article by Avan and White (2001) [5]. The authors have indicated the significance of managing clinical research, as much as the attention is provided in managing science. Effective research depends first upon the purpose and objectives and related research questions. These essential steps drive key decisions on the study type, the analysis procedures, and the timeline and budget to execute the study. The research effort emerges by selecting appropriate study types based on coherent research questions and appropriate data collection and ultimately concludes by drawing inferences on the basis of properly planned analyses. Current clinical research is based on these steps.

Although all the areas involved with clinical research management are important, treatment evaluation is one of the most important tasks in medical research. The standards that are used in evaluating efficacy/effectiveness of treatments, like the inclusion of control groups to assess the treatment, have been developed during the last few decades. To guarantee the internal validity of the comparisons between treated and control groups, the groups have to be comparable at the beginning of the study. This can be achieved by randomized allocation of patients to treatment. Furthermore, as far as possible patient and physician should be blinded to treatment in order to avoid subjective influences on treatment results. Groups should still be comparable when analyzing the trial, thus an analysis according to the principle of intention-to-treat ("as randomized") should be performed. These indispensable principles for design, conduct, and analysis of clinical trials are widely accepted and contribute to reliable and credible results.

Blume and Peipert (2004) [6] have addressed the significance and challenge of using randomization in controlled clinical trials. The authors confirmed that the randomized controlled trial (RCT) is considered the highest level of medical evidence. In their brief overview, the authors discussed several key principles of the RCT. First, balance is paramount. Comparison groups must have similar proportions of participants with "important" prognostic and confounding factors. Randomization may or may not achieve this balance; if it does not, statistical adjustments should be used. Second, a statistical analysis should emphasize comparability and not mask dissimilarity. If the trial was indeed randomized, certain analysis techniques, such as an intention-to-treat analysis, should always be presented. Third, additional bias-reducing techniques, such as concealing treatment assignments from treating

TABLE 10.1 Summary of Clinical Stages in Drug Development.

Study Phase	Goal	Type of Subjects	Size of Study	Average Study Duration
Phase 1	First in human (FIH)	Healthy volunteers*	50–100	< 1 year
Phase 2	Proof-of-concept dose response	Patients	100–300	~ 2 years
Phase 3	Efficacy	Target patient population	1000–5000	~ 3 years
Phase 4	Post approval	Patients	Variable	Variable

*Exception in specific oncology treatment.

physicians and participants (i.e., masking) and using clearly defined exclusion and inclusion criteria, should be used wherever possible.

To manage and reasonably handle the complexity in setting clinical trials among those involved, such as sponsors and regulatory authorities, human clinical investigations have been divided into four phases. These are summarized in Tables 10.1 and 10.2 and briefly reviewed here:

Phase 1 studies are conducted to assess the safety of a drug or device. This initial phase of testing, which can take several months to complete, usually includes a small number of healthy volunteers (20–100) who are generally paid for participating in the study. The study is designed to determine the effects of the drug or device on humans, including how it is absorbed, metabolized, and excreted. This phase also investigates the side effects that occur as dosage levels are increased. About 70% of experimental drugs pass this phase of testing.

Phase 2 studies are conducted to test the efficacy of a drug or device. This second phase of testing can last from several months to 2 years and involves up to several hundred patients. Most Phase 2 studies are randomized trials where one group of patients receives the experimental drug, while a second "control" group receives a standard treatment or placebo. Often these studies are "blinded," which means that neither the patients nor the researchers know who has received the experimental drug. This allows investigators to provide the pharmaceutical company and the FDA with comparative information about the relative safety and effectiveness of the new drug. About one-third of experimental drugs successfully complete both Phase 1 and Phase 2 studies. It sometimes involves one investigator center or more than one depending on the effectiveness of enrollment.

Phase 3 studies involve randomized and blind testing in several hundred to several thousand patients within multicenters of investigation. This large-scale testing, which can last several years, provides the pharmaceutical company and the FDA with a more thorough understanding of the effectiveness of the drug or device, the benefits, and the range of possible adverse reactions. 70–90% of drugs that enter Phase 3 studies successfully complete this phase of testing. Once Phase 3 is complete, a pharmaceutical company can request FDA approval for marketing the drug.

TABLE 10.2(A) Studies Conducted in Phase 1 Clinical Studies.

Pharmacokinetics	Safety	Pharmacodynamics and Others
Bioavailability	Tolerability and safety (short-term safety of max tolerated dose)	PK/PD model and simulation
C_{max}	Drug–drug interaction	CP on patient volunteers not responsive to conventional therapy
AUC	Food effect	Dosage form
T_{max}	Effect on vital sign and determine AE or ADR	Protein binding
Half-life		
Clearance		
Radiolabelled ADME: route of elimination, metabolite ID		One center-uncontrolled—no control group-trial

TABLE 10.2(B) Studies Conducted in Phase 2 and Design.

Proof-of-concept

1. Safety profile in patients
2. Primary aim: drug therapeutic effect, dose range
3. Frequency of dosing
4. Safety and tolerability: drug consequent adverse effect
5. Benefit to treatment
6. One or multicenter study: uncontrolled—no control group, or controlled group

TABLE 10.2(C) Studies Conducted in Phase 3 and Design.

Large-scale efficacy clinical trials

Multicenter trials
Special population safety and efficacy
Long-term safety
Comparative studies against placebo
Controlled, blinded, randomized design

TABLE 10.2(D) Studies Conducted in Phase 4/postmarket Approval.

Postapproval, therapeutic use in patients

More diversified safety data
Long-term chronic use
Comparative to other drugs in class
Cost-effective

Phase 4 studies, often called postmarketing surveillance trials, are conducted after a drug or device has been approved and is on the market for consumer sale. Pharmaceutical companies have several objectives at this stage: (1) to compare a drug with other drugs already in the market; (2) to monitor a drug's long-term effectiveness and impact on a patient's quality of life; and (3) to determine the cost-effectiveness of a drug therapy relative to other traditional and new therapies. Phase 4 studies can result in a drug or device being taken off the market or restrictions of use placed on the product depending on the findings in the study.

10.2 GENERAL CLINICAL STUDY DESIGN: ENROLLMENT, RESPONSIBILITIES, AND DOCUMENTATION

As mentioned earlier, clinical trials are conducted to collect data regarding the safety and efficacy of new drug and device development. There are several steps and stages of approval in the clinical trials process before a drug or device can be sold in the consumer market, if ever. The processes that are initiated—from protocol approval, selection of team and site, approval of process, enrollment of patients, generating data and entry, statistical analysis to determine the significance of treatment, then reporting until drug registration for public use—require several meetings with regulatory authorities during the progress from investigational new drug (IND) filing and approval and Phase 1 to Phase 3 clinical investigations, as indicated in Figure 10.1.

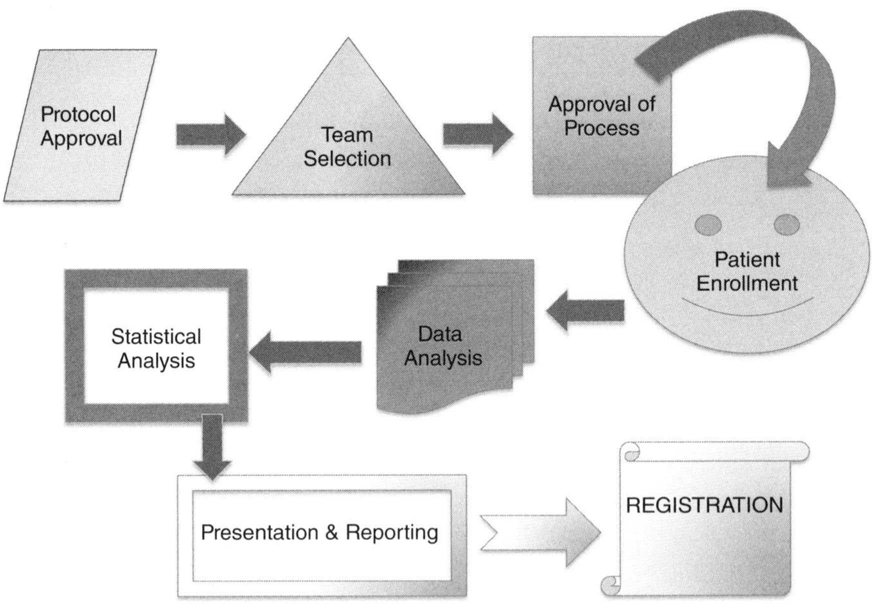

Figure 10.1 Sequential procedures in establishing and completing clinical investigations.

To simplify the process, there are three components that any clinical investigation is based on:

1. *Experimental unit*: represents the subject patient from target population under study
2. *Treatment:* can be either placebo or combination of a new pharmaceutical entity, new diet, surgical procedure, medical device, diagnostic test, health education program, or no treatment
3. *Evaluation*: lists assessment of laboratory parameters or physician examination that reveals the safety (adverse events) and efficacy of treatment
   a. Includes assessing the quality of life
   b. Assesses pharmacogenomics, utility of biomarkers, pharmacoeconomics (cost-minimization/cost-effectiveness)

There are two initial elements in any clinical study of new medicine: first, its safety in target patients; second, its efficacy (clinical endpoints) taken in consideration with the identification of the measured level of therapeutic effectiveness of (1) reduction of mortality (live longer), (2) reduction of disability (do more), and (3) reduction of suffering (feel better).

Both safety and efficacy are evaluated relative to how the treatment is intended to be used. Also, a study needs to consider what other treatments are available and the severity of the disease or condition. In general, the benefits must outweigh the risks; for example, several anticancer drugs have severe side effects that would not be acceptable for an over-the-counter pain medication, yet the cancer drugs have been approved since they are used under a physician's care and are used for a life-threatening condition.

10.2.1 Clinical Study Protocol

To start designing a clinical investigation, it is essential to initiate the setting of study protocol, or the clinical research plan. A principal investigator (PI), who is often a doctor, leads a clinical study. Members of the research team regularly monitor the participants' health to determine the study's safety and effectiveness. Both the PI and the selected study team set up the protocol that is carefully designed to safeguard the participants' health and provide descriptions and answers for the specific research questions, such as:

- Who is eligible to participate in the trial: study population, the inclusion and exclusion criteria, number of subjects (enrollment process)
- Details about tests, procedures and endpoints (e.g., clinical effect with few biases), treatment methods (randomization and blinding/masking), medications and choice of comparator, and dosages
- Where, when, and how the study will be conducted
- The length of the study and what information will be gathered

The sponsor designs the trial in coordination with a panel of expert clinical investigators, including what alternative/existing treatments to compare to the new agent and what type(s) of patients might benefit. If the sponsor cannot obtain enough subjects at one location, investigators at other locations are recruited to join the study. During the trial, investigators recruit patients with the predetermined characteristics, administer the treatment(s), and collect data on the patients' health for a defined time period.

10.2.2 Patient Selection and Eligibility Criteria

Clinical trials are not carried out in random samples of the general population. In a variety of ways, those who design and carry out trials (a PI and study team) seek to identify a population of patients in whom the effect of a drug, if present, can be demonstrated. In successful recruitment, some critical basic procedures can be considered:

1. When a heterogeneous pool of patients is selected, that can cause a significant decrease in accuracy and reliabilit,y as it raises bias/variability issues. Obviously homogeneous selection will reduce bias and minimize variability.
2. Demographic differences among patients such as age, sex, height, and weight are expected, but it is not advised to recruit patients with differences based on disease state and concomitant therapies.
3. Optimum selection of patients is for those that will most likely demonstrate the effect of the drug.
4. As recommended by Temple (2010, 2008, 1994) [7–9], recruitment based on "enrichment" of clinical study population fits well into the growing interest in individualization of drug therapy, creating tension with another approach, which fits with "real-world studies." There are three types of enrichment:
   a. Practical: reduce "noise" by excluding subjects who can't show effect, who likely stay in trial, who don't have other diseases that interfere with the assessment, OR whose disease does not disappear
   b. Prognostic: include patients for whom the treatment decreases the rate of an event (heart attack, stroke, tumor reoccurrence)
   c. Predictive: identify patients who are more likely to respond to treatment for mechanistic reasons (express an appropriate receptor in anticancer).

Enrichment specifies the inclusion and exclusion criteria and can allow realistic smaller size, reduce cost, and improve the risk-benefit relationship.

Some of the considerations here are shared under the more general topic of design of experiments, but there can be others, in particular related to patient confidentiality and ethics.

In the United States, the geriatric populations constitute only 14% of the population, while they consume over one-third of drugs. They are often excluded from trials because their greater health issues and drug use complicate data interpretation. Furthermore, women (e.g., pregnant women), children, and individuals with unrelated medical conditions are also frequently excluded.

In general, careful planning, realistic recruitment, and effective strategy can lead to successful trials without delay. As stated by Temple (2010) [7], enrichment designs hold great promise for increasing the power of studies to detect the real effect of a treatment.

10.2.3 Typical Study Design Features

When designing a clinical study, there are four major tasks the study design is based on: (1) treatment sequences (single, parallel, or crossover; (2) method of assigning treatment (randomized/nonrandomized or stratification); (3) selection of control (noncontrolled, controlled, concurrent control); and (4) blinding and masking (open unblinded, single blinded, double dummy).

10.2.3.1 Randomized Clinical Trials For its significance, the randomized clinical trial will be discussed here in some detail. RCTs constitute a powerful methodology in "evidence-based medicine" that has provided the evidentiary basis for many of the interventions of modern medicine [10].

An RTC is a type of medical experiment where the people being studied are randomly allocated one or another of the different treatments under study (either the new tested drug, reference drug, or placebo). The RCT is the most prefered method of investigating the efficacy or effectiveness in clinical studies of various types of medical intervenion and might provide information on adverse drug reactions. It is conducted by assigning the new intervention in a random fashion to subjects that are enrolled after they are found to be eligible for the study (but before the study begins).

Random allocation of treatments among subjects in real trials is a complex process, but as mentioned by Temple (2010) [7], conceptually the process is like tossing a coin. After randomization, the two (or more) treatment groups of subjects are followed in exactly the same way, and the only differences between the care they receive—for example, procedures, tests, outpatient visits, and follow-up calls—should be those intrinsic to the treatments being compared. Again, as mentioned above, the most important advantage of proper randomization is that it minimizes allocation bias in the assignment of treatments by balancing both known and unknown prognostic factors. RCT is the name only for trials that contain control groups, in which groups receiving the experimental treatment are compared with control groups receiving no treatment, receving excepient, in which case the study is called a placebo-controlled study, or receiving previously tested treatment, called a positive-control study. There is another type of randomized study called "randomized trials" without the mention of controls, which describes studies that compare multiple treatment groups with each other (in the absence of a control group) [11]. Not all randomized trials are RCT and some of them could never be, in cases where controls would be impractical or unethical to institute.

Despite the powerful utility of RCT, it has certain limitations [12]. RCTs can be used only to answer one or a few questions per trial while still preserving randomization [13]. They may not address the utility of clinical studies for health-care providers, patients, and other stakeholders because of limited generalizability in many cases. There is an unlimited number of questions regarding clinical utility in personalized medicine, and new questions continue to emerge as clinical practice experience with different interventions in pragmatic settings continues to increase over time. Because of cost, time, and, in some cases, ethical burdens to performing RCTs in all situations, alternative and credible study designs are needed for generating evidence of clinical utility [14].

A combination of prospective clinical trials and observational studies is needed to speed the translation of clinical research to medical practice. It has been suggested that methodologically robust observational studies, pragmatic clinical trials, and real-time surveillance of individual patient outcomes can play a vital role in providing the evidence of clinical utility for a genetic test intended to reduce the overall risk of medicines [15]. Large clinical-outcomes databases, such as those of pharmacy benefit managers or those likely to emerge with the dawning of the electronic health record era [16], and clinical trials in real-world settings can represent effective and efficient approaches to generate evidence of clinical utility [17]. This can be especially true in looking at clinical outcomes in a large number of patient subgroups encountered in everyday clinical practice. In these cases, an effective medical record system with accurate phenotype information is critical [18]. It would seem that having a portfolio of acceptable approaches for evidence generation would be the most robust and efficient method for translating biomarker test information (genetic or other types) to the clinic [19]. This would mean that the evidence needed for demonstration of clinical utility could acceptably derive from a range of study designs depending on such variables as the significance of the clinical outcome of interest (e.g., mortality vs. blood pressure reduction), the background noise or extraneous variability (e.g., inherent in multisite clinical trials and lack of centralized biomarker measurements), the level of acceptable uncertainty (e.g., in benefits and risks relative to alternative treatments), extent of corroborating or supportive evidence, the magnitude of the intervention effect, and the extent of replication and/or trends in results. Achieving consensus on when RCTs are not required, what constitutes overwhelming or mechanistically supportable evidence of clinical utility, and when well-designed observational trials would suffice for establishing clinical utility would advance the field of personalized medicine [20]. As described by Woodcock, there are many complementary sources of evidence (mechanistic, pharmacological, and observational studies) that can all contribute to establishing clinical utility [21].

10.2.3.2 Blinding versus Masking Blinded studies are also called masked studies. A blinded study is one in which information about the test that might lead to bias in the results is concealed from the tester, the subject, or both until after the test. Bias may be intentional or unconscious. In single studies, also called single-masked studies, the participants do not know which medicine is being used, so they can describe what happens without bias. "Blind" (or "masked") studies are designed to prevent members of the research team or study participants from influencing the results. This allows scientifically accurate conclusions. In single-blind ("single-masked") studies, only the patient is not told what is being administered. In a double-blind study, only the pharmacist knows; members of the research team are not told which patients are getting which medication, so that their observations will not be biased. If medically necessary, however, it is always possible to find out what the patient is taking. A blinding trial is used wherever new drugs, known drugs, placebo, or no treatment are compared without influences from testers' preferences or expectations. In some disciplines, such as drug testing, blind experiments are considered essential. In developmental psychology, while blind experiments would be useful, they are impractical or unethical. Although it would be informative to raise children under arbitrary experimental conditions, it is obviously a violation of ethics and human rights.

In contrast to blinding imposed on researchers, technicians, subjects, and funders, open trial is a study in which both the researchers and participants know which treatment is being administered. Open-label trials may be appropriate for comparing two very similar treatments to determine which is most effective. An open-label trial may be unavoidable under some circumstances, such as comparing the effectiveness of a medication to intensive physical therapy sessions. An open-label trial may still be randomized and uncontrolled, with all participants receiving the same treatment.

10.2.4 Responsibilities: IRBs, Regulatory Authorities, Sponsor, PI, Patients

10.2.4.1 Institutional Review Boards In medicinal clinical trials, approval of many things by an IRB, or ethics board—the protocol and study plan, recruiting and consent(s), the data collection forms and supporting documentation, PI and team, and study objective—is necessary before all informal research can begin. The study protocol is not approved by an IRB before the sponsor recruits sites to conduct the trial. IRB offices can be found in state universities and most hospitals. However, other researchers (such as in walk-in clinics) use independent IRBs. The IRB scrutinizes the study for both medical safety and protection of the patients involved in the study and before it allows the researcher to begin the study.

10.2.4.2 Role of Regulatory Agencies Regulatory requirements and enforcement abilities are different in different countries. Currently, about 50% of all clinical trials are carried out in Eastern Europe and Central and South America as well as in Asia (especially in China for oncology drug therapy). No obligatory registration system governs clinical trials in these countries, and many do not follow European instructions in their operations.

To make uniform clinical investigations and public health around the globe, an initial step was taken to harmonize clinical trial protocols across countries of the European Union in 1980. After 1990, coordination between Europe, Japan, and the United States led to a joint regulatory-industry initiative under the International Conference on Harmonization of Technical Requirements for Registration of Pharmaceuticals for Human Use (ICH). ICH, while involving multiple participants, primarily comprises experts from regulatory authorities and pharmaceutical manufacturer associations from the United States, Europe, and Japan. The organization develops guidelines for pharmaceutical development and registration under four categories: safety, quality, efficacy, and multidisciplinary.

Unfortunately, only 50% of foreign clinical study sites were audited by the FDA compared to 100% of U.S clinical investigations. Criticism has been raised with regard to foreign clinical studies, especially in developing countries, in relation to the rights and welfare of study participants, the integrity of study data, and the relevance of data to the U.S. population. In the United States, the FDA reviews the files of local site investigators after they have finished participating in a study in order to determinee if they were correctly following study procedures.

In addition to the random audit of PI sites, the FDA and the EMA are reviewing the preclinical data filing before clinical investigation start (IND application) and after completion of Phase 3 clinical investigation (new drug application [NDA]) before launching the drug for public use. The schedule of IND and NDA along with other meetings with FDA during the clinical investigations is shown in Figure 10.2.

An article by Ledwith and DeGeorge from Merck & Co. Safety Assessment & Laboratory Animal Resources in 2011 [22] described the major changes made to ICH—specifically ICH M3, which is a multidisciplinary guidance that recommends international standards for nonclinical safety studies to support clinical studies and marketing authorization. The recommendations in ICH M3 are intended to facilitate the timely conduct of clinical trials, reduce the use of animals in accordance with the "3R"—reduce/refine/replace—principles, and promote the safe, ethical development and availability of new pharmaceuticals. However, unlike other ICH guidelines at the time, there were still differences between the regions on certain points denoted

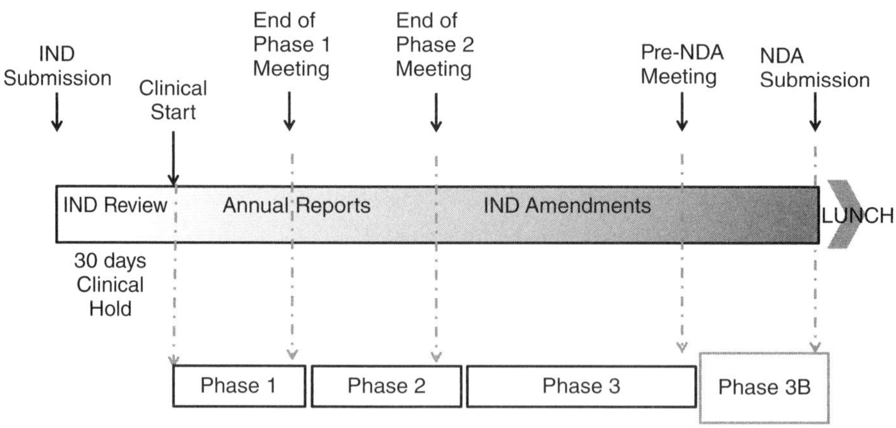

Figure 10.2 Clinical drug development, phases, timeline, and regulatory submission timeline.

within the ICH M3 guidance (e.g., duration of nonrodent chronic studies and requirements for reproductive toxicity testing). These regional differences have become increasingly problematic because, for the purpose of supporting global marketing applications, clinical trials are more frequently being conducted as simultaneous, multiregional trials. In addition, new approaches and regional recommendations have been developed for areas of toxicology (e.g., phototoxicity, abuse liability, and combinations) and technical aspects of toxicity testing (e.g., criteria for high-dose selection) that were not covered by ICH M3; in many cases, there is conflicting guidance between the regions. Furthermore, a new approach to early drug development, namely, exploratory clinical trials, has been developed. This is a consequence of the high rate of drug development failure and has been enabled by the increasing availability of biomarkers and their utility in improving the understanding of drug activity. In 2006, based on these drivers, ICH initiated the multistep process of revising the M3 guidance, employing a rigorous data-driven approach to resolve certain aspects (e.g., evaluating the impact of the duration of chronic studies on identification of new toxicities and comparing data from exploratory vs. definitive embryo-fetal toxicity studies to determine whether the significant risks for early clinical development were identified in the exploratory studies). The newly revised guidance, ICH M3(R2), was adopted in 2009. Highlights of the changes include new sections on exploratory clinical studies, combination products, and various types of specific toxicity tests performed on a case-by-case basis. In addition, the revised guidance contains additional details and changes pertaining to aspects of the previous recommendations for nonclinical testing. As stated by the authors, these major changes may facilitate drug development and translate into clinical practice.

10.2.4.3 Responsibility of Sponsor In general, the sponsor is the pharmaceutical or biotech company that leads the development of new medicine under clinical investigation. The sponsor is responsible for covering all the administration and financial burdens of the clinical investigation, which may vary depending on many factors, such as the number of patients required, number of clinical sites, and number of staff that operate these sites. Pharmaceutical companies also have to cover the cost of manufacturing the drug(s)/device(s) tested, salaries for PIs and study teams, payments to the contract research organization (CRO) involved in conducting the studies, and any site management organization, as well as any outside consultants. In addition, pharmaceutical companies cover all the financial needs for the company's staff responsible for the research and development processes during drug discovery and preclinical drug development, including toxicity studies; clinical staff involved in recruiting patients, collecting data for the sponsor study materials, and communication with the local researchers, including on-site monitoring by the CRO before and during the study; and any payments to patients enrolled in the trial. These costs are incurred over several years. In the United States, sponsors may receive a 50% tax credit for clinical trials of drugs being developed for the treatment of rare (orphan) diseases.

10.3 INTEGRATION OF CLINICAL TRIALS WITH PRECLINICAL ABSORPTION, DISTRIBUTION, METABOLISM, AND EXCRETION (ADME), DRUG–DRUG INTERACTION (DDI), AND PHARMACOGENOMICS IN INVESTIGATING

10.3.1 Assessment of DDI and Disposition

In a recent study to review the data on the drug disposition and DDI listed in FDA 2013 NDAs, Yu and colleagues [23] analyzed 27 approved drugs. In 2013, a total of 25 NDAs and 2 biologic license applications (BLAs) were approved by the FDA. Eight of these (30%) were cancer treatments, including four kinase inhibitors, making oncology the most represented therapeutic area. Among the 27 new molecular entities (NMEs) approved in 2013, 22 (81%) had drug metabolism or transporter data available, and therefore were fully analyzed in this review (18 [67%] provided hepatic impairment [HI] and/or renal impairment [RI] studies). The NDAs without those studies were not evaluated in this review and comprised radioactive diagnostic or therapeutic agents, as well as a cytolytic antibody.

The authors in this review aimed to demonstrate:

- Integration of in vitro data with clinical data can be necessary to identify every drug metabolism-mediated and drug transporter-mediated DDI.
- The correlation between in vitro and in vivo DDI using the scientific approaches and probes recommended by the regulatory authorities and guidance.
- The comparison between the disposition of these drugs that will characterized with regard to drug metabolism and transport in impaired hepatic and renal organs with the disposition in healthy patients.
- The frequency of drugs that were identified as substrates for drug-metabolizing enzymes (DMEs) and drug transporters in comparison to those that were identified as inhibitors in preclinical and clinical testing in accordance with the 2012 FDA drug interaction guidance.

The authors presented the significance of integrating the evaluation of DDI studies and their mechanisms during the drug development process. Their statement has been supported by the results generated from the detailed analysis of NDA reviews for drugs that were approved by the FDA in 2013 and were generally consistent with current regulatory recommendations. As indicated in this report, the drug interaction profiles of these drugs were well characterized using probe markers and known inhibitors and inducers of DMEs. Furthermore, there were substantial scientific efforts to elucidate the mechanisms and clinical significance of drug transporters in the disposition transporter-based DDIs of these drugs. Because of the need to address dose adjustment in hepatic- and renal-impaired patients, the review assessed the disposition and potential DDI in these conditions compared to healthy individuals. By using the University of Washington Drug Interaction Database, the results of these analyses indicated that among 27 NMEs approved, 22 (81%) were well characterized with regard to drug metabolism, transport, or organ impairment, in accordance with the 2012 FDA drug interaction guidance, and were fully analyzed in the review [23]. In the in vitro testing, the majority of drugs were found to be substrates or inhibitors/inducers of at least one DME or transporter. However, in vivo, 50% of these drugs (n = 11) showed clinical relevance in DDIs, with most related to the NMEs as victim drugs and CYP3A being the most affected enzyme. As perpetrators, the overall effects for NMEs were much less pronounced, compared with when they served as victims. The pharmacokinetic (PK) evaluation in patients with HI or RI provided further understanding on the disposition of these drugs. To evaluate the extent of DDI in clinical studies, the metric application using the area under the curve (AUC) ratio of inhibited versus control, ($AUC_{inhibited\ or\ induced}/AUC_{control}$), was applied. A clinically significant interaction was revealed when the resulting AUC ratio ≥ 2. In addition, important or significant labeling modifications or recommendations were also noted.

The study showed that CYP3A was involved in the metabolism of the most NMEs in vitro (17 of 22), and 10 were further confirmed to be substrates of CYP3A in vivo. In addition, the largest DDI observed in vivo was caused by CYP3A inhibition, with ibrutinib being the victim drug. As perpetrators, 17 drugs (77%) showed positive inhibition or induction toward at least one enzyme in vitro. In contrast, in vivo, only three NMEs (simeprevir, alogliptin, and eslicarbazepine acetate) were found to be enzyme inhibitors and two NMEs (dabrafenib and eslicarbazepine acetate) to be enzyme inducers, highlighting the challenge of translating inhibition and induction data from in vitro to in vivo. The overall in vivo effect of NMEs as perpetrators was much less pronounced, with the largest AUC ratio less than 2, compared with when the NMEs served as victim drugs, where the largest AUC ratio observed was greater than 20. Regarding clinical induction data, significant inductions were mainly related to NMEs as victim drugs and, in most cases, involved induction of CYP3A by known inducers. When NMEs were considered as perpetrators, two compounds were found to significantly induce CYP3A, dabrafenib and eslicarbazepine acetate. Dabrafenib significantly reduced the AUC of the sensitive CYP3A substrate midazolam by over 70%, whereas eslicarbazepine acetate decreased simvastatin exposure by 50%. Both drugs have recommendations in their labels regarding their possible inductive effect on coadministered drugs.

With regard to drug transporters, three NMEs were shown to be in vivo inhibitors of P-glycoprotein (P-gp): alogliptin, canagliflozin, and simeprevir, with simeprevir also inhibiting OATP1B1 in vivo. In vitro, in contrast, 85% of NMEs showed a positive interaction with at least one transporter, either as a substrate or inhibitor. Only three NMEs showed no interaction with any transporter, and in all three cases, P-gp was the only transporter tested. These data indicate that when tested in vitro, many NMEs appear to be substrates or inhibitors of at least one transporter. These results do not necessarily translate to in vivo, and the clinical relevance of the transporter-mediated DDI may be minimal, especially when compared with DME-mediated DDI. Reasons for this may include extensive protein binding of drugs, resulting in low free circulating concentrations, as well as interplay between DMEs and transporters [24]. For example, canagliflozin and simeprevir are both greater than 98% protein bound, whereas simeprevir is a substrate and inhibitor of several P450s, all of which may contribute to the moderate effects observed in the digoxin (a P-gp substrate) interaction studies. Less agreement was observed for probe substrate or inhibitor of other transporters with translation of in vitro data to the clinical setting. Very few purely transporter-based drug interactions with over 2-fold changes in substrates exposure were detected. Only the interaction between sofosbuvir (400 mg single dose) and cyclosporine (administered as a high single dose of 600 mg) was related to inhibition of P-gp and breast cancer resistance protein (BCRP) and yielded an increase in sofosbuvir AUC of almost 4-fold. However, the exposure of the predominant circulating inactive metabolite (GS-331007) was unchanged, and considering sofosbuvir safety margins, the effect of cyclosporine on sofosbuvir PK was not considered clinically significant by the sponsor and no dose adjustment was required. Also of note, sofosbuvir plasma exposure was also increased by coadministration of simeprevir (150 mg once daily for 12 or 24 weeks), an inhibitor of P-gp (AUC ratio of 3.2).

For the impact of pharmacogenetics on clinical data among the 27 tested drugs, no genotyping data were available and no specific pharmacogenetic analyses were performed in the NDA reviews. However, it was discovered that in Phase 3 clinical trials subjects of East Asian ancestry treated with simeprevir ($n = 14$) had 3.4-fold higher exposure of simeprevir than the pooled Phase 3 population, and this higher exposure was associated with an increased frequency of adverse reactions. The increase in simeprevir exposure may be clinically relevant and is likely due to some as-yet-unidentified genetic variation, therefore the product label states, "There are insufficient safety data to recommend an appropriate dose for patients of East Asian ancestry. The potential risks and benefits should be carefully considered prior to use in patients of East Asian ancestry."

Among the 15 NMEs evaluated for HI studies, 4 demonstrated an AUC ratio greater than 2 in the HI patients versus normal controls. The highest AUC ratio (6.0) was observed for ibrutinib in the moderate HI population (Child-Pugh B). However, the ibrutinib label states that there was "insufficient data to recommend a dose in patients with baseline HI" and the recommendation is that ibrutinib should be avoided in these patients. The next largest change in AUC was observed for simeprevir, showing 2.4- and 5.2-fold increases in AUC for moderate and severe HI patients, respectively. Although no dose recommendations have been provided for simeprevir in these patients, the label states that the potential risks and benefits of simeprevir should be carefully considered before use in patients with moderate or severe cases. Furthermore, bazedoxifene (with conjugated estrogens) demonstrated AUC ratios of 3.6, 2.1, and 4.3 in patients with mild, moderate, and severe impairment, respectively, and hence has been contraindicated in women with any known HI or disease. The increases in AUC observed in HI patients may be attributed to the fact that these compounds all undergo extensive hepatic metabolism and have also shown high biliary excretion, with > 80% being eliminated in the feces [23]. Finally, no dose adjustment has been recommended for any HI patients treated with sofosbuvir that showed AUC ratios greater than 2 in patients with moderate and severe impairment.

In the case of RI, 4 out of 17 NMEs demonstrated an AUC ratio greater than 2 in RI patients versus normal controls. Gadoterate meglumine showed the largest effect in

RI patients, with 3.5- and 9.3-fold increases in AUC and 61.3 and 87.3% decreases in clearance in moderate and severe RI patients, respectively. These extensive changes in exposure and clearance in RI compared to control resulted from the predomination of renal clearance of gadoterate meglumine; however, no dose adjustment was suggested for RI populations, but the product label contains a black box warning for the risk of the life-threatening adverse event nephrogenic systemic fibrosis in patients with chronic severe kidney disease. For another drug, alogliptin, of which 80% of the elimination is renal, the changes in AUC ratios in moderate RI, severe RI, and the end-stage renal disease (ESRD) populations of 2.0, 3.6, and 4.7 were observed, respectively. In this case, a dose adjustment is recommended for moderate RI, severe RI, and ESRD patients. Similarly, the prodrug eslicarbazepine acetate, which is 90% eliminated by renal excretion as parent and its glucuronide conjugate, its clearance decreased by 37.9, 52.6, 60.6, and 28.9% in mild RI, moderate RI, severe RI, and ESRD, respectively. Hence, a dose reduction has been recommended in the product label for patients with moderate and severe RI.

As recommended by the authors in this review article [23], it is important to study drug PK in impaired populations because the AUC ratios observed in HI or RI patients, or pharmacogentic-modified patients (although no clear example was listed here), may be on the same order of magnitude as those observed in clinical drug interaction studies.

In another review article by the FDA Office of Clinical Pharmacology and Office of Translational Sciences, Huang et al. (2007) [25] stated that issuing guidance or guidelines is considered as one of the most effective ways that the regulatory agencies are able to communicate with sponsors to ensure that the drug development processes are well guided and regulated. In the past 10 years, clinical pharmacology guidances covering important areas have been issued, including PK data in patients with RI and HI, dose-response studies, and DDIs as those that has been followed in the previous review article by Yu et al. (2014) [23, 26–28]. The authors indicated that guidance reflects the agency's current view in the evaluation of DDIs during drug development and includes the following principles:

1. Understanding of metabolism, transport, and DDI is critical to the benefit/risk assessment of a drug.
2. Using an integrated approach incorporating in vitro and in vivo metabolism and transport studies to elucidate the underlying mechanisms and to evaluate the potential for drug interactions can reduce the number of studies needed and optimize our knowledge.
3. Appropriately designed studies are critical to providing important information for proper labeling.
4. The clinical significance of DDIs should be interpreted based on well-defined exposure-response data and analyses.
5. Classification of CYP inhibitors and substrates can aid in the study design and cross-labeling of drugs.
6. DDI information should be appropriately placed in the labeling.

By following these principles, Katzenmaier et al. (2011) [29] established a new clinical study with limited sampling strategy to assess CYP3A activity and evaluated the time course of reversible (voriconazole) and irreversible (ritonavir) CYP3A inhibition. The clinical study is based on randomized trial of two groups, each with eight healthy participants, that received CYP3A inhibitors voriconazole or ritonavir orally for 9 days, with 3 mg midazolam (MDZ) administered before the inhibitor treatment, on days 1, 2, 3, 5, 8, and 9 during inhibitor treatment, and on days 10, 11, and 12 (3 days) after discontinuation. Plasma AUC between 2 and 4 h after oral administration of MDZ, in a solution form, strongly correlated with MDZ clearance. Using this parameter, maximum inhibition of voriconazole and ritonavir was calculated, and it occurred only 48 h after starting of the inhibitor (percentage of baseline MDZ clearance, voriconazole: 10.6%; ritonavir: 8.4%). Recovery of CYP3A activity occurred with a half-life of 24 h after voriconazole discontinuation, while in the case of ritonavir, the inhibition was maintained strong 3 days after discontinuation. These findings underscore the substantial and gradual alterations in dose requirements in the first days of and after such combination therapies.

As stated by the authors, this method of limited sampling might facilitate the quantification of CYP3A activity by measuring the partial AUC between 2 and 4 h after oral MDZ administration. The accuracy over the full range of enzymatic activity was observed and results were able to distinguish the differences in the time courses of a reversible inhibitor and an irreversible inhibitor. With both ritonavir and voriconazole, the maximum inhibition of CYP3A activity occurred in 2 days after administration of the inhibitors. However, major differences were seen between the voriconazole and ritonavir groups after discontinuation, showing that the inhibitory effect did not disappear within hours; instead, 3 or more days may be required for restoration of normal CYP3A activity.

In general, as demonstrated in chapter 5, clinical DDI study can be set up in several designs [30]. The most common design is the randomized crossover design, where subjects are randomized to one of two sequences and subjects randomized to one sequence where a single test dose of the object drug for PK assessment in one period (1) is administered, followed by administration of the object drug for PK during multiple dosing of the precipitant (at steady state) in a second period to assess the effect of the precipitant on the PK of the object drug. Dosing of the precipitant drug is continued through the period of PK assessment of the object drug. In the second sequence, the order of treatments is reversed such that the single-dose PK of the object drug is evaluated at the steady state of precipitant that was coadministered in period 1, followed by evaluation of the single-dose PK of the object drug in the absence of the precipitant in period 2. A long washout period separates periods 1 and 2, ensuring complete washout of both the object drug and the precipitant drug to reach baseline of metabolic activity at the start of the second period. These DDI study designs are depicted in Figure 5.4(A). Other study designs for DDI investigations are shown in Figure 5.4(B). In reality, the fixed-sequence design was the most common, with $\sim$ 70% of all DDI studies evaluated. If neither the fixed-sequence design nor randomized crossover design is feasible, then an alternate design that may be used is the randomized parallel group design with a placebo reference [as depicted in

Figure 5.4(C)]. In this design, subjects are randomized into one of two groups. One group of subjects undergo PK characterization of the object drug alone in period 1, followed by PK characterization of the object drug in the setting of coadministration with the precipitant drug in period 2, and a second group of subjects undergo PK characterization of the object drug alone in period 1 and of the object drug following blinded administration of a placebo in period 2. Such a design provides the ability to separate out the effect of a potential DDI from any confounding period effects.

10.3.2 Mechanism Underlying Drug Therapy (Aromatase Inhibitors) for Breast Cancer

Aromatase, also called estrogen synthetase or estrogen synthase, is an enzyme responsible for a key step in the biosynthesis of estrogens. It is a member of the cytochrome P450 superfamily, which is a monooxygenase that catalyzes many reactions involved in steroidogenesis. In particular, aromatase is responsible for the aromatization of androgens into estrogens, as indicated in Figure 10.3. The aromatase enzyme can be found in many tissues including gonads, brain, adipose tissue, placenta, blood vessels, skin, and bone, as well as in tissue of endometriosis, uterine fibroids, breast cancer, and endometrial cancer. It is an important factor in sexual development. Aromatase is localized in the endoplasmic reticulum, where it

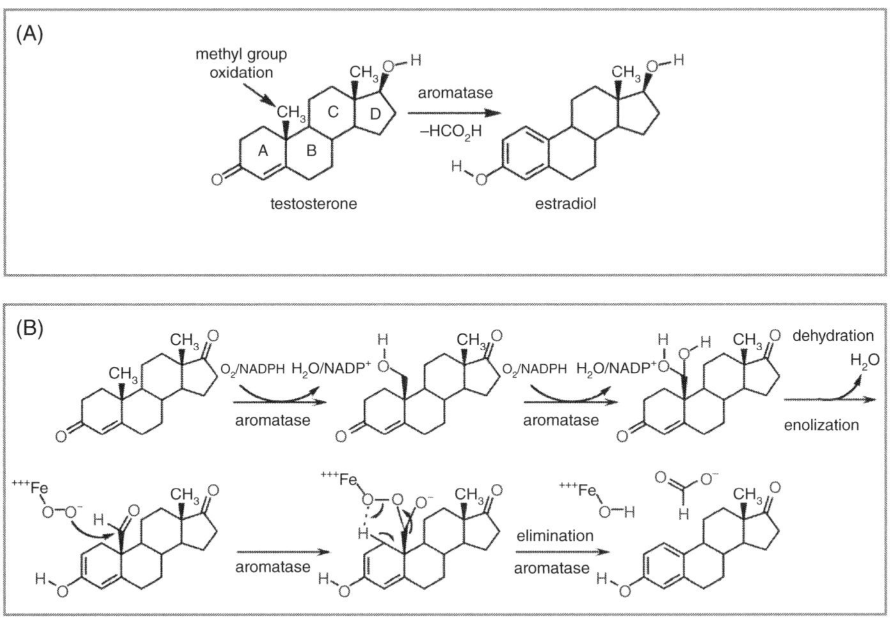

Figure 10.3 General reaction for the conversion of testosterone to estradiol catalyzed by aromatase. The enzyme aromatase (P450 19) converts the ring A to aromatic form (A) by the detailed reaction in (B).

is regulated by tissue-specific promoters that are in turn controlled by hormones, cytokines, and other factors. It catalyzes the last steps of estrogen biosynthesis from androgens (specifically, it transforms androstenedione to estrone and testosterone to estradiol). These steps include three successive hydroxylations of the 19-methyl group of androgens, followed by simultaneous elimination of the methyl group and aromatization of the ring [31]. Aromatase (cytochrome P450 19) is the rate-limiting enzyme in the biosynthesis of estrogens from androgens.

The gene expresses two transcript variants. In humans, the gene CYP19, located on chromosome 15q21.1, encodes the aromatase enzyme. The gene has nine coding exons and a number of alternative noncoding first exons that regulate tissue-specific expression. There are known factors that increase aromatase activity including age, obesity, insulin, gonadotropins, and alcohol. Aromatase activity appears to be enhanced in certain estrogen-dependent local tissue next to breast tissue, endometrial cancer, endometriosis, and uterine fibroids.

Aromatase inhibitors stop the production of estrogen in postmenopausal women. Aromatase inhibitors work by blocking the enzyme aromatase, which turns the hormone androgen into small amounts of estrogen in the body. This means that less estrogen is available to stimulate the growth of hormone-receptor-positive breast cancer cells. Aromatase inhibitors can't stop the ovaries from making estrogen, so aromatase inhibitors only work in postmenopausal women. There are three aromatase inhibitors that are approved for use by the FDA: Arimidex (chemical name anastrozole); Aromasin (chemical name exemestane); and Femara (chemical name letrozole).

Desta and colleagues (2011) [32] summarized their results generated from a multicenter, open-label prospective clinical trial in randomly assigned women ($n \approx 250$ in each arm) to receive 2 years of treatment with either oral letrozole (2.5 mg/day) or oral exemestane (25 mg/day). As has been reported, letrozole is a nonsteroidal triazole derivative, one of the third-generation potent and selective competitive inhibitors of aromatase, which effectively depletes plasma and tissue estrogen [33]. Letrozole is now widely used as adjuvant therapy in postmenopausal women with hormone receptor–positive early stage breast cancer and for extended therapy in patients with early stage breast cancer who have already received 5 years of adjuvant tamoxifen therapy [34]. Also, despite its proven efficacy, the studies showed that letrozole does not benefit all patients, and a significant proportion of patients are at increased risk for serious long-term adverse effects, notably musculoskeletal symptoms, bone loss, and fractures [34]. The interpatient variability in letrozole response remains unknown, and there are no identified biomarkers that help in determining the female patients who will derive benefit from letrozole treatment. It has been identified, though, that the variability in metabolism (as in the case of tamoxifen) can be one of the factors that contribute to variability in drug response [35]. Letrozole is cleared predominantly by metabolism through the CYP enzyme system to pharmacologically inactive metabolite (carbinol) and probably to ketone metabolites [36]. The carbinol metabolite is rapidly excreted, primarily in urine, in the form of glucuronide conjugates [36]. The PK of letrozole varies widely among patients, thus it may contribute to differences in biochemical and clinical responses to this drug. Consequently, identifying the mechanisms and factors that are responsible for the interindividual variability in the PK of letrozole can be important so that better and individualized drug therapy for patients will be ensured.

As reported, letrozole metabolism involved P450 enzymes, therefore, the PK of the drug is most likely to be controlled by the P450 enzyme activity, and it is important to identify well the specific P450 enzymes that catalyze letrozole metabolism. In an in vitro study, letrozole was shown to be a competitive inhibitor of CYP2A6, with marginal effects of other CYPs tested [37,38], with reported data on the contribution of CYP3A4. However, there are no published data that systematically address the contribution of these or other enzymes in vitro. More importantly, the value of these in vitro data in predicting in vivo letrozole exposure has not been studied. The authors then intended to investigate the association between plasma letrozole concentrations and genetic variants of CYP2A6 and CYP3A5 in DNA samples from patients of this study. From the analysis of letrozole plasma concentrations, the results showed high interpatient variability ($> 10-$fold) and were associated significantly with CYP2A6 genotypes ($P < 0.0001$), body mass index (BMI) ($P < 0.0001$), and age ($P = 0.0035$). However, CYP3A5 genotypes showed no association with plasma letrozole concentrations. The data of this study strongly suggest that CYP2A6 predominantly mediated the clearance of letrozole in vivo, and CYP2A6 metabolic status, along with BMI and age, may serve as a biomarker of the efficacy of letrozole treatment or a predictor of adverse effects in patients with breast cancer.

10.3.3 Mechanism Underlying Drug Therapy (Metformin) for Type 2 Diabetes

The kidneys play an important role in the elimination of many endogenous compounds and drugs from the body. Elimination processes in the kidney are driven, in part, by membrane transporters expressed on the basolateral and apical membranes of renal epithelial cells, working in sequential steps to mediate tubular secretion. Human H+/organic cation antiporter multidrug and toxin extrusion 1 (MATE1) and paralog, MATE2 were identified [39–41]. Functional changes in the activity or expression level of MATE2-K caused by genetic variants in the coding or basal promoter regions of this transporter may result in changes in the expression and activity levels of drugs that are substrates for the transporter. By applying the pharmacogenetic practice in drug therapy during the clinical investigation, Choi et al. (2011) [42] looked into the impact of genetic variants of MATE1 and found it to be associated with metformin response in patients with diabetes [43–45].

This study was conducted to identify and functionally characterize genetic variants in the coding and basal promoter regions of MATE2-K in populations constituting four major ethnic groups, hence to determine the impact of these variants on metformin response. Four nonsynonymous variants and four variants in the MATE2-K basal promoter region were identified from ethnically diverse populations. Two nonsynonymous variants, c.485C > T and c.1177G > A, were shown to be associated with significantly lower metformin uptake and reduction in protein expression levels. MATE2-K basal promoter haplotypes containing the most common variant, g.-130G > A (> 26% allele frequency), were associated with a significant increase in luciferase activities and reduced binding to the transcriptional repressor myeloid zinc finger 1 (MZF-1). Patients with diabetes who were homozygous for g.-130A had a significantly poorer response to metformin treatment, assessed as relative change in glycated hemoglobin (HbA1c), as compared with carriers of the reference allele ($P = 0.002$), which concludes that MATE2-K plays a role in the antidiabetes response to metformin. This study as well others demonstrated that the use of genetic information in pharmacogentic research did and will improve outcomes for patients and lead to personalized medicine in clinical practice.

10.4 CLINICAL PHARMACOLOGY STUDIES OF SPECIAL POPULATIONS

The last 3 years have seen outstanding growth in the personalized medicine and translating the concept from dream to practice despite the ethical challenges and difficulties in applying this approach in drug therapy. Examples of the progress have been discussed in chapter 9 and in section 10.3 of this chapter. Looking at mean differences resulting from drug therapy in clinical trials and comparing with placebo has been the standard for more than 4 decades. Defining subpopulations with specific genetic profiles, age, or impaired organ functions making recruitment for clinical trials complex. Advancements in translational research and availability of tools and technologies, especially for genetic testing, should facilitate the recruitment efforts to target subpopulations who are more likely to benefit from drug therapy and to distinguish them from those that are more likely to have adverse events. An article by Reynolds (2012) [46] provided an insight into the practice of tailoring therapeutic intervention to an individual patient's disease, demographic characteristics, genetics, environment, lifestyle, and health status, because many tools are now available that allow translational scientists and clinicians to determine factors that distinguish between individual patients, although there are medical communities who believe that individualization benefits only outliers (as the in case of warfarin). However, the proportion of patients who do not display the average response varies across drugs and diseases [47]. The authors summarize tools used in individualized therapy—such as individualization based on pharmacogenomics, the use of population imaging, and the use of cell-based models, such as Epstein-Barr virus–transformed lymphoblastoid cell lines—which serves as a system for studying the genetics of drug response, as indicated by Cox et al. (2012) [48] who indicated that this model is a cost-effective way to identify genetic variants in drug transporter and pharmacodynamic genes.

Dose adjustment in general can also be possible for subpopulations that are outside the common age brackets or with impairments in liver or renal function to ensure drug efficacy without adverse events. Examples of these subpopulations are described below.

10.4.1 Pediatrics and Geriatrics

An understanding of dose adjustment is needed at both ends of the age spectrum. For example, Leeder and Kearns (2012) [49] indicated that individualization based on genotype-phenotype relationship may not be sufficient in newborns and infants because a specific genotype may not predict the same phenotype in all individuals, as in the case of CYP2C19-mediated pantoprazole clearance, where the number of functional CYP2C19 alleles may not predict drug clearance in neonates compared to children and adults. One reason that explains the lack of a genotype-phenotype relationship in neonates is because CYP2C19 is just beginning to be expressed in these young subjects. Hence, in neonates it is difficult to individualize therapy based on CYP2C19 genotype.

Traditionally, pediatric populations are dosed with drugs that are studied in adults with an adult-adjusted dose in some nonspecific way, either based on body weight or body surface area ratio. Since the FDA Modernization Act (FDAMA) from 1997 provides an additional 6 months of marketing exclusivity, the pharmaceutical industry has been encouraged to carry out studies in children. Several translational research approaches are conducted to enhance the knowledge of how children can be included in clinical investigations and in determination of dosing regimens. Those approaches have been reported in a review article by Yanni and Allegaert (2012) [50]. In addition, in a recent review article by Szychot at al. (2014) [51], new molecular methods were reported that specifically allow the evaluation of leukemia and provide treatment of necessary intensity. Also, this review aimed to provide insight into the progress in leukemia treatment that had been possible due to advances in molecular genetics over the last few decades. The new diagnostic methods predict patients' prognosis and also provide a specific therapy depending on the molecular and genetic characteristics of patients. In the review, 25 articles regarding novel diagnostic and therapeutic methods as well as prognostic factors published between 1992 and 2011 were included. Those articles focus mostly on molecular and cytogenetic testing allowing revolutionary methods of patient classification and individual therapy for this highly heterogeneous group of disorders. The authors stated that implementation of molecular genetic testing to evaluate the type of leukemia allowed pediatric oncologists and hematologists to adjust the intensity of treatment, improve outcome, minimize toxicity of therapies, and considerably lower the risk of side effects. In the last few decades there has been great improvement in survival among children suffering from hematopoietic malignancies. Progress made in molecular genetics allowed the creation of new treatment protocols that are designed to maintain a high cure rate for children with leukemia, while reducing toxicity.

As individuals get older their renal function deteriorates, leading to a need for specific dosing guidance. Groups of interest are subjects with ages from 65 to 75 years, 75 to 85 years, and > 85 years [52]. One of the first drug problems in the elderly occurred with the drug benoxaprofen, a nonsteroidal anti-inflammatory agent. Benoxaprofen was initially approved at a single dose of 600 mg/day but turned out to have a much longer half-life in the elderly population, particularly in subjects over 80 years of age. Subjects, all aged over 80 years, had cholestatic jaundice and died probably related to the excessive benoxaprofen serum levels seen in this population, suggesting that the drug was being overdosed particularly in the elderly [53]. This episode led to the requirement for PK studies in the elderly to ensure the dosing is consistent with the younger population and adverse events are kept to a minimum.

Therapeutic success of acute myeloid leukemia (AML) was reported to be less common in older patients (> 65 years) than in younger patients with AML [54]. Cytarabine in combination with an anthracycline has been the mainstay of AML therapy for many years and continues to serve as the foundation for the current standard therapeutic regimen. The recent review by Kaur et al. (2014) [54] discussed the PK and metabolic and toxicological issues associated with antileukemic agents used to treat elderly patients (> 60 years) with AML. This expert opinion review indicated that a profound and predictable change often occurs with age and can have effects on drug metabolism, PK, and toxicity with consequences bearing on overall efficacy. Therapeutic options for the treatment of AML have remained static for many years, but it has become clear that among elderly patients with AML, improved antileukemic therapy is greatly needed.

10.4.2 Renal Impaired

As discussed in previous chapters, particularly in chapters 4 and 9, if a drug is excreted by kidney, either as the unchanged parent drug or as major metabolites, the plasma, blood, and tissue levels are likely to rise in the presence of increasing impairment of renal functions. The FDA requires an estimation of RI subjects that chronically use drugs whether they are renally or nonrenally excreted [55]. The results of these small PK studies are then reported in the label with statements such as "use with caution in subjects with renal impairment."

However, there are some drugs that become toxic as the serum levels increase, and this has led to various methods for monitoring and adjusting dose in response to RI. Aminoglycoside antibiotics (e.g., gentamicin) have traditionally been used in RI even though they cause both ototoxicity and nephrotoxicity at high systemic levels. Dose modification to maintain efficacy in all degrees of RI and patients undergoing hemodialysis ensured very successful therapeutic strategy in terms of outcome. Because of the general concerns about using drugs in RI, efforts have been made to offer guidelines for such prescribing that go beyond the approved label. Physicians seeking clarity on the appropriate dosing guidance in RI may find confusion in the various official and secondary sources of information [56]. Nephrologists are comfortable using a limited number of drugs in subjects with severe renal failure or on dialysis, but there is some confusion in the general population, who may incidentally have mild or moderate RI. Although there are theoretical concerns, it does not appear that there is a major safety issue for the majority of drugs, and those drugs that are well known to cause problems have appropriate dose adjustment guidelines.

10.4.3 Hepatic Impaired

Like RI, it is recommended that caution be considered in mild, moderate, and severe liver function as measured by the Child-Pugh classification when treatment with drugs that are metabolized by the liver is made. Unlike renal disease, there is no simple factor that determines the degree of liver function. The study design recommended by the FDA [57] involves studying the drug in at least eight usable subjects with the relevant amount of hepatic dysfunction and an age- and sex-matched control group. This provides PK information but does not really address whether or not there are consequences of chronic use in such subjects and whether or not the dose needs adjusting. There are some drugs that may be needed to modulate the hepatic disease, such as prednisolone and other drugs that are needed for chronic or acute therapy in subjects who happen to have some degree of HI, but in general, it is easier to avoid or reduce the use of therapeutic drugs in subjects with marked HI. Dose adjustments in hepatic dysfunction and safety considerations may be intensified in subjects with HI. There is no clarity in dose adjustments for particular drugs even though the PK changes resulting from a clinical dosage trial in each subject may be clear.

10.4.4 Genetic Polymorphic Populations

Impairment in renal and hepatic functions can be a result of lack of expression of certain DMEs or drug transporters or sometimes the expression of nonfunctional enzymes or transporters. Variations in P450 enzymes or transporters are modified by polymorphic genetic variability. The genetic polymorphism changes the disposition, PK, and clearance of drugs, thus modification of the drug dose (or individualized drug therapy) is essential to reduce adverse events and maintain efficacy. Although the determination of the implication of genetic polymorphism in the disposition of drugs has been evaluated to reduce the adverse drug reaction, but the practical implications are not contained in current labeling, these findings were reported and argue that science should be used prospectively in drug research [58].

In the case of tramadol drug therapy, which is centrally acting as an active analgesic and also metabolized to O-desmethyl-tramadol, is an active (M1), the O-demethylation is catalyzed by cytochrome P450 2D6 and the N-demethylation is catalyzed by CYP2B6 and CYP3A4. There is a wide variability in PK because of CYP genetic polymorphism [59]. The tramadol label insert states: "The formation of the active metabolite, M1, is mediated by CYP2D6. Approximately 7% of the population has reduced activity of the CYP2D6 isoenzyme of cytochrome P-450. Based on a population PK analysis of Phase I studies with immediate-release tablets in healthy subjects, concentrations of tramadol were approximately 20% higher in 'poor metabolizers' versus 'extensive metabolizers,' while M1 concentrations were 40% lower. In vitro drug interaction studies in human liver microsomes indicate that inhibitors of CYP2D6 (fluoxetine, norfluoxetine, amitriptyline, and quinidine) inhibit the metabolism of tramadol to various degrees, suggesting that concomitant administration of these compounds could result in increases in tramadol concentrations and decreased concentrations of M1. The full pharmacological impact of these alterations in terms of either efficacy or safety is unknown."

It would appear that the science continues to improve to help understand how genetic polymorphism explains differences in toxicity and to some extent efficacy. It is still not possible to adequately label the proposed change in dose with any precision.

10.4.5 Different Ethnic Populations

It is hard to tease out the effects of ethnic factors from genetic polymorphism as discussed in the last section. Ethnic factors include intrinsic factors such as genetics and metabolism and extrinsic factors such as diet, medical practice, and use of concomitant drugs, tobacco, or alcohol. There are probably no pure ethnic factors, but the proportion of the population with a particular genetic polymorphism varies in different ethnic populations. Ethnic factors have become an issue in drug development with the ICH guideline [60,61]. It is clearly desirable to include as diverse a population as possible in clinical trials to help understand and anticipate any ethnic differences, but the practical labeling consequences are less clear. It is apparent that some drugs have been dosed lower in Japan than the United States, but there is greater concordance in dosing between Europe and the United States. A full review of the literature underpinning ethnic differences has been carried out,and includes questions on the precision of the ethnic group being tested as well as differences in methodology used in different studies [35]. Modern drug development incorporates a diverse population with multicentered multinational studies becoming the norm in Phase 3. However, the interpretation of this diverse data does not lead to individual drug labeling advice.

REFERENCES

1. Meinert CL, Tonascia S. Clinical trials: design, conduct, and analysis. Oxford: Oxford University Press, 1986, 3.
2. Metcalfe NH. Sir Geoffrey Marshall (1887–1982): respiratory physician, catalyst for anaesthesia development, doctor to both Prime Minster and King, and World War I Barge Commander. J Med Biogr 2011;19(1):10–14.
3. Zweig MH, Robertson EA. Why we need better test evaluations. Clin Chem 1982;28(6):1272–1276.
4. Farrington P, Miller E. Clinical trials. Methods Mol Med 1996;4:251–268.
5. Avan BI, White F. Why a research study needs a design? J Pak Med Assoc 2001;51(5):197–201.
6. Blume J, Peipert JF. Randomization in controlled clinical trials: why the flip of a coin is so important. J Am Assoc Gynecol Laparosc 2004;11(3):320–325.
7. Temple R. Enrichment of clinical study populations. Clin Pharmacol Ther 2010;88(6):774–778.
8. Temple R. Complexities in drug trials: enrichment, biomarkers and surrogates. Interview with Robert Temple. Biomark Med 2008;2(2):109–112.
9. Temple RJ. Special study designs: early escape, enrichment, studies in non-responders, commun. Stat Theory Methods 1994;23:499–531.
10. Holliday EG, Scott R, Attia J. Evidence-based medicine in the era of biomarkers: teaching a new dog old tricks? Clin Pharmacol Ther 2010;88:740–742.
11. Ranjith G. Interferon-alpha-induced depression: when a randomized trial is not a randomized controlled trial. Psychother Psychosom 2005;74(6):387–388.
12. Woodcock J. Paving the critical path of drug development: the CDER perspective. Nat Rev Drug Discov 2014;13(11):783–784.

13. Lesko LJ, Zineh I, Huang S-M. What is clinical utility and why should we care? Clinical Pharmacol Therap 2010;88:729–733.
14. Woodcock J. The prospects for personalized medicine in drug development and drug therapy. Clin Pharmacol Ther 2007;81:164–169.
15. Zerhouni E. Medicine: the NIH roadmap. Science 2003;302:63–72.
16. Roses A. Personalized medicine: elusive dream or imminent reality—a commentary. Clin Pharmacol Ther 2007;81:801–805.
17. Epstein RS. Pharmacy benefit managers: evaluating clinical utility in the real world. Clin Pharmacol Ther 2010;88:880–882.
18. March JS, Silva SG, Compton S, Shapiro M, Califf R, Krishnan R. The case for practical clinical trials in psychiatry. Am J Psychiatr 2005;162:836–846.
19. Carleton, B. Demonstrating utility of pharmacogenetics in pediatric populations: methodological considerations. Clin Pharmacol Ther 2010;88:757–759.
20. Lesko LJ. Personalized medicine: elusive dream or imminent reality. Clin Pharmacol Ther 2007;81:807–816.
21. Woodcock J. Assessing the clinical utility of diagnostics used in drug therapy. Clin Pharmacol Ther 2010;88:765–773.
22. Ledwith BJ, DeGeorge JJ. Changes to ICH Guideline M3: New and Revised Guidance on Nonclinical Safety Studies to Support Human Clinical Trials and Marketing Authorization. Clin Pharmacol Ther 2011;89(2):295–299.
23. Yu J, Ritchie TK, Mulgaonkar A, Ragueneau-Majlessi I. Drug disposition and drug–drug interaction data in 2013 FDA new drug applications: a systematic review. Drug Metab Dispos 2014;42(12):1991–2001.
24. Chu X, Korzekwa K, Elsby R, Fenner K, Galetin A, Lai Y, Matsson P, Moss A, Nagar S, Rosania GR, et al., and International Transporter Consortium. Intracellular drug concentrations and transporters: measurement, modeling, and implications for the liver. Clin Pharmacol Ther 2013; 94:126–141.
25. Huang S-M, Temple R, Throckmorton DC, Lesko LJ. Drug interaction studies: study design, data analysis, and implications for dosing and labeling. Clin Pharmacol Ther 2007;81:298–304.
26. Guidance for Industry Drug Interaction Studies—Study Design, Data Analysis, Implications for Dosing, and Labeling Recommendations, 2012. [Online]. Available: http://www.fda.gov.libproxy.lib.unc.edu/downloads/Drugs/GuidanceComplianceRegulatoryInformation/Guidances/UCM292362.pdf.
27. EMA. Draft Guideline on the Investigation of Drug Interactions, 2010. [Online]. Available: http://www.ema.europa.eu/docs/en_GB/document_library/Scientific_guideline/2010/05/WC500090112.pdf.
28. EMA. Guideline on the investigation of drug interactions. EMA, 2012.
29. Katzenmaier S, Markert C, Riedel KD, Burhenne J, Haefeli WE, Mikus G. Determining the time course of CYP3A inhibition by potent reversible and irreversible CYP3A inhibitors using a limited sampling strategy. Clin Pharmacol Ther 2011;90(5):666–673.
30. Venkatakrishnan K, Pickard MD, von Moltke LL. A quantitative framework and strategies for management and evaluation of metabolic drug–drug interactions in oncology drug development: new molecular entities as object drugs. Clin Pharmacokinet 2010;49(11):703–727.

31. Czajka-Oraniec I, Simpson ER. Aromatase research and its clinical significance. Endokrynol Pol 2010;61(1):126–134.
32. Desta Z, Kreutz Y, Nguyen AT, Li L, Skaar T, Kamdem LK, Henry NL, Hayes DF, Storniolo AM, Stearns V, Hoffmann E, Tyndale RF, Flockhart DA. Plasma letrozole concentrations in postmenopausal women With breast cancer are associated with CYP2A6 genetic variants, body mass index, and age. Clin Pharmacol Ther 2011;90(5):693–700.
33. Geisler J, Haynes B, Anker G, Dowsett M, Lønning PE. Influence of letrozole and anastrozole on total body aromatization and plasma estrogen levels in postmenopausal breast cancer patients evaluated in a randomized, cross-over study. J Clin Oncol 2002;20:751–757.
34. Coates AS, et al. Five years of letrozole compared with tamoxifen as initial adjuvant therapy for postmenopausal women with endocrine-responsive early breast cancer: update of study BIG 1-98. J Clin Oncol 2007;25:486–492.
35. Goetz MP, et al. Pharmacogenetics of tamoxifen biotransformation is associated with clinical outcomes of efficacy and hot flashes. J Clin Oncol 2005;23:9312–9318.
36. Pfister CU, et al. Effect of age and single versus multiple dose pharmacokinetics of letrozole (Femara) in breast cancer patients. Biopharm Drug Dispos 2001;22:191–197.
37. Jeong S, Woo MM, Flockhart DA, Desta Z. Inhibition of drug metabolizing cytochrome P450s by the aromatase inhibitor drug letrozole and its major oxidative metabolite 4,4′-methanol-bisbenzonitrile in vitro. Cancer Chemother Pharmacol 2009;64:867–875.
38. Murai K, Yamazaki H, Nakagawa K, Kawai R, Kamataki T. Deactivation of anti-cancer drug letrozole to a carbinol metabolite by polymorphic cytochrome P450 2A6 in human liver microsomes. Xenobiotica 2009;39:795–802.
39. Otsuka M, Matsumoto T, Morimoto R, Arioka S, Omote H, Moriyama YA. Human transporter protein that mediates the final excretion step for toxic organic cations. Proc Natl Acad Sci USA 2005;102:17923–17928.
40. Masuda S, et al. Identification and functional characterization of a new human kidney-specific H+/organic cation antiporter, kidney-specific multidrug and toxin extrusion 2. J Am Soc Nephrol 2006;17:2127–2135.
41. Tanihara Y, Masuda S, Sato T, Katsura T. Ogawa O, Inui K. Substrate specificity of MATE1 and MATE2-K, human multidrug and toxin extrusions/H(+)-organic cation antiporters. Biochem Pharmacol 2007;74:359–371.
42. Choi JH, Yee SW, Ramirez AH, Morrissey KM, Jang GH, Joski PJ, Mefford JA, Hesselson SE, Schlessinger A, Jenkins G, Castro RA, Johns SJ, Stryke D, Sali A, Ferrin TE, Witte JS, Kwok PY, Roden DM, Wilke RA, McCarty CA, Davis RL, Giacomini KM. A common 5′-UTR variant in MATE2-K is associated with poor response to metformin. Clin Pharmacol Ther 2011;90(5):674–684.
43. Choi JH, et al. Identification and characterization of novel polymorphisms in the basal promoter of the human transporter, MATE1. Pharmacogenet Genomics 2009;19:770–780.
44. Meyer ZU, Schwabedissen HE, Verstuyft C, Kroemer HK, Becquemont L, Kim RB. Human multidrug and toxin extrusion 1 (MATE1/SLC47A1) transporter: functional characterization, interaction with OCT2 (SLC22A2), and single nucleotide polymorphisms. Am J Physiol Renal Physiol 2010;298:F997–F1005.
45. Jablonski KA, et al. Common variants in 40 genes assessed for diabetes incidence and response to metformin and lifestyle intervention in the diabetes prevention program. Diabetes 2010;59:2672–2681.
46. Reynolds KS. Achieving the promise of personalized medicine. Clin Pharmacol Ther 2012;92(4):401–405.

47. Woodcock J, Lesko LJ. Pharmacogenetics—tailoring treatment for the outliers. N Engl J Med 2009;360:811–813.
48. Cox NJ, Gamazon ER, Wheeler HE, Dolan ME. Clinical translation of cell-based pharmacogenomic discovery. Clin Pharmacol Ther 2012;92:425–427.
49. Leeder JS, Kearns GL. Interpreting pharmacogenetic data in the developing neonate: the challenge of hitting a moving target. Clin Pharmacol Ther 2012;92:434–436.
50. Yanni S, Allegaert K. Drug metabolism and interactions in pediatric patients. In: Encyclopedia of drug metabolism and interactions, ed. Lyubimov AV. Hoboken, NJ: Wiley, 2012, 11:1–44.
51. Szychot E, Brodkiewicz A, Peregud-Pogorzelski J. How have advances in our understanding of the molecular genetics of paediatric leukaemia led to improved targeted therapies for these diseases? Adv Clin Exp Med 2014;23(3):469–474.
52. Shah RR. Drug development and use in the elderly: search for the right dose and dosing regimen. Br J Clin Pharmacol 2004;58(5):452–469.
53. Halsey JP, Cardoe N. Benoxaprofen side-effect profile in 300 patients. BMJ 1982;284:1365–1368.
54. Kaur I, Constance JE, Kosak KM, Spigarelli MG, Sherwin CM. An extensive pharmacokinetic, metabolic and toxicological review of elderly patients under intensive chemotherapy for acute myeloid leukemia. Expert Opin Drug Metab Toxicol 2014;9:1–13.
55. U.S. Department of Health and Human Services. Food and Drug Administration. Pharmacokinetics in patients with impaired renal function—study design, data analysis, and impact on dosing and labeling. FDA, 2010.
56. Vidal L, Shavit M, Fraser A, et al. Systemic comparison of four sources of drug information regarding adjustment of dose for renal function. BMJ 2005;331(7511):263.
57. U.S. Department of Health and Human Services. Food and Drug Administration Center for Drug Evaluation and Research (CDER) Center for Biologics Evaluation and Research (CBER). Pharmacokinetics in patients with impaired hepatic function—study design, data analysis, and impact on dosing and labeling. FDA, 2003.
58. Kirchheiner J, Fuhr U, Brockmöller J. Pharmacogenetics-based therapeutic recommendations—ready for clinical practice? Nat Rev Drug Discov 2005;4:639–647.
59. Grond S, Sablotzki A. Clinical pharmacology of tramadol. Clin Pharmacokinet 2004;43(13):879–923.
60. Huang S-M, Temple R. Is this the drug or dose for you? Impact and consideration of ethnic factors in global drug development, regulatory review, and clinical practice. Clin Pharmacol Ther 2008;84:287–294.
61. Bjornsson TD, Wagner JA, Donahue SR, et al. A review and assessment of potential sources of ethnic differences in drug responsiveness. J Clin Pharmacol 2003;43:943–967.

11

REGULATORY SUBMISSION: MIST AND DRUG SAFETY ASSESSMENT

11.1 DRUG DEVELOPMENT AND APPROVAL PROCESSES ACCORDING TO THE FOOD AND DRUG ADMINISTRATION (FDA), EUROPEAN MEDICINES AGENCY (EMA), AND OTHER REGULATORY AUTHORITIES

Any drug product requires the submission of an investigational new drug (IND) application, followed by a review process and the regulatory authorization for marketing in the United States. The IND is the vehicle through which a sponsor advances to the next stage of drug development known as clinical trials (human trials). Registration of any new therapeutic products in the United States is as follows:

- Sponsor seeks permission to test a new drug or biologic in humans with the primary goal to determine if the product is reasonably safe for initial use in humans, and if the therapeutic product exhibits pharmacological activity that justifies commercial development.
- The process usually begins in Phase 1; however it can begin in Phase 2 or 3 if there is adequate human experience.
- FDA reviews the IND application (contained in Title 21, CFR, Part 312) focusing on:
  - Patient safety within Phase 1.
  - Assessment of the scientific quality of the clinical evaluation included in Phase 2/3.

Translational ADMET for Drug Therapy: Principles, Methods, and Pharmaceutical Applications, First Edition. Souzan B. Yanni.

When a product is identified as a viable candidate for further development, the sponsor's collecting of the data and information necessary to establish that the product will not expose humans to unreasonable risks when used in early stage clinical studies occurs during a new drug's early preclinical development.

In Europe, the Committee for Medicinal Products for Human Use (CHMP) under the EMA requires the following steps to be taken before drug registration:

1. Step I (120 days)—presubmission to application by which the CHMP provides the following:
    1. Early advice
    2. Assessment team
    3. Application
    4. Validation
2. Step II (210 days)—scientific evaluation:
    1. Assessment reports
    2. List of questions (and when the process can stop)
    3. CHMP opinion
    4. Possibility to appeal
    5. Transfer to EU Commission
3. Step III (67 days)—decision-making process after validation by the EMA and review of the application. The EMA will ensure that the opinion of the CHMP is given within a maximum of 210 days. However, within the 210 days the opinion can be positive, conditional, or negative. The opinion itself is not legally binding; only the EU Commission can turn this into a legally binding decision across the 15 EU member states, and hence the opinion must then enter the commission decision-making process.

In the United States, the investigational new drug/new drug application (IND/NDA) corresponds with investigational medicinal product dossiers/marketing application authorizations (IMPD/MAA) in Europe and centralised procedure/ decentralised procedure (CP/DCP) in Japan and other health authorities across geographies.

As discussed in chapter 10, the International Conference on Harmonization of Technical Requirements for Registration of Pharmaceuticals for Human Use (ICH) was set up in 1995 to regulate clinical investigations around the pharmaceutical industry in the EU, Japan, and the United States. The goal is to improve (through harmonization) efficiency of development/registration of new medicinal products based on scientific consensus. These countries via their regulatory agencies—the FDA, the EMA, and the Japan Medical Association (JMA)—are making a solid commitment to implement ICH recommendations.

In the United States, the regulation and control of new drugs has been based on the NDA before U.S. commercialization. The NDA application is the vehicle through which drug sponsors formally propose that the FDA approve a new pharmaceutical for sale and marketing in the United States. The data gathered during the animal studies and human clinical trials of an IND become part of the NDA.

11.2 STUDIES REQUIRED FOR IND AND NDA

11.2.1 Types of INDs, Types of Information, and Timelines

There are two types of INDs, commercial INDs, which are the applications submitted primarily by companies whose ultimate goal is to obtain marketing approval for a new product, and a broad category known as noncommercial INDs. The vast majority of INDs are filed for noncommercial research. The types of data needed for the IND/IMPD applications in general include information covering three broad areas:

1. Animal pharmacology and toxicology studies: preclinical data to permit an assessment as to whether the product is reasonably safe for initial testing in humans.
2. Manufacturing information: information pertaining to the composition, manufacture, stability, and controls used for manufacturing the drug substance and the drug product. This information is assessed to ensure the company can adequately produce and supply consistent batches of the drug.
3. Clinical protocols and investigator information as well as detailed protocols for proposed clinical studies: information necessary to assess whether the initial-phase trials will expose subjects to unnecessary risks, as well as to determine the qualifications of clinical investigators who oversee the administration of the experimental compound and hence to assess whether they are qualified to fulfill their clinical trial duties.

Under these three broad areas, the IND applications are filed and reviewed for the following information:

- Chemistry
- Pharmacology/toxicology
- Pharmacology and drug distribution
- Toxicology data
- Medical review
- Safety review
- Statistical review

11.2.1.1 Chemistry and Manufacturing Control The chemistry sections of drug applications address issues related to drug identity, manufacturing control, and analysis. The manufacturing and processing procedures for drug information are filed and reviewed to ensure that the compound is adequately reproducible and stable. If the drug is either unstable or not reproducible, then the validity of any clinical testing would be undermined because one would not know what was really being used in the patients, and, more importantly, the studies may pose significant risks to participants. Furthermore, in this section the drug sponsor states the quality of chemistry of either the drug substance or the drug product, or the manufacturing of either the drug substance or the drug product, and the presence of any signals of potential human risk. In case of the presence of such risks, the sponsor proposes how to monitor such risks. Also, the sponsor states any chemistry and manufacturing differences between the drug product proposed for clinical use and the drug product used in the animal toxicology trials, as any of these differences may affect the safety profile of the drug product.

11.2.1.2 Pharmacology/Toxicology The pharmacology/toxicology evaluation is for the results of animal testing and an attempt to relate animal drug effects to potential effects in humans.

11.2.1.3 Pharmacology and Drug Distribution (21 CFR 312.23(a)(8)(I)) This section of the application should contain, if known: (1) a description of the pharmacologic effects and mechanism(s) of action of the drug in animals, and (2) information on the absorption, distribution, metabolism, and excretion of the drug. The regulations do not further describe the presentation of these data, in contrast to the more detailed description of how to submit toxicological data. To the extent that such studies may be important to address safety issues, or to assist in the evaluation of toxicology data, they may be necessary; however, lack of this potential effectiveness may not put the Clinical Phase 1 study on hold.

11.2.1.4 Toxicology Data Present Regulations (21 CFR 312.23(a)(8)(ii)(a)) Regulations require an integrated summary of the toxicological effects of the drug in animals and in vitro. The particular studies needed depend on the nature of the drug and the phase of human investigation. When species specificity, immunogenicity, or other considerations appear to make many or all of the toxicological models irrelevant, sponsors are encouraged to contact the agency to discuss toxicological testing. In the United States, pivotal toxicological studies are done in two species: one rodent (i.e., rats) and one nonrodent (i.e., dogs). The safe starting dose is based on toxicological findings observed in the most sensitive species.

11.2.1.5 Medical Review Medical/clinical reviewers, often called medical officers, are almost exclusively physicians. Medical reviewers are responsible for evaluating the clinical sections of submissions, such as the safety of the clinical protocols in an IND or the results of this testing as submitted in the NDA. Within most divisions, clinical reviewers take the lead role in the IND or NDA review and are responsible for synthesizing the results of the animal toxicology, human pharmacology, and clinical reviews to formulate the overall basis for a recommended agency action on the application.

During the IND review process, the medical reviewer evaluates the clinical trial protocol to determine: (1) if the participants will be protected from unnecessary risks; and (2) if the study design will provide data relevant to the safety and effectiveness of the drug. Under federal regulations, proposed Phase 1 studies are evaluated almost exclusively for safety reasons. Since the late 1980s, FDA reviewers have been instructed to provide drug sponsors with greater freedom during Phase 1, as long as the investigations do not expose participants to undue risks. In evaluating

Phase 2 and 3 investigations, however, FDA reviewers also must ensure that these studies are of sufficient scientific quality to be capable of yielding data that can support marketing approval.

11.2.1.6 Safety Review Following review of an initial IND submission, the Center for Drug Evaluation and Research (CDER) has 30 calendar days in which to decide if a clinical hold is necessary (i.e., if patients would be at an unacceptable risk or if CDER doesn't have the data to make such a determination). Generally, drug review divisions do not contact the sponsor if no concerns arise with drug safety and the proposed clinical trials. If the sponsor hears nothing from CDER, then on day 31 after submission of the IND, the study may proceed as submitted.

11.2.1.7 Statistical Review Statisticians evaluate the statistical relevance of the data in the NDA with the main tasks of evaluating the methods used to conduct studies and the various methods used to analyze the data. The purpose of these evaluations is to give the medical officers a better idea of the power of the findings to be extrapolated to the larger patient population in the country.

11.2.1.8 Timelines and Clinical Hold Decision A clinical hold is the mechanism that CDER uses when it does not believe, or cannot confirm, that the study can be conducted without unreasonable risk to the subjects/patients. If this occurs, the center will contact the sponsor within the 30-day initial review period to stop the clinical trial. CDER may either delay the start of an early phase trial on the basis of information submitted in the IND or stop an ongoing study based on a review of newly submitted clinical protocols, safety reports, protocol amendments, or other information. When a clinical hold is issued, a sponsor addresses the issue causing the hold before the order is removed.

11.2.1.9 Notify Sponsor Once a clinical hold is placed on a commercial IND, the sponsor will be notified immediately. For both individual and commercial INDs, the division is required to send a letter within 5 working days describing the reasons for the clinical hold, and the sponsor may then respond. The division then reviews the sponsor's response and decides within 30 days whether the hold should be lifted. If the decision is made to lift the hold, the sponsor will be informed and the trial may begin.

11.2.2 Metabolites in Safety Testing (MIST) Regulation—Safety Assessments in Humans

The most difficult issues to solve in an IND are chemistry, manufacturing and control information, and pharmacology and toxicology. In response to recent regulatory guidance on the safety evaluation of metabolites of candidate drugs in man, novel approaches have been adopted to define the identities and levels of circulating metabolites in the plasma of both human subjects and the animal species employed in safety assessment programs. Through these analyses, exposure margins are established for human drug metabolites that circulate above a certain threshold in order to provide assurance that the preclinical toxicology program takes into account metabolites of the drug candidate as well as the parent molecule. Metabolites judged to be disproportionate or "unique" by regulatory criteria may require more detailed evaluation. In general, a drug can yield several metabolites, and measuring the exposure of these metabolites in toxicology studies conducted early in the drug development process is not the common practice. In addition, in the majority of early drug development programs, the identities of most of the metabolites may not even be known [1]. In 2002, a group of scientists from the pharmaceutical industry proposed a guideline for assessing the contributions of metabolites to toxicity termed "metabolites in safety testing" or "MIST" [2]. This document attempted to define those situations in which metabolites should be further studied to help define risk assessment for the parent drug. If the metabolite is present in humans at 25% or more of the total drug-related material in circulation, it merits further investigation as a potential contributor to safety findings. As mentioned in section 7.5, the assessment was based on a quantitative testing as a percentage of total radioactivity, but not in absolute concentration terms. However, Smith and Obach (2005) [1] aimed to build on MIST and propose a set of criteria to be used to determine whether a metabolite should be more extensively studied, focusing less on relative abundance (proportion, percentage), as has been suggested in the MIST guideline [2], and more on absolute abundance (concentration, mass), also taking into consideration the structure of the metabolite relative to the parent drug and the potential toxic mechanisms of metabolites.

When assessing the potential toxicity of a drug, what is actually being examined is a complex mixture of chemicals: drug plus all metabolites and impurities. Developed policies have been described for assessing the safety of impurities in drugs [3]. These policies are not directly applicable for metabolites. Chemical structures of impurities can be substantially different from those of the parent drug, since they may derive from the chemical process used to synthesize the drug. As completely unrelated structures, impurities can elicit toxic responses very different from those of the parent drug; thus, the examination of the safety of impurities can have stringent criteria. Metabolites are generated from the parent drug; therefore, in most cases they bear structural similarity to the parent drug.

Traditionally, the primary focus of safety assessment paradigms in pharmaceutical development centered on measuring exposure to the drug candidate itself, first preclinically in animal models to support first in human studies and, subsequently, during clinical trials to establish appropriate safety margins for continued clinical development. The regulatory requirements for animal metabolism and toxicity as well as pharmacokinetics (PK) studies are somewhat vague. When reviewing the FDA IND requirement, the pharmacology and drug disposition section stipulates that the sponsor must provide in the IND application "A section describing the pharmacological effects and mechanisms of action of the drug in animals; and information on the absorption, distribution, metabolism, and excretion of the drug, if known." While this regulation makes no explicit reference to the use of radiolabel [4], the broad consensus of sponsors and regulators is that the only practical means of obtaining the full information specified is a preclinical radiolabel-based absorption, distribution, metabolism, and excretion (ADME) study [5]. The phrase in the regulation "if known" is usually interpreted to mean that a complete understanding of animal ADME is not a requirement for approval of the IND application but suggests that such information must be provided at some point, though an IND application lacking such data would undoubtedly generate questions from regulatory reviewers.

In 2008, the FDA published the MIST guidance [6], which in some respects paralleled the recommendations of the 2002 paper [2] but differed significantly in other respects, notably in defining the threshold for human metabolites of interest as those whose AUC_p values exceeded 10% of that of the parent drug when measured under steady-state dosing conditions. Moreover, as described by Baillie (2012) [7], it was recommended that metabolites falling into this category should be demonstrated to be present in the plasma of at least one of the animal species selected for preclinical safety assessment at levels at least as great as those observed (or predicted) in human subjects at the highest clinical dose. Should this requirement for exposure margins (i.e., animal $AUC_p \geq$ human AUC_p) not be met, then the metabolite in question is designated "disproportionate," and additional safety studies may be required to assure its human safety. While in some cases metabolite exposure in animals can be increased by administering progressively higher doses of the parent drug, this approach is not always successful because of a number of factors, for example, limited solubility in the dosing vehicle, saturation of drug absorption, and species differences in metabolic pathways. In this case, the FDA guidance recommends that separate animal toxicology studies with the preformed (synthetic) metabolite should be considered. Finally, the guidance emphasizes the need for information on human drug metabolism to be acquired as early as possible in the drug development process so that issues related to disproportionate human metabolites may be recognized and addressed before the initiation of large-scale (Phase 3) clinical trials.

In the 2012 American Association of Pharmaceutical Scientists (AAPS) MIST Symposium held in Chicago on October 15, 2012, regulatory experts and industrial scientists joined together to discuss their perspectives and strategies in addressing contemporary MIST recommendations [8]. As reported by Baillie (2012) [7], the regulatory guidance indicates that metabolites identified in human plasma should circulate at similar or greater concentrations in at least one of the animal species used in nonclinical safety assessment of the parent drug. However, synthetic standards for the metabolites often do not exist or they are intractable to synthesize, thus introducing multiple challenges in drug development for the quantitative comparison of metabolites between humans and animals. The final guidance from FDA in 2008 [6], which ultimately culminated with recommendations from ICH in 2010 [9], laid out criteria describing when a human metabolite needs to be measured in laboratory animal species and defined the circumstances under which direct testing of a metabolite in animal toxicology studies would be needed to provide a risk assessment.

After consensus was reached regarding the 10% of total drug-related material threshold over which metabolites might be considered important, drug metabolism and bioanalysis specialists began to focus on how to measure metabolite exposures in laboratory animals and humans. Developing and validating standard bioanalytical methods for metabolites were feasible [10], although there can be challenges in obtaining authentic standards of metabolites that are difficult to synthesize and measure multiple analytes simultaneously. In response to the bioanalytical challenges posed by these expectations, investigators have proactively developed alternate approaches that can offer assurance of relative exposures to metabolites in animals and humans. These include the application of radiometrically calibrated metabolite standards [11,12]; creating metabolite standards from biological sources that are quantitated using nuclear magnetic resonance (NMR) spectroscopy [13–15]; or demonstrating the relative exposures to metabolites across species using HPLC-MS peak area comparisons from animal and human plasma extracts [16–18].

11.2.3 Highlights of the AAPS 2013 MIST Symposium

11.2.3.1 ICH M3(R2) and Metabolite Issues Comments and recommendations were generated by Abigail Jacobs, who indicated that sometimes a metabolite that's "major" in humans may be "minor" in animal test species. Such metabolites could be at greater risk of being disproportionate metabolites. But nonclinical studies in animals are often conducted at a dose level that is far greater (in milligrams per kilogram) than in humans when normalized by weight; these major human metabolites(s) may still have been acceptably evaluated in the animals with dosing of the parent drug, and no further toxicology studies on the metabolites are warranted.

Before clinical trials, in vitro metabolic and plasma protein binding data for animals and humans and preclinical systemic exposure data are usually available. Subsequently, before large numbers of subjects are exposed to an investigational drug, nonclinical absorption, distribution, metabolism, and excretion results and in vitro biochemical information relevant to potential drug interactions are submitted to the FDA per ICH M3(R2).

The revised ICH M3(R2) contains a new section on potential metabolite issues. Per ICH M3(R2), evaluation of metabolites is warranted when the metabolites are observed at exposures > 10% of total drug-related material exposure and at significantly greater levels in humans than the maximum exposure seen in the toxicity studies. Not all disproportionate metabolites are of concern, including most glutathione conjugates, most glucuronide metabolites, and metabolites with an additional hydroxyl group.

Clarity on some issues was also discussed and included in the final review of revised ICH M3(R2). These issues are:

1. The definition and calculation method of the 10% threshold. The "10% threshold" refers to when a human metabolite comprises greater than 10% of the measured total exposure to drug and metabolites, usually based on group mean area under the curve (e.g., AUC $0-\infty$).
2. The guidance states that the metabolite needs to be at "significantly" greater exposure in humans to merit concern. The use of the term "significantly" does not refer to statistical significance. Characterization of metabolite toxicity

is generally considered adequate when animal exposure is at least 50% the exposure seen in humans at the "to be marketed" dose. However, when a metabolite comprises more than 50% of the total human exposure, it is appropriate for exposure to the metabolite in animals to exceed that in humans at the to be marketed dose.

3. It is important to have adequate exposure to the metabolite in one species used in general toxicity evaluation, in one species used in a carcinogenicity study (when carcinogenicity evaluation is warranted or an in vivo micronucleus study when carcinogenicity evaluation is not warranted), and in one species used in an embryo–fetal development study.
4. The evaluation can be based on single-dose radiolabeled ADME data in humans. However, if exposure data later collected from multiple-dose human studies indicate that steady-state levels of a metabolite exceed 10% of total drug-related material, then additional nonclinical evaluation of metabolite should be considered.
5. Because parent drug and metabolites contribute to target organ toxicity in animals at the maximum tolerated dose (MTD), exposure comparisons across species should be conducted at the MTD in animals compared to the maximum exposure in humans at the therapeutic dose, assuming the toxicity of concern can be adequately monitored in humans and does not pose an unacceptable risk. However, if toxicity at the MTD cannot be monitored in humans or poses an unacceptable risk, then the exposure comparison should be conducted at the no adverse effect level for the toxicity of concern.
6. Some metabolite types may not be of toxicological concern. With rare exceptions, most glutathione conjugates are formed by conjugation with reactive metabolites to form excretory metabolites that are not of toxicological concern. Most glucuronides are not of concern, except those that undergo chemical rearrangement (e.g., reactive acyl glucuronides). It was noted that generally it is not feasible to test highly reactive metabolites independently because of their instability, but they are assumed to contribute to the overall nonclinical toxicity of the drug.
7. Clinical studies assessing safety pharmacology endpoints are conducted during Phase 1 and will have already been assessed in humans before a full characterization of the human metabolites is conducted. Therefore, nonclinical safety pharmacology studies are generally not warranted for the characterization of metabolites. However, if a safety pharmacology signal is seen in humans not predicted by nonclinical studies with the parent, then additional safety pharmacology studies of these human metabolites can be considered to better understand the mechanism.
8. The guidance does not specifically address prodrugs. If the animal species converts the prodrug to the active metabolite similarly to humans, then a standard testing approach as recommended in ICH M3(R2) can be used. Further, if the active metabolite is not adequately produced in the animal species, then the

target molecule for toxicologic evaluation is the active metabolite, and therefore additional testing beyond that recommended for metabolites can be appropriate.

11.2.3.2 Early Assessment of MIST Liability of a Clinical Drug Candidate without the Use of Radiolabel As commented by Ronald White, a consequence of MIST is that analytical chemists have been challenged to find and quantitate human metabolites in plasma during early clinical trials without using a radiolabeled drug. This is currently resolved by applying modern start-of-the-art mass spectrometers, which have the ability to accurately determine molecular weights of ions with mass/charge measurement error < 5 ppm. When combined with sophisticated data-processing algorithms such as mass defect filtering, isotope pattern filtering, and background subtraction, these methods allow us to discern drug metabolites among the many background compounds present in biofluids. The currently available techniques may be reliably integrated to identify the structure of metabolites in the plasma and urine samples collected in typical Phase 1 clinical trials. Comparison between animal and human exposures to the metabolites can be achieved by semiquantitation of metabolites using liquid chromatography (LC-UV, LC-MS/MS), peak area ratio comparison with radiolabeled calibrant, and quantitative NMR standards. This allows a sponsor to comply with regulatory expectations for metabolite safety assessment without the need to wait for the conventional ^{14}C–ADME studies that are usually conducted in Phase 2 or 3.

11.2.3.3 MIST: How Do We Deal with Surprises? As commented by Brian Booth, when novel metabolites are discovered in humans at significantly higher concentrations than those observed in nonclinical studies, several safety-related questions will be raised. For example, is the metabolite active (e.g., SN-38 for irinotecan)? Is there sufficient nonclinical coverage to assure safety? Will we need to monitor this metabolite? If so, how and when will this issue be addressed? These data from humans are used to develop the exposure-response (PK/PD) relationships that support approval and guide dose adjustments in different patient settings. However, analytical methods take time to develop and validate, and there is considerable anxiety over when these methodologies are needed. In this case, a fit-for-purpose approach should be used. Frequently, initial Phase 1 data are not used to determine the approvability or the labeling for a novel product. Often, the PK behavior of the drug and any metabolites is characterized throughout drug development. This fact allows time for a method for an unexpected metabolite to be validated such that the metabolite can be characterized in later trials. The general rule is full validation of bioanalytical methods of parent and metabolites should be available for drug final approval and labeling [19]. This rule allows flexibility in the time frame for generating the data.

11.2.3.4 A Simple LC-MS/MS Method for Evaluating MIST Coverage A simple HPLC-MS/MS method was presented by Hongying Gao whereby quantitative comparisons of exposures to metabolites were made between animals and humans in the absence of authentic standards of the metabolites, calibration curves, and other attributes of standard bioanalytical methods. A statistical analysis showed that if the experimentally determined MS response ratio of animal/human was ≥ 2.0, then the actual exposure ratio is unity or greater ($p < 0.01$). The confidence level in such a ratio increases exponentially with the measured MS response ratio of animal/human. This method offers time- and resource-sparing advantages to ascertaining metabolite exposure comparisons between humans and laboratory animal species. Gao indicated that it is important to note that pharmacologically active metabolites offer an exception to this approach, as quantitative exposure data for such metabolites are critical to establishing PK/PD relationships. Pharmacologically active metabolites require standard bioanalytical methods using authentic standards and should be included in the bioanalytical method as soon as possible in the drug development timeline. Most metabolites are nonactive metabolites, and often, a simple LC-MS/MS measurement is sufficient to demonstrate that an animal toxicology study with the parent drug has covered the safety of the human metabolites, since animals are dosed at a higher level when corrected by body weight. This data-driven bioanalysis strategy would increase the rigor of the bioanalysis accordingly based on the results of the animal-to-human MS response ratio measurements rather than the stage of the drug in the development. A validated or qualified method using synthetic standard will be needed only in rare cases where an actual exposure measurement of a metabolite is needed [20].

11.3 DRUG LABELING AND BLACK BOX WARNING

The main principle of drug labeling is to provide the prescriber with sufficient objective information to make informed prescribing decisions. Drug labeling refers to all of the printed information that accompanies a drug, including the label, the wrapping, and the package insert. Drug labeling is regulated by the FDA Division of Drug Marketing, Advertising and Communications. These regulations apply to prescription drugs, over-the-counter (nonprescription) drugs, and dietary supplements. The FDA requires that drug labeling be balanced and not misleading, and most important, that it be scientifically accurate so it provides clear instruction to health-care physicians (and pharmaceutical researchers) for prescription drugs and to consumers for over-the-counter drugs and supplements. Labeling regulations require that the statement of ingredients must include all ingredients, in the order in which they are used in the drug. These ingredients must also be identified by their established name.

The structure of FDA drug labels has progressed over time, and in 2006 new regulations were put in place to make the organization and chapter headings of drug labels consistent and easier to understand for both the patient and physician. Over 1,116 of the most commonly prescribed drugs have been now covered with the full label information in the *Physicians' Desk Reference* (PDR) or at the FDA-CDER website. Issues of dosing to target patients and those under specific population categories, drug metabolism, and clinical pharmacology have to be carefully reviewed for any drug prescribed to avoid unwanted consequences. The label should cover any special instructions such as taking the medicine with food or water. In addition, the label should spell out storage instructions and general instructions, such as discussing questions with a health professional, because the label is not intended to be all-inclusive. Important issues are addressed as "black box warnings" although there is some dispute about whether or not this information is used in a consistent manner [21]. The length of the label may also reflect a defensive legal side as well as the more helpful informational aspect. Despite the carefully crafted labeling information, drugs are increasingly prescribed off-label, which means using approved drugs in unapproved situations [22]. The drug manufacturer must spell out the symptoms of any adverse reactions to the drug. The FDA requires the manufacturer to collect this information. If the patient faces any risk of drug tolerance or dependency while taking the drug, the label must contain a warning.

Modern drugs are fully investigated before approval and marketing and tend to be very robust in dose and rarely need dose modification. The recommended drug doses are currently based on a full development program providing appropriate doses to maximize efficacy and minimize safety issues. Before considering the impact of drug metabolism, it is worth considering how recommended drug doses are decided and the factors that may influence that decision both before and after regulatory approval. For new drugs or old drugs for new indications, the process of drug development leads to a choice of dose, dosing interval, and method of administration. Appropriate formulations are then manufactured to satisfy those needs and the definitive clinical trials are usually carried out using those formulations, which are then taken through to marketing.

In addition, a full drug development program uses clinical pharmacology to investigate PK and pharmacodynamics, including potential metabolic variability and drug–drug interactions (DDIs). The full description of the efficacy, safety, and clinical pharmacology information collected is then presented in the drug labeling for the use of physicians and patients. Whether or not this information leads physicians to make dosing modifications for individual subjects is discussed. Examples from the past of drugs that were inappropriately dosed or that required dose modification based on variations in drug metabolism or excretion are listed below.

As shown in Figure 11.1, selection of new drugs usually involves a choice from a range of chemicals with different physical and chemical properties, leading to potential benefits in terms of absorption, distribution, and elimination. Once selected, the chemical is tested in animals to determine the nature of any toxicity at high doses and the serum levels of drug that are associated with those toxicities. This allows the first use in man using appropriate algorithms to compare kinetics in different species and choosing a dose with an adequate safety margin [23]. From that point clinical pharmacology guides the process, but the minimum dose is considered from observation of efficacy and the maximum dose from the appearance of toxicity or maximum tolerability. Thus, at the end of the development program, a recommended dose range is chosen together with appropriate formulations and a full label to guide prescribing.

The basis for the dose range is supported by clinical trials in predominantly healthy subjects, that is, those with the disease of interest but otherwise healthy. Additional studies provide information on absorption, distribution, and elimination; potential interaction with food; DDIs; and PK in the presence of renal and hepatic impairment. The product is then launched and subsequent information becomes available through

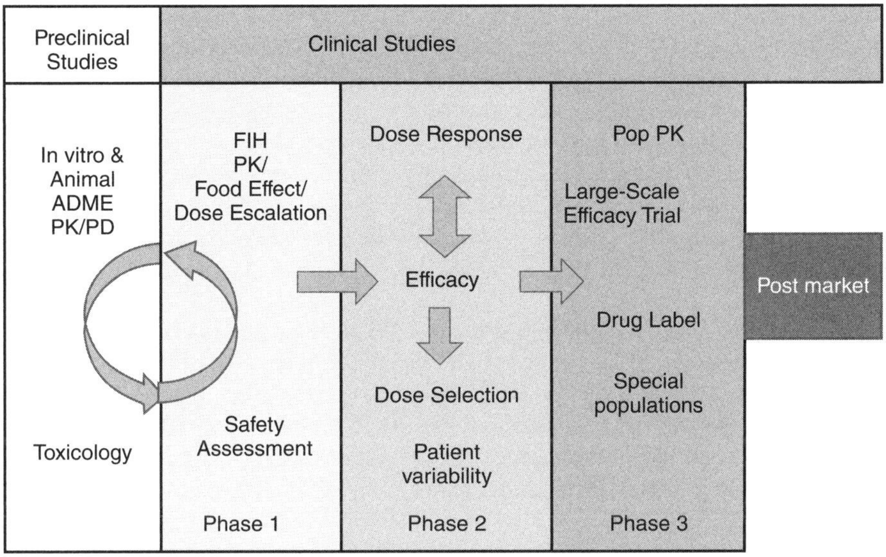

Figure 11.1 Drug development processes: from preclinical to market.

postmarketing surveillance, Phase 4 clinical trials, and other sources of information. This ultimately leads to changes in the label, providing guidance to physicians and users. Thus, information from real-world use leads to continuing updates to prescribing information as the product life cycle continues.

11.3.1 Sections Included in Drug Label

11.3.1.1 Drug Dosing Regulators have expanded the clinical pharmacology requirements for new drugs and new formulations so that a complete assessment of desirable dosing to target patient population and general recommendation to most patient populations. That type of information is included in the label for drugs that is designed to help good prescribing.

11.3.1.2 Age in Drug Labeling Dosing estimation at both ends of the age spectrum, pediatric and geriatric, is needed. Traditional, pediatric populations are not included in clinical investigation and the doses are allometrically scaled based on adult doses and body weight, which leads to underdosing and thus poor therapeutic results, or body surface area, which causes overdosing and drug-related toxicity as discussed in chapter 9. Since 1997, the pharmaceutical industry has been encouraged to carry out studies in pediatrics because of the FDA Modernization Act (FDAMA), which provides an additional 6 months of marketing exclusivity by including pediatrics in clinical investigation. Thus the off-label use of drugs in pediatric populations decreased. This has provided more appropriate information to prescribe in children based on actual pediatric drug exposure and clearance in the disease of interest [24,25].

As individuals get older their renal function deteriorates, leading to a need for specific dosing guidance. Groups of interest are subjects with ages from 65 to 75, 75 to 85, and older than 85 [26]. Therapeutic problems were observed in geriatric patients with > 80 years of age after treatment with benoxaprofen when it had a much longer half-life compared to younger patients. Extensive elevation of benoxaprofen in patients' serum caused cholestatic jaundice, and the patients died [27]. This incident pointed out the need to optimize drug dosing by conducting PK studies in the elderly, thus ensuring the dosing is consistent with the younger population and adverse events are kept to a minimum [28].

11.3.1.3 Renal and Hepatic Impairment When renal and hepatic functions are impaired, drugs that depend on these functions for elimination can build up in blood circulation, thus causing drug toxicity. Consequently, PK studies in patients with renal and hepatic impaired functions are required by the FDA to estimate the change in exposure of chronic use of drugs that depend on these organs in elimination [29,30]. The results of these small PK studies are then reported in the label with statements such as "use with caution in subjects with renal impairment or use with caution in subjects with hepatic impairment." Dose adjustment is usually recommended, or another choice of therapeutic agents when it is possible. However, there is no clarity in dose adjustments for particular drugs even though the PK changes may be clear, resulting in a clinical dosage trial in each subject as suggested by some authors [31].

11.3.1.4 Drug Metabolism In this section of a drug label, a specification on all aspects of in vitro and in vivo drug metabolism and pathways is included. In addition, studies that describe the role of metabolism resulting in DDIs are presented in this section. The interaction potential, either if the drug is a victim or if the drug is a perpetrator, is fully defined in this label section with statements about the PK changes.

For example, in the case of a pretty commonly used drug, Lipitor (atorvastatin), under the heading of drug interactions, the following statements are shown:

- The risk of myopathy during treatment with statins is increased with concurrent administration of fibric acid derivatives, lipid-modifying doses of niacin, cyclosporine, or strong CYP3A4 inhibitors (e.g., clarithromycin, HIV protease inhibitors, and itraconazole). Lipitor is metabolized by cytochrome P450 3A4. Concomitant administration of Lipitor with strong inhibitors of CYP3A4 can lead to increase in plasma concentrations of atorvastatin. The extent of interaction and potentiation of effects depend on the variability of effect on CYP3A4.
- Atorvastatin AUC was significantly increased with concomitant administration of Lipitor 80 mg with clarithromycin (500 mg twice daily) compared to that of Lipitor alone. Therefore, in patients taking clarithromycin, caution should be used when the Lipitor dose exceeds 20 mg.
- Atorvastatin AUC was significantly increased with concomitant administration of Lipitor 40 mg with ritonavir plus saquinavir (400 mg twice daily) or Lipitor 20 mg with lopinavir plus ritonavir (400 + 100 mg twice daily) compared to that of Lipitor alone. Therefore, in patients taking HIV protease inhibitors,

caution should be used when the Lipitor dose exceeds 20 mg. Atorvastatin AUC was significantly increased with concomitant administration of Lipitor 40 mg and itraconazole 200 mg. Therefore, in patients taking itraconazole, caution should be used when the Lipitor dose exceeds 20 mg.

- Interaction of Lipitor and grapefruit that contains one or more components that inhibit CYP3A4 is pronounced. Grapefruit can increase plasma concentrations of atorvastatin, especially with excessive grapefruit juice consumption (> 1.2 L/day).

Although sometimes interaction may result in inhibition of the metabolism of a certain drug by other drugs, thus increasing its bioavailability, exposure, and hence its efficacy (which is not the case with grapefruit), its consumption may result in inhibition of Lipitor-hepatic elimination, thus its plasma accumulation and toxicity. The statement in the label section seems to cause concern for consumption of grapefruit while taking atorvastatin.

Also, atorvastatin and its metabolites are substrates of the OATP1B1 transporters. Inhibitors of the OATP1B1 (e.g., cyclosporine) can increase the bioavailability of atorvastatin.

- Atorvastatin AUC was significantly increased with concomitant administration of Lipitor 10 mg and cyclosporine 5.2 mg/kg/day compared to that of Lipitor alone. In cases where coadministration of Lipitor with cyclosporine is necessary, the dose of Lipitor should not exceed 10 mg.
- Concomitant administration of Lipitor with inducers of cytochrome P450 3A4 (e.g., efavirenz and rifampin) can lead to variable reductions in plasma concentrations of atorvastatin. Owing to the dual interaction mechanism of rifampin, simultaneous coadministration of Lipitor with rifampin is recommended, as delayed administration of Lipitor after administration of rifampin has been associated with a significant reduction in atorvastatin plasma concentrations.

Lipitor as perpetrator was specified as follows: when multiple doses of Lipitor and digoxin were coadministered, steady-state plasma digoxin concentrations increased by ~ 20%.

- Patients taking digoxin should be monitored appropriately. This advice refers to digoxin rather than Lipitor and clearly implies that the dose of digoxin may need to be lowered, but no further clinical advice is listed of the most safe but effective dose in this case, though it has been suggested that digoxin dosing should not be modified unless there was an evidence of toxicity.

Another case of Lipitor as perpetrator is when it is coadministred with oral contraceptives such as norethindrone and ethinyl estradiol.

- Lipitor in this case increased AUC values for norethindrone and ethinyl estradiol. Label information stated that these increases should be considered when selecting an oral contraceptive for a woman taking Lipitor, which may indicate a lower dose oral contraceptives should be chosen, but no comments are made on whether the efficacy of those lower doses will be sustained.

In these cases when the efficacy of oral contraceptives is critical and needs to be maximized, physicians may have the option of prescribing other statins that are not modulated through cytochrome P450 3A4 and that may be an alternative solution, although that is not specifically mentioned in the label for atorvastatin.

- No clinically significant effect on prothrombin time when administered to patients receiving chronic warfarin treatment while treated with Lipitor.

This is an important labeling statement because it discusses the outcome, that is, prothrombin time, rather than the usual presentation of PK drug levels.

The conclusion is that a good deal of clinical pharmacology takes place to help understand the metabolism of the drug of interest. Although label information does not specify the clinical advice, the physician must then decide on how to deal with each DDI issue.

11.3.1.5 Genetic Polymorphism, Ethnic Differences Variations in cytochrome P450 metabolic pathways are modified by polymorphic genetic variability. As an example is the case of tramadol, a centrally acting analgesic that is by itself active and has an active metabolite O-desmethyl-tramadol (M1). The O-demethylation is catalyzed by cytochrome P450 2D6 and the N-demethylation is catalyzed by CYP2B6 and CYP3A4. There is a wide variability in PK because of CYP genetic polymorphism [32], which is reflected in the current tramadol label, which reads as follows: "The formation of the active metabolite, M1, is mediated by CYP2D6. Approximately 7% of the population has reduced activity of the CYP2D6 isoenzyme of cytochrome P-450. Based on a population PK analysis of Phase I studies with immediate-release tablets in healthy subjects, concentrations of tramadol were approximately 20% higher in 'poor metabolizers' versus 'extensive metabolizers, while M1 concentrations were 40% lower. In vitro drug interaction studies in human liver microsomes indicate that inhibitors of CYP2D6 (fluoxetine, norfluoxetine, amitriptyline, and quinidine) inhibit the metabolism of tramadol to various degrees, suggesting that concomitant administration of these compounds could result in increases in tramadol concentrations and decreased concentrations of M1. The full pharmacological impact of these alterations in terms of either efficacy or safety is unknown."

It would appear that the science continues to improve to help understand how genetic polymorphism explains differences in toxicity and to some extent efficacy. However, it is still not possible to adequately label the proposed change in dose with any precision.

In addition to genetic polymorphism, variability in the PK due to ethnic factors has become an issue in drug development, and it is hard to differentiate it from the variability due to genetic polymorphism [33]. Ethnic factors include genetics, metabolism, diet, medical practice, concomitant drugs, tobacco/alcohol, and environment. Because genetic polymorphism varies in different ethnic populations, it is desirable to include as diverse a population as possible in clinical trials to help understand and anticipate any ethnic differences. When it comes to the practicality of drug labeling, that may become challenging and less clear. For drugs such as the antifungal voriconazole, which depends on CYP2C19 in its metabolism and clearance, a lower dose was recommended in Japan than the United States [24], but there is greater concordance in dosing between Europe and the United States although there are some exceptions [34].

Modern drug development incorporates a diverse population with multicentered multinational studies becoming the normal practice in Phase 3 clinical investigations. However, the interpretation of this diverse data does not lead to individual drug labeling advice.

REFERENCES

1. Smith DA, Obach RS. Seeing through the MIST: abundance versus percentage. Commentary on metabolites in safety testing. Drug Metab Dispos 2005;33:1409–1417.
2. Baillie TA, Cayen MN, Fouda H, Gerson RJ, Green JD, Grossman SJ, Klunk LJ, LeBlanc B, Perkins DG, Shipley LA. Drug metabolites in safety testing. Toxicol Appl Pharmacol 2002;182:188–196.
3. U.S. Department of Health and Human Services. Food and Drug Administration Guidance for Industry Q3B. Impurities in new drug products. FDA, 1997. Available: http://www.fda.gov.libproxy.lib.unc.edu/cder/guidance/1317fnl.htm#ATTACHMENT %20II (accessed March 28, 2005).
4. White RE, Evans DC, Hop CE, Moore DJ, Prakash C, Surapaneni S, Tse FL. Radiolabeled mass-balance excretion and metabolism studies in laboratory animals: a commentary on why they are still necessary. Xenobiotica 2013;43(2):219–225.
5. Penner N, Xu L, Prakash C. Radiolabeled absorption, distribution, metabolism, and excretion studies in drug development: why, when, and how? Chem Res Toxicol 2012; 25:513–531.
6. U.S. Department of Health and Human Services. Food and Drug Administration. Guidance for industry: safety testing of drug metabolites. Available: www.fda.gov/cder/guidance (accessed November 15, 2011).
7. Baillie TA. Drug metabolism in drug safety evaluation. In: Encyclopedia of drug metabolism and interactions, ed. Lyubimov AV. Hoboken, NJ: Wiley, 2012, 5:1–24.
8. Gao H, Jacobs A, White RE, Booth BP, Obach RS. Meeting report: Metabolites in Safety Testing (MIST) Symposium—safety assessment of human metabolites: what's REALLY necessary to ascertain exposure coverage in safety tests? AAPS J 2013;15(4):970–973.
9. ICH M3(R2). 2010. Available: http://www.fda.gov/Drugs/GuidanceCompliance-Regulatory Information/Guidances/ucm065006.htm (accessed April 2013).
10. Timmerman P, Kall MA, Gordon B, Laakso S, Freisleben A, Hucker R. Best practices in a tiered approach to metabolite quantification: views and recommendations of the European Bioanalytical Forum. Bioanalysis 2010;2:1185–1194.

11. Yu CP, et al. A rapid method for quantitatively estimating metabolites in human plasma in the absence of synthetic standards using a combination of LC/MS and radiometric detection. Rapid Commun Mass Spectrom 2007;21:497–502.
12. Yi P, Luffer-Atlas D. A radiocalibration method with pseudo internal standard to estimate circulating metabolite concentrations. Bioanalysis 2010;2:1195–1210.
13. Zhang D, Raghavan N, Chando T, Gambardella J, Fu Y, Zhang D, et al. Humphreys WG LC-MS/MS-based approach for obtaining exposure estimates of metabolites in early clinical trials using radioactive metabolites as reference standards. Drug Metab Lett 2007;1:293–298.
14. Vishwanathan K, Babalola K, Wang J, Espina R, Yu L, Adedoyin A, Talaat R, Mutlib A, Scatina J. Obtaining exposures of metabolites in preclinical species through plasma pooling and quantitative NMR: addressing metabolites in safety testing (MIST) guidance without using radiolabeled compounds and chemically synthesized metabolite standards. Chem Res Toxicol 2009;22:311–322.
15. Walker GS, Ryder TF, Sharma R, Smith EB, Freund A. Validation of isolated metabolites from drug metabolism studies as analytical standards by quantitative NMR. Drug Metab Disp 2011;39(3):433–440.
16. Gao H, Deng S, Obach RS. A simple liquid chromatography-tandem mass spectrometry method to determine relative plasma exposures of drug metabolites across species for metabolite safety assessments. Drug Metab Dispos 2010;38(12):2147–2156.
17. Gao H, Obach RS. A simple liquid chromatography-tandem mass spectrometry method to determine relative plasma exposures of drug metabolites across species for metabolite safety assessments. II. Application to unstable metabolites. Drug Metab Dispos 2012; 40(7):1290–1296.
18. Ma S, Li Z, Lee K-J, Chowdhury SK. Determination of exposure multiples of human metabolites for MIST assessment in preclinical safety species without using reference standards or radiolabeled compounds. Chem Res Toxicol 2010;23(12):1871–1873.
19. U.S. Department of Health and Human Services. Food and Drug Administration. Guidance for industry: bioanalytical method validation. FDA, 2001. Available: http://www.fda.gov/downloads/Drugs/GuidanceCompliance Regulatory-Information/Guidances/UCM070107.pdf (accessed June 2013).
20. Gao H, Obach RS. Addressing MIST (metabolites in safety testing): bioanalytical approaches to address metabolite exposures in humans and animals. Curr Drug Metab 2011;12(6):578–86.
21. Wagner AK, Chan KA, Dashevsky I, et al. FDA drug prescribing warnings: is the black box half empty or half full? Pharmacoepidemiol Drug Saf 2006;16(6):369–386.
22. Stafford RS. Regulating off-label drug use rethinking the role of the FDA. N Engl J Med 2008;358:1427–1429.
23. U.S. Department of Health and Human Services. Food and Drug Administration, Center for Drug Evaluation and Research (CDER). Estimating the maximum safe starting dose in initial clinical trials for therapeutics in adult healthy volunteers. FDA, 2005.
24. Yanni SB, Annaert PP, Augustijns P, Ibrahim JG, Benjamin DK, Thakker DR. In vitro hepatic metabolism explains higher clearance of voriconazole in children versus adults: role of CYP2C19 and FMO. Drug Metab Dispos 2010;38;45–51.
25. Yanni SB, Smith PB, Benjamin DK Jr, Augustijns PF, Thakker DR, Annaert PP. Higher clearance of micafungin in neonates compared with adults: role of age-dependent micafungin serum binding. Biopharm Drug Dispos 2011;32(4):222–232.

26. Shah RR. Drug development and use in the elderly: search for the right dose and dosing regimen. Br J Clin Pharmacol 2004;58(5):452–469.
27. Taggart HM, Alderdice JM. Fatal cholestatic jaundice in elderly patients taking benoxaprofen. Br Med J 1982;284(6326):1372.
28. Routledge PA, O'Mahony MS, Woodhouse KW. Adverse drug reactions in elderly patients. Br J Clin Pharmacol 2004;57(2):121–126.
29. U.S. Department of Health and Human Services. Food and Drug Administration, Center for Drug Evaluation and Research (CDER). Pharmacokinetics in patients with impaired renal function—study design, data analysis, and impact on dosing and labeling. FDA, 2010.
30. Verbeeck RK. Pharmacokinetics and dosage adjustment in patients with hepatic dysfunction. Eur J Clin Pharmacol 2008;64:1147–1161.
31. Delco F, Tchambaz L, Schlienger R, et al. Dose adjustment in patients with liver disease. Drug Saf 2005;28(6):529–545.
32. Dombey S. Clinical pharmacology in drug labeling. The impact of drug metabolism and clinical pharmacology on recommended dose of drugs. In: Encyclopedia of drug metabolism and interactions, ed. Lyubimov AV.Hoboken, NJ: Wiley, 2012, 11:1–12.
33. ICH. Harmonised tripartite guideline ethnic factors in the acceptability of foreign clinical data E5(R1). Current Step 4, version dated February 5, 1998.
34. Huang S-M, Temple R. Is this the drug or dose for you? Impact and consideration of ethnic factors in global drug development, regulatory review, and clinical practice. Clin Pharmacol Ther 2008;84:287–294.

INDEX

Translational ADMET for Drug Therapy: Principles, Methods, and Pharmaceutical Applications, First Edition. Souzan B. Yanni.